Advances in Design for Cross-Cultural Activities Part I

Advances in Human Factors and Ergonomics Series

Series Editors

Gavriel Salvendy
Professor Emeritus
School of Industrial Engineering
Purdue University

Chair Professor & Head
Dept. of Industrial Engineering
Tsinghua Univ., P.R. China

Waldemar Karwowski
Professor & Chair
Industrial Engineering and
Management Systems
University of Central Florida
Orlando, Florida, U.S.A.

3rd International Conference on Applied Human Factors and Ergonomics (AHFE) 2010

Advances in Applied Digital Human Modeling
Vincent G. Duffy

Advances in Cognitive Ergonomics
David Kaber and Guy Boy

Advances in Cross-Cultural Decision Making
Dylan D. Schmorrow and Denise M. Nicholson

Advances in Ergonomics Modeling and Usability Evaluation
Halimahtun Khalid, Alan Hedge, and Tareq Z. Ahram

Advances in Human Factors and Ergonomics in Healthcare
Vincent G. Duffy

Advances in Human Factors, Ergonomics, and Safety in Manufacturing and Service Industries
Waldemar Karwowski and Gavriel Salvendy

Advances in Occupational, Social, and Organizational Ergonomics
Peter Vink and Jussi Kantola

Advances in Understanding Human Performance: Neuroergonomics, Human Factors Design, and Special Populations
Tadeusz Marek, Waldemar Karwowski, and Valerie Rice

4th International Conference on Applied Human Factors and Ergonomics (AHFE) 2012

Advances in Affective and Pleasurable Design
Yong Gu Ji

Advances in Applied Human Modeling and Simulation
Vincent G. Duffy

Advances in Cognitive Engineering and Neuroergonomics
Kay M. Stanney and Kelly S. Hale

Advances in Design for Cross-Cultural Activities Part I
Dylan D. Schmorrow and Denise M. Nicholson

Advances in Design for Cross-Cultural Activities Part II
Denise M. Nicholson and Dylan D. Schmorrow

Advances in Ergonomics in Manufacturing
Stefan Trzcielinski and Waldemar Karwowski

Advances in Human Aspects of Aviation
Steven J. Landry

Advances in Human Aspects of Healthcare
Vincent G. Duffy

Advances in Human Aspects of Road and Rail Transportation
Neville A. Stanton

Advances in Human Factors and Ergonomics, 2012-14 Volume Set: Proceedings of the 4th AHFE Conference 21-25 July 2012
Gavriel Salvendy and Waldemar Karwowski

Advances in the Human Side of Service Engineering
James C. Spohrer and Louis E. Freund

Advances in Physical Ergonomics and Safety
Tareq Z. Ahram and Waldemar Karwowski

Advances in Social and Organizational Factors
Peter Vink

Advances in Usability Evaluation Part I
Marcelo M. Soares and Francisco Rebelo

Advances in Usability Evaluation Part II
Francisco Rebelo and Marcelo M. Soares

Advances in Design for Cross-Cultural Activities Part I

Edited by

Dylan D. Schmorrow
and
Denise M. Nicholson

CRC Press
Taylor & Francis Group
Boca Raton London New York

CRC Press is an imprint of the
Taylor & Francis Group, an **informa** business

CRC Press
Taylor & Francis Group
6000 Broken Sound Parkway NW, Suite 300
Boca Raton, FL 33487-2742

Printed in the United States of America on acid-free paper
Version Date: 20120529

International Standard Book Number: 978-1-4398-7028-0 (Hardback)

Visit the Taylor & Francis Web site at
http://www.taylorandfrancis.com

and the CRC Press Web site at
http://www.crcpress.com

Table of Contents

Section I: Perceptual Training for Cross Cultural Decision Making

Section II: Tactical Culture Training: Narrative, Personality, and Decision Making

Section III: Application of Human, Social, Culture Behavioral Modeling Technology

Section IV: Architecture for Socio-Cultural Modeling

Section V: Strategic and Tactical Considerations for Cross-Cultural Competency

Section VII: ICEWS and SAA: Early Warnings

Preface

We are excited to introduce this two-volume collection of papers presented at the 2012 Cross Cultural Decision Making (CCDM) Conference in San Francisco, the second international gathering of its kind. Following the inaugural CCDM conference in 2010, 64 presentations were collected in one volume; this year more than 175 abstracts were submitted to CCDM, and the best of those presentations have been collected in two volumes; this tremendous growth reflects the increasing relevance of research in this area.

The CCDM conference focuses on improved decision making across a variety of cultural constructs, including geographical, historical, sociological, organizational, team, and technology interactions. This conference includes the research of experts and industry practitioners from multidisciplinary backgrounds, including sociology, linguistics, human-computer interaction, human factors engineering, systems engineering, military science, psychology, neuroscience, instructional design, and education, who showcase the latest advances in our understanding of the role of culture on decision making in numerous settings. Improved decision making among members of diverse teams and within organizational systems, and innovative ways to measure and assess that process, comprise the foundation for many projects discussed in these volumes.

Part of the growth in this year's CCDM conference can be attributed to the expanding sociocultural behavioral modeling community: scientists, engineers, and practitioners who resonate with the focus of CCDM, many of whom previously participated in the Human Social Culture Behavior (HSCB) Focus conference series. Since 2008, significant advances in applied research and technology development have improved our ability to understand sociocultural behavior, detect relevant sociocultural signals, forecast through persistent sensing of the environment, and mitigate with measurable courses of action grounded in the social and behavioral sciences. Through innovative and rigorous applied research, advanced technology development, and prototypes, HSCB research has helped build the sociocultural behavior science base.

For example, this volume showcases a variety of discussions on the increasingly influential impact of social media as a driver of social change and as a bridge for cultural differences. Technology's role in modeling and mediating cultural and social differences is also discussed in several chapters, from a wide array of perspectives and applications. The implications of culture for analysis and decision making are explored from divergent angles, and the conclusions presented will guide future research efforts.

The influence of culture on decision making is pervasive, as reflected in the diverse disciplines represented by those individuals and entities involved in sociocultural research and engineering. The two-volume CCDM collection features papers that discuss emerging concepts, theories, and applications of cross-cultural decision making knowledge. The work described in these chapters reflects dedicated research by a wide range of expert academics and practitioners from around the world.

The chapters in these volumes are grouped in the following broad categories:

Advances in Design for Cross-Cultural Activities: Part I

- Perceptual Training for Cross Cultural Decision Making
- Tactical Culture Training: Narrative, Personality, and Decision Making
- Applications of Human, Social, Culture Behavioral Modeling Technology
- Architecture for Socio-Cultural Modeling
- Strategic and Tactical Considerations for Cross-Cultural Competency
- Commercial Research and Applications of Social-Cultural Science
- ICEWS and SAA: Early Warnings

Advances in Design for Cross-Cultural Part: II

- Multifarious Modeling Discussions
- Verification, Validation, and Assessment
- Language, Trust, and Culture
- Social Media and Culture
- Social Science and Culture

Each of the chapters of this book were either reviewed or contributed by the members of Editorial Board. For this, our sincere thanks and appreciation goes to the Board members listed below:

This book will be of special value to professionals, researchers, and students in the broad field of human social and cultural behavior and decision making. As our performers continue to develop their tools, especially in support of US and coalition forces, we must always remain aware of the limits of our understanding and keep our research grounded in good social and behavioral science.

We hope this book is informative, but even more, that it is thought-provoking. We hope it inspires, leading the reader to contemplate other questions, applications, and potential solutions to the challenges of intercultural collaboration, decision making, and innovation.

April 2012

Dylan Schmorrow
Office of the Secretary of the Defense
Washington, DC, USA

Denise Nicholson
MESH Solutions LLC, a DSCI Company
Orlando, Florida, USA

Editors

Section I

Perceptual Training for Cross Cultural Decision Making

CHAPTER 1

Perceptual Training for Cross Cultural Decision Making (Session Overview)

Sae Schatz and Denise Nicholson

MESH Solutions, LLC
Orlando, FL
sschatz@dsci.com, dnicholson@dsci.com

ABSTRACT

Today's battlefield is a dynamic environment, filled with subtle physical, psychological, and social threats. Modern military personnel must not only possess typical warfighting abilities, but they must also be able to rapidly perceive, understand, and then respond to a range of ambiguous stimuli, including nuanced social and cultural indicators. In order to help military personnel—particularly enlisted ground forces—manage such threats, we are exploring advanced perceptual training for complex urban spaces. Broadly, the goals of this research effort are twofold: First, we seek to address immediate gaps in contemporary Marine Corps training by developing, testing, and ultimately delivering a suite of instructional methodologies and technologies that enhance the training of advanced perceptual skills. Second, we are investigating emerging Science and Technology (S&T) challenges, including the categorization of militarily relevant culturally situated cues, the modeling of sociocultural patterns of life, and development of associated instructional methods and performance metrics. This session comprises representatives from the multidisciplinary Perceptual Training Systems and Tools (PercepTS) team, and this paper establishes the context and rationale for the combined effort. It discuss the status quo of Marine Corps perceptual training, offers general guidance about perceptual training drawn from the literature, and outlines our team's unified approach to addressing the these research challenges.

Keywords: Perception, Combat Hunter, patterns of life, human behavior pattern recognition, cognitive readiness, training simulation

1 BACKGROUND

Today's battlefield is filled with subtle physical, psychological, and social threats. Modern military personnel must not only possess typical warfighting abilities, but they must also be able to rapidly perceive, understand, and then respond to a range of ambiguous stimuli, including nuanced social and cultural cues. These stimuli may include immediate threats, such as snipers or suicide bombers, and they may also include more subtle, behavioral indicators. For instance, warfighters may need to interpret cues that imply a foreign national is lying, signs that point to an insurgent network, or macro-cultural patterns that reveal the collective outlook of the denizens in a particular region.

In order to help military personnel—particularly enlisted ground forces—excel at such tasks, we are exploring advanced perceptual training methodologies and technologies as part of the Perceptual Training Systems and Tools (PercepTS) effort. Broadly, the goals of this research effort are twofold. First, we seek to synthesize and apply existing research in order to address immediate needs in United States Marine Corps (USMC) perceptual training. Second, we intend to investigate emerging Science and Technology (S&T) challenges, in order to advance the scientific body of knowledge in this topical area.

1.1 Contemporary Marine Corps Perceptual Training

Presently, facets of perceptual training are incorporated throughout the enlisted Marine Corps training continuum. Most notably, perceptual skills are explicitly trained to Marine ground forces in three programs of instruction (for more detail on these curricula see Schatz, Dolletski-Lazar, et al., 2011):

Marksmanship. Rifle marksmanship is a core Marine Corps competency that relies upon visual perception, including visual search, target detection, and range estimation (Chung et al., 2004).

IED Detection. Improvised Explosive Device (IED) detection instruction typically involves training in the recognition of various types of explosive devices as well as practice-focused field evolutions, in which trainees are exposed to practice environments containing mock IEDs (e.g., Hess & Sharps, 2008, Murphy, 2009).

Both the marksmanship and the IED detection courses engender critical knowledge, skills, and attitudes. However, neither of these programs of instruction includes the perceptual training of social, cultural, and behavioral cues.

Combat Hunter. Combat Hunter serves as the primary Marine Corps program of instruction for learning sustained observation of social patterns of life and for enhancing personnel's social, cultural, and behavioral sensemaking skills. Through the Combat Hunter course, personnel learn how to evaluate people's biometric signs, read human behavior, identify geographic indicators (like footprints), and objectively analyze the "atmosphere" of a locale. Broadly speaking, Combat Hunter trains situation awareness, sensemaking, mental simulation, and dynamic decision-making for urban operational environments (Schatz, Reitz, Fautua, & Nicholson, 2010). Figure 1 depicts Marines in the Combat Hunter course viewing a mock village (not shown) and attempting to make sense of the location by conducting sustained observation and creating mental "baselines" of the local behavioral, social, and cultural norms.

Figure 1 Marines in the Combat Hunter course viewing a mock village (not shown) and attempting to make sense of the location by conducting sustained observation and creating mental "baselines" of the local behavioral, social, and cultural norms (photo by the first author).

1.2 PercepTS and Combat Hunter Course

From 2005–2007, the Juba Sniper terrorized American warfighters in Baghdad. "Like a wartime boogeyman, Juba had twice been reported captured, only to spring up again, Freddie Krueger-style, in Jihadist propaganda and U.S. military patrol briefings" (Morin, 2007). At the same time, casualties from IEDs and small arms fire were on the rise (US casualty report). In January 2007, the USMC sought a novel solution. With assistance from the Marine Corps Warfighting Lab (MCWL), a diverse group of subject matter experts (SMEs) were assembled with a common goal: Turn Marines from the hunted into the hunters. From this meeting, Combat Hunter was formed (Gideons, Padilla, and Lethin, 2008).

MCWL and the Office of Naval Research developed Combat Hunter as rapidly as possible, in order to meet the urgent needs described above. As a result, certain aspects of the course initially lacked rigor (for a full description, see Schatz, Reitz, et al., 2010), including incomplete documentation and validation of the course content, limited availabilities of assessment apparatus beyond subjective instructor ratings, inadequate throughput, lack of support for in-unit training, and high course costs, particularly for the role-player support. The Office of Naval Research created the PercepTS project, in part, to address gaps in the current Combat Hunter program of instruction and related Marine Corps training. PercepTS will also expand the body of science and technology in this area, conducing advanced research into perceptual training, patterns of life, dynamic tailoring of simulations, and human behavior modeling.

1.2.2 PercepTS Science and Technology Objectives

The PercepTS project formally addresses the following Marine Corps Training and Education (T&E) Science and Technology Objectives (STOs). Brief descriptions, in our own words, of the most relevant aspects of each STO are listed below (TECOM S&T Working Group, 2012):

T&E STO-1: Warrior Decision Making: Develop knowledge products and technology capabilities that enhance cognitive, relational, and perceptual skills for small unit leaders, better preparing them to make decisions in ambiguous and dangerous conditions, operate from a commander's intent, and act with minimal supervision.

T&E STO-4: Experiential Learning Technologies and Methodologies: Investigate scenario-based instructional methods and technologies that optimize experiential learning for topics such as cultural, social, and behavioral interactions. Build rapid and accurate assessments into these instructional systems.

T&E STO-5: Adaptive Simulated Entities: Develop simulation-based training systems that include behaviorally accurate virtual role-players that can represent a variety of cultures, populations, and domains. Demonstrate the ability to adapt (or "tune") these entities' behaviors to respond to changes in the simulation, including dynamic tailoring of a scenario for instructional purposes.

T&E STO-9: Cultural and Language Proficiency Tools: Develop training technologies and instructional scenarios that focus on culture-general skills and cross-cultural competence to prepare Marines to interact in wide-ranging and rapidly changing cross-cultural situations.

2 PERCEPTUAL TRAINING SYSTEMS AND TOOLS

PercepTS is a five-year S&T project, currently in its second year. The subsections below describe outcomes from the completed first-year cycle and plans for the future phases. The completed work and future plans will be conducted by the entire PercepTS team, meaning that a range of scientists and developers (in addition to the authors) have contributed or will contribute to each of the following. The broad PercepTS team is represented in the other papers in this session.

2.1 Year 1: Baselining (Completed)

The PercepTS project recently completed its first year, which focused on empirical and theoretical baselining. During this baselining phase, we conducted theoretical research into perceptual–cognitive skills, empirical research on maintaining perceptual performance under stress, testing and observation of real-world Marine Corps perceptual training, and ongoing requirements generation based on these analyses and interviews with Navy and Marine Corps stakeholders. Through this effort, the team discovered more specific gaps and opportunities involving Marine Corps perceptual training, and we devised a corresponding research and development plan to address these needs. Bolton and Squire's presentation, in this session, describes these research challenges in more detail. See their chapter, on "Warfighters' Perception and Human Behavior."

2.1.1 Theoretical Research: Perceptual–Cognitive Skills

Research on perception has a long history (e.g., Gibson & Gibson, 1955; Gibson, 1969; Goldstone, 1998). Yet, despite ongoing attention, an actionable, broad-spectrum categorization for perceptual skills has not been developed. Toward that end, our team has created a comprehensive taxonomy for perceptual–cognitive skills, designed specifically to support their training and education.

The PercepTS taxonomy is a two-dimensional model in which levels of operational control (i.e., amount of conscious thought or control) interconnect with temporal properties (i.e., features of time) to generate broad categories of skill characteristics. Under each category, or taxon, lies a representative compilation of specific skills. These skills will be the focus of the training and education products created under PercepTS.

2.1.2 Empirical Research: Perceptual Resilience

Perceptual skills are among the first to fail under stressful conditions. Hence, PercepTS is investigating training approaches that will better engender enduring perceptual skills and facilitate enhanced performance under chaotic operational conditions. The PercepTS approach is unique in that it combines intentional perceptual instruction with explicit stress resilience instruction, adapted for maintenance of perceptual performance under stress.

Initial academic review and an experimental trial (N = 60) were completed in FY11. Results suggest that trainees who completed the experimental perceptual resilience training outperformed those who received only the experimental perceptual training (i.e., without the resilience component); further, both of these experimental cohorts significantly outperformed the control group, who received "placebo" training, which included a currently used computer-based trainer on military vehicle discrimination. These laboratory-based results (with civilian participants) are promising, and they suggest that further inquiry with the military community may be warranted.

2.1.3 Field Research: Combat Hunter Data Collection

The PercepTS team collected data on learning outcomes and training enablers within the USMC Combat Hunter program of instruction. These analyses directly informed the PercepTS requirements, and the data collected during baselining support content development for, and downstream impact assessment of, future PercepTS products. PercepTS team members observed and collected data at five Combat Hunter or Combat Hunter-like courses.

Detailed measures of declarative, procedural, and applied knowledge were collected in each venue. Data were collected at the USMC Schools of Infantry East and West, as well as at the Joint and Coalition Bold Quest combat identification event. Finally, results previously collected from the Joint Border Hunter exercise were in a comparable format, allowing much of these data to be directly compared to the baselining results.

2.1.4 Stakeholder Interviews

The PercepTS development effort employs a Human–Systems Integration (HSI) process that incorporates best practices of human factors and systems engineering. These best practices include frequent interaction with stakeholders, soliciting their opinions and collecting their feedback on prototypes. In the first year of the PercepTS effort, numerous informal and semi-formal interviews were conducted with Combat Hunter trainees and instructors, as well as decision makers with the USMC Training and Education Command, USMC Program Manager for Training Systems, and Office of Naval Research. Advice from subject matter experts was also considered.

Nicholson and Schatz discuss the PercepTS HSI approach in their presentation, in this session. See their chapter on "Bridging the Gap Between Humans and Systems."

2.1.5 Requirements Generation

Through theoretical inquiry, empirical baselining, and stakeholder knowledge elicitation, the team has devised an approach to address current and future Marine Corps training needs. The team plans to deliver a range of perceptual–cognitive training products, from low-tech knowledge products to immersive technology solutions such as an adaptive "Virtual Ville" that makes use of dynamic tailoring, immersive stimulation, and advanced human behavioral modeling techniques.

Figure 2 depicts a high-level road map of the technology deliverables and the Fiscal Years (FYs) during which each will be emphasized. Roughly, one major technology deliverable is featured in each fiscal cycle; initial deliverables address novice-level knowledge, skills, and attitudes (KSAs) and later deliverables target progressively more advanced KSAs.

The following subsections describe the vision for future PercepTS deliverables, based upon lessons learned from the baseline data collection. The operational requirements for each component will be shaped through ongoing research, stakeholder input, and feasibility analyses. In other words, facets of the general plans outlined below may change (particularly for the deliverables associated with FY14 and FY15).

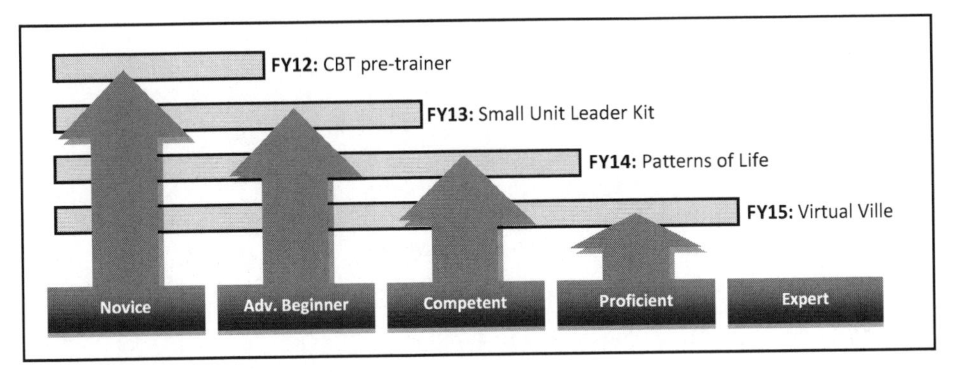

Figure 2 High level PercepTS roadmap of major deliverables

2.2 Year 2: Combat Hunter CBT (Currently In Progress)

Previous research on Combat Hunter identified the need to deliver pre-training content and establish course prerequisites (Fautua et al., 2010; Kobus et al., 2009). Our own baselining investigations also confirmed these recommendations. The Combat Hunter course includes specialized language and a substantial amount of declarative and procedural knowledge that students must memorize before they can develop more meaningful, higher-level Combat Hunter skills. This means that Marine Corps instructors spend considerable time on this low-level instruction; however, the lower-level content could be more efficiently taught using instructional technology.

In the second year of the project, the team intends to deliver a Combat Hunter Computer-Based Trainer (CBT). This introductory course will provide instruction on relevant declarative and procedural Combat Hunter materials, which should better prepare trainees to attend the existing face-to-face Combat Hunter courses. In this way, we believe that instructors in the face-to-face course can use their valuable time to focus on the "hard" skills, as well as hands-on demonstration, in-class exercises, and practical applications.

PercepTS team members, Tarr et al., will offer more information on the CBT and its relevant instructional approaches in their presentation in this session, called "Instructional Strategies for Cross Cultural Perceptual Training."

2.3 Year 3: Perceptual Training Kit

In their 2010 study of Combat Hunter, Spiker and Johnston recommended that developers create a suite of game-based training tools to support ongoing instruction once students exit the course. Our team similarly endorses the development of post-course materials. We add that the supplementary instructional materials should be designed for small-unit leaders to support their in-garrison and in-theater training. Small-unit leaders, such as Marine squad and platoon leaders, act as the primary mentors and instructors of Marine Corps enlisted personnel. These leaders often train their personnel during breaks in their regular activities. As such, the majority of the components in any supplementary instructional package should function in field settings without requiring significant technology support.

The PercepTS Small Unit Leader Perceptual Training Kit will be a comprehensive set of knowledge products, intended to emphasize intermediate practice and skill sustainment. Colloquially, the Kit may be considered a "lite" train-the-trainer package coupled with resources that small-unit leaders can leverage for flexible education, training, and integrated practice. With the Kit, small-unit leaders can grow their Marines' skills beyond the advanced-beginner levels with which they graduate the face-to-face Combat Hunter course. The Kit can also facilitate ongoing sustainment training.

The Kit will contain a variety of instructional materials, such as recommended training exercises and photos that can be used for Combat Profiling exercises. It will also include a developmental model of expert "perceivers" at the novice, intermediate, and expert levels. This model will be translated into assessment products, such as Behaviorally Anchored Rating Scales (BARSs), Situational Judgment Tests (SJTs), and Tactical Decision Games (TDGs).

PercepTS team members, Ross and Militello, will offer more information on Combat Hunter expert models and the research supporting the PercepTS Small Unit Leader Perceptual Training Kit in their presentation in this session. See their chapter, "Expert Models of Perception in Cross Cultural Contexts."

2.4 Year 4: Patterns of Life

The Marine Corps STOs advise that additional research should focus on social, cultural, and behavioral simulation-based trainers, with a special emphasis on their human behavior models. These STOs coincide with the simulation-related recommendations we identified through the baselining phase. However, before such sophisticated simulations can be developed, additional investigation into patterns of life principles, related behavioral models, and associated adaptive simulation-based instructional strategies must be developed.

In the fourth year of the PercepTS effort, the team will deliver research findings related to patterns of life. This work will synthesize theories from a variety of social and computational sciences, incorporate advanced behavioral modeling techniques, and integrate adaptive instructional strategies that uniquely enhance this training. The team is also investigating vicarious learning strategies for patterns of life instruction (where secondary trainees benefit from observing and interacting with other trainees' simulation-based training playbacks).

In his presentation, PercepTS team member, Wray, will offer more information on some of the research supporting the patterns of life inquiry, including the dynamic tailoring approaches and vicarious learning strategies to be explored. See his chapter, entitled "Tailoring Culturally Situated Simulation for Perceptual Training."

2.5 Year 5: Virtual Ville Simulation

The culminating PercepTS deliverable will be the Virtual Ville, an immersive team trainer designed to support experiential learning of perceptual–cognitive skills, including patterns-of-life sensemaking and related communication competencies, within the context of sustained observation (see Figure 3). The Virtual Ville is intended to be a scenario-based instructional simulation with Virtual Reality (VR) optics, such as binoculars and night-vision goggles. In the scenarios, Marines will observe a location (e.g., a small town) from an Observation Post or Combat Outpost, located between 300–1000 meters away. They will have to identify the patterns of activity within the region to establish a baseline, identify anomalies, and, ultimately, to predict deleterious events before they occur (i.e., "left of bang").

Figure 3 Artist's conception of the Virtual Ville

Work conducted during the patterns of life project phase (described in 2.4) will inform the Virtual Ville, and the simulator will incorporate instructional principles and assessment metrics created during development of the CBT and Small-Unit Leader Kit. The Virtual Ville will also feature dynamic tailoring, i.e., the manipulation of scenario features and extrinsic feedback for instructional purposes. Some of these adaptive features include the manipulation of communications, adjustments of the saliency of patterns of life cues, and variations on the ratio of signal-to-noise in the scenario.

PercepTS team member, Wray, will offer more information on the Virtual Ville during his presentation, and Bartlett et al. further describe the relationship between communication and perception in their presentation, "Can You Hear Me Now?"

ACKNOWLEDGMENTS

This work is supported in part by the Office of Naval Research. The views and conclusions contained in this document are those of the authors and should not be interpreted as representing the official policies, either expressed or implied, of the Department of Defense or Office of Naval Research. The U.S. Government is authorized to reproduce and distribute reprints for Government purposes notwithstanding any copyright notation hereon. This work was supported, in part, by the Office of Naval Research project N00014-11-C-0193, Perceptual Training Systems and Tools (PercepTS).

REFERENCES

Chung, G. K. W. K., S. O. Nagashima, G. C. Delacruz, J. J., Lee, R. Wainess, and E. L. Baker. 2011, January. "Review of Rifle Marksmanship Training Research." Retrieved from http://www.cse.ucla.edu/products/reports/R783.pdf

Fautua, D., S. Schatz, D. Kobus, V. A. Spiker, W. Ross, J. H. Johnston, D. Nicholson, and E. A. Reitz. 2010. *Border Hunter Research Technical Report* (IST-TR-10-01). Orlando, FL: University of Central Florida, Institute for Simulation & Training.

Gideons, C. D., F. M. Padilla, and C. R. Lethin. 2008. Combat Hunter: The training continues. *Marine Corps Gazette* September 2008: 79–84.

Hess, A. B. and M. J. Sharps. 2006,. *Identification and interpretation of peripheral sources of hazard in complex crime situations.* Western Psychological Association, Palm Springs, CA, April.

Kobus, D. A., Palmer, E. D., Kobus, J. M., and Ostertag, J. R. 2009. *Assessment of the Combat Hunter Trainer Course (CHTC): Lessons Learned* (PSE Report 09-08). San Diego, CA: Pacific Sciences and Engineering Group.

Morin, M. 2007. "Juba the Sniper Legend haunting troops in Iraq." *Stars and Stripes.* Retrieved February 8, 2012, from http://www.stripes.com/news/juba-the-sniper-legend-haunting-troops-in-iraq-1.63062

Murphy, J. S., ed. 2009. *Identifying experts in the detection of improvised explosive devices (IED2).* Unpublished manuscript.

Schatz, S., E. Reitz, D. Nicholson, and D. Fautua. 2010. Expanding Combat Hunter: The science and metrics of Border Hunter. In *Proceedings of the Interservice/Industry Training, Simulation and Education Conference (I/ITSEC)*, November 28–December 2, 2010, Orlando, FL.

Spiker, V. A. and J. H. Johnston. 2010. *Limited Objective Evaluation of Combat Profiling Training for Small Units.* Suffolk, VA: US Joint Forces Command.

CHAPTER 2

Methodology for Developing the Expert Model: The Case of the Combat Hunter

William A. Ross, Laura G. Militello

Cognitive Performance Group
Orlando, FL
bill@CognitivePerformanceGroup.com

ABSTRACT

Members of small tactical units interact with patterns of life, where they are immersed in complex adaptive environments often including novel problems. Simply stated, the cognitive demands for Marine infantrymen are growing because the adversaries they face have become so adaptive. To succeed, the infantryman must learn to apply keen ***intuition***, leverage individual ***tacit knowledge*** and lived ***experiences***, ***recognize*** complex inter-dependencies in situations, and ***act decisively***. In order to understand why some small unit leaders have an exceptional knack for using these abilities, we interviewed Combat Hunters particularly proficient in using their cognitive abilities and perceptual skills to anticipate enemy actions and operate inside the enemy's decision cycle. We applied a number of cognitive task analysis (CTA) methods to understand the nature of the expertise and to transform our understanding into executable learning strategies. This research has significant implications for understanding how small unit leader learn to think as well as assessment of the cognitive abilities during training before the first day of combat.

Keywords: Combat Hunter, expert model, cognitive task analysis, mental model

1. INTRODUCTION

Small unit leaders report that current operations require them to make sense of complex patterns of life where they must solve problems and make decisions. These operational requirements ask them to draw on abilities that are acquired experientially and not through practice or classroom instruction. In the narratives used to re-construct combat engagements, the actions and decisions illustrated that among today's Combat Hunters there was considerable variation regarding how they responded and what factors were most significant. Within these differences, we recognized that some leaders possessed a knack for deciding and acting because they anticipated threat actions before they were obvious to others. When we interviewed this group, they reported the importance of actively learning *how* to think and not *what* to think when dealing with ill-structured problems under tactical conditions where situations do not progress according to a "script". In order to deepen our understanding of this expertise, we conducted field research to learn more about the skills possessed by expert Combat Hunters, who were conducting a training course. The training audience included individuals from the U.S. Army, U.S. Border Patrol, and other law enforcement agencies (Ross et al., 2010).

2. PURPOSE

The purpose of this research was to demonstrate a methodology to construct and verify an expert model of the Combat Hunter archetype. Expert Combat Hunters are individuals who consistently demonstrate an exceptional ability to interact with dynamically complex of patterns of life through enhanced observation, abductive reasoning and visual interpretation skills. When combined, these abilities work to improve predictions about the actions and intentions of the adversary and enable timely decision making and problem solving under stress. In the parlance of the expert Combat Hunter, this allows the warfighter to operate "left of bang". Based on these findings, the resulting Combat Hunter Expert Model can inform the design of training solutions and a program of instruction.

The expertise we sought to capture included individual strategic knowledge, lived experiences and best practices that consistently resulted in the Combat Hunter operating proactively (see Figure 1).

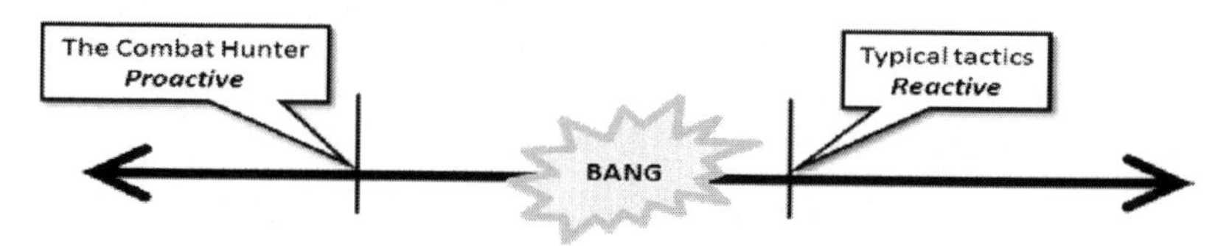

Figure 1. Taking actions that pre-empt the attacker by contacting, capturing or killing him before his attack can take place is the goal of the Combat Hunter.

3. METHODOLOGY - APPLIED COGNITIVE TASK ANALYSIS METHODS

Our research methodology was built upon several terms of reference adopted from our earlier research into decision making and expertise development (Klein and Hoffman, 1993; see Table 1). Specifically, for this project, we interviewed Combat Hunters who possessed two types of expertise: 1) Human Trackers, (N = 6) and 2) Combat Profilers, (N = 9). Each cohort member was highly experienced in law enforcement or military applications of their expertise. The participants were part of an instructional team that was assembled to provide classroom and practical lessons for the training audience. Each participant voluntarily participated on the condition of anonymity.

Table 1. Reference Terms for Combat Hunter Task Analysis

Reference Term	Definition
Mental Model	An individual resource that provides a schema for organizing knowledge, information and processes into patterns and relationships that are typical of an operational context and can be extended to specific situations
Decision Requirements Table	A tabular representation of decisions or dilemmas that serves as a tool for synthesizing the components of an individual's lived experience
Expert Model	A common set of best practices that consistently result in an enhanced ability to perform complex tasks efficiently and to adapt to novel or non-routine situations
Competency Model	A graphical representation of knowledge, skills, and abilities (KSAs) across jobs, functional areas, and job levels. It establishes direct line-of-sight between individual competency requirements and the broader organizational goals with a strategic- and future-orientation

The Human Trackers' backgrounds included military and law enforcement experience spanning decades. Trackers were accomplished at perceiving cues, pursuing quarry and interpreting evidence to determine number, rate of movement, direction, and intentions on all types of terrain, under any condition. The Trackers specialized in applying their observation skills and cognitive abilities to assume the mindset of the quarry and in converting their understanding into plausible narratives that could lead to a capture (Militello et al., 2011).

The Combat Profilers' backgrounds also included military and law enforcement experience spanning decades; however, they were significantly younger than the trackers in this study. Profilers were accomplished at perceiving baseline conditions and detecting anomalies within a pattern of life that indicated a requirement to decide or act. The Profilers specialized in applying their observation skills and cognitive abilities to detect anomalies, assess risks in complex, high risk situations and convert their understanding into a decision or actions that would defeat the threat or disrupt its decision making.

A two-part interview protocol was developed. Part one of the interview included Demographics, a Task Diagram (Militello and Hutton, 1998), and Critical Decision Method (Crandall et al., 2006). Part two of the interview was a team ranking activity. The same semi-structured interview format was used for all interviews.

Cognitive task analysis (CTA) tools (Hoffman and Militello, 2008) were used to conduct this research. For this project, two experienced qualitative researchers observed 18 days of classroom and field instruction. In addition, a suite of three complementary interview techniques was used including the Task Diagram, Concept Maps, and Critical Incident Method to interview course instructors.

Interviews were conducted individually. One to two interviewers were present for each interview, which lasted approximately two hours. Each interview was audio recorded and transcribed to facilitate qualitative data analysis. Permission to record the interviews was requested of each interviewee, and recorders were occasionally when requested. Interviewees were assured that all data would be treated as confidential.

Data were analyzed iteratively and individual narratives were transformed into Decision Requirements Tables (DRTs). Integrated concept maps were built based on themes found across the incidents and highlighted in the DRTs. Clustered decisions and dilemmas were noted and used to identify relevant mental models. The concept maps were also analyzed to define the macro-cognitive components and the underlying performance requirements. These requirements were assessed and compared to the results of the team ranking scales to produce a competency model that included knowledge, skills, abilities and aptitudes common to highly skilled Combat Hunters (Militello et al., 2011). Through these analyses, expert models for tracking and combat profiling expertise were represented (see Figure 2).

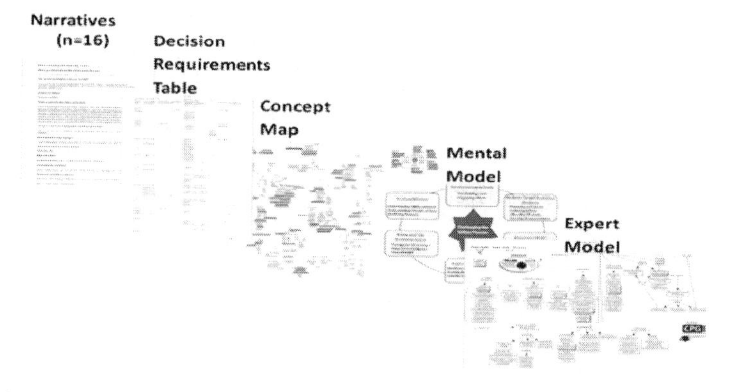

Figure 2. Iterative analyses of the narrative data provided a traceable means for constructing an Expert Model of the Combat Hunter.

The expert models are the building blocks for a Combat Hunter's decisionmaking. The expert models allowed us to align decisions, problem solving strategies and novice errors. Individual groupings became the learning cases for a program of instruction and lesson materials.

4. RESULTS – WHAT DID THE INTERVIEW DATA REVEAL

The principal elements of the daily activities of a Combat Hunter are illustrated below. The activities illustrate a surface level understanding of the domain and represent the application of strategic knowledge and problem solving strategies that distinguish expert performance. They do not, however, describe how a Combat Hunter thinks (see Figure 3).

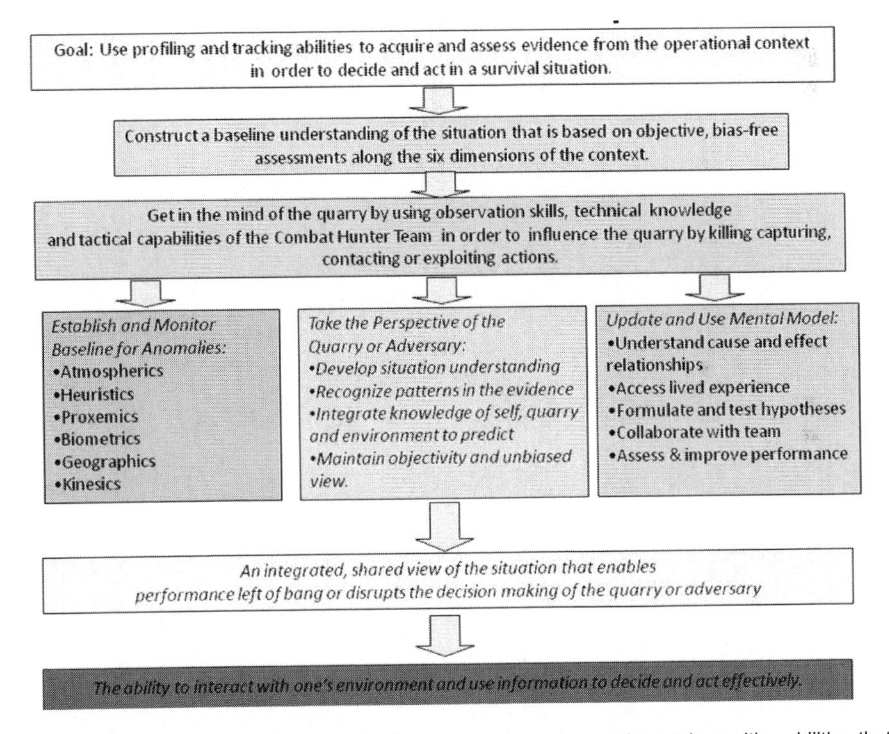

Figure 3. Combat Hunter skills and behaviors are based on perception and cognitive abilities that increase survivability and effectiveness.

Although activities define the behavioral model that is trained, they do not reveal the underlying mental models that are activated by highly experienced Combat Hunters to support decision making and problem solving. Consequently, the next step was to define the underlying mental models.

5. FRAMEWORK – BUILDING BLOCKS FOR EXPERTISE

The narrative data that were analyzed as part of the CTA described more than 200 years' cumulative experience. The experiences were rich in detail and provided a comprehensive understanding of performance and expertise. Highly effective Combat Hunters perceived the problem, adopted a mindset, and applied problem solving strategies to achieve success faster and more precisely than their less experienced counterparts and adversaries. This emerging framework highlighted the importance of understanding cultures and patterns of life in relation to a perceived baseline. The framework also suggested that these mental models are updated and applied in critical decision contexts. The framework for the Combat Hunter Expert Model is illustrated in Figure 4.

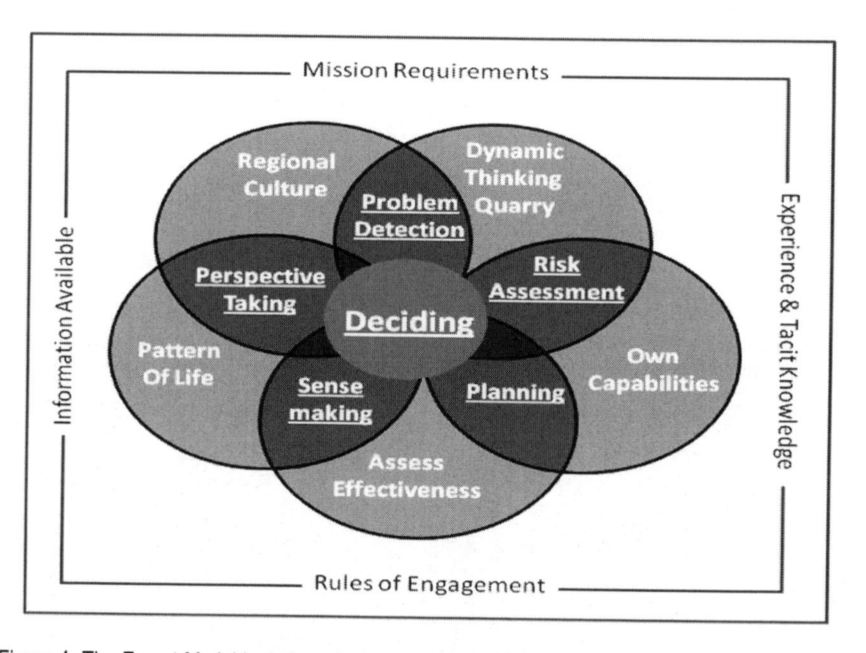

Figure 4. The Expert Model includes mission considerations, mental models and macro-cognitive abilities that describe the Combat Hunter practice.

The representation of the Combat Hunter Expert Model illustrates three inter-related elements of expert performance: 1) Mission Considerations, 2) Mental Models, and 3) Macro-cognitive Abilities.

Mission Considerations. We identified factors that influence how the Combat Hunter filters information or perceives the situation. These include a focus on the mission requirements and the rules of engagement that must be considered. He/she also weighs information that he/she senses and compares that to his/her lived experiences and tacit knowledge. These factors enable the Combat Hunter to frame the problem context and facilitate decisions and actions.

Mental Models. Combat Hunters are always vigilant and aware of their surroundings. Five mental models are accessed whenever a Combat Hunter is engaged with the operational environment: 1) Regional Culture, 2) Patterns of Life, 3) Dynamic Thinking Enemy, 4) Own Capabilities, and 5) Effectiveness. These mental models are the primary means that he/she uses to identify courses of actions and determines whether to decide or act. These models provide the strategic knowledge employed. The Combat Hunter's mental models mean that he/she,

- possesses a working knowledge of the Regional Culture. This mental model helps to explain political, military, economic and physical characteristics of the operating environment. An understanding of these elements provides a values-based context for explaining behaviors of large groups and factions.
- possesses an in-depth knowledge of the Pattern of Life. This mental model allows him/her to apply dimensions of human terrain analysis at the individual level. It also provides the ability to establish and maintain a baseline, from which anomalies can be detected and assessed.
- attempts to take on the mindset of the quarry. He/she knows the quarry's tactics, his/her weapons and his/her center of gravity. He/she understands a quarry's operating norms and uses that mental model to anticipate how the adversary or quarry will decide and act in response to his/her actions.
- is introspective and understands his/her own capabilities, when dealing with a specific threat or quarry. These insights include his/her model of how weapons and tactics are employed, as well as intrapersonal intelligence about his/her strengths, weaknesses and biases.
- combines information, knowledge and intelligence in a manner that informs his/her judgment about how well he/she is executing his plan and the need to adapt to change.

Macro-cognitive Abilities. The Combat Hunters we interviewed for this study described in the narratives an exceptional ability to make timely decisions about the quarry that proved accurate. These cognitive strengths distinguished this group's ability to leverage their mental models to solve problems especially when conditions of stress, high risk, uncertainty, and multi-tasking were present. Each of these abilities was enhanced by the quality of observation and communication that were used to make sense of ill-structured and novel problems. Six macro-cognitive abilities were routinely practiced: 1) Sensemaking 2) Problem Detection, 3) Perspective Taking, 4) Planning, 5) Risk Assessment, and 6) Decision Making (Klein et al., 2006).

- Sensemaking is a cognitive function performed by Combat Hunters to assess the atmospherics and determine the normal baseline for a problem context. This baseline is not fixed; it requires re-calibrating when mission considerations and pattern of life are considered and changed.
- Problem Detection is a continuous cognitive process used by Combat

Hunters to scan the baseline and look for anomalies within the pattern of life and regional culture. In the Combat Hunter's world anomaly detection and action equate to survival.

- Perspective-Taking is a cognitive function performed by Combat Hunters to perceive the problem and mission requirements from alternative perspectives. This enables the Combat Hunter to construct a personal mindset, the quarry and others so that his/her decisions and actions are framed to support proactive interactions.
- Planning includes real-time re-planning based on the current situation. This is a process for adapting to change and producing solutions that work. Planning involves mental simulation that facilitates the allocation and prioritization of resources to accomplish mission requirements. It also provides a context for anticipating 2nd and 3rd order consequences and identifying usable courses of actions.
- Risk Assessment involves a continuous re-assessment of potential threats as the mission changes based on actions of the quarry or as dictated by the pattern of operations. Risks are identified and categorized to inform judgment and actions. Risk management leverages the Combat Hunter's experience and strategic knowledge.
- Decisionmaking is supported by strategic knowledge, mental models and cognitive abilities. Combat Hunters select an option resulting in a moral, ethical and legal decision to contact, capture, or kill the quarry. Deciding also depends on the Combat Hunter's ability to mentally simulate futures and shape the situation to achieve that end-state.

6. IMPLICATIONS AND CONCLUSIONS

We identified and studied a small group of Combat Hunters who possess specialized skills for sensemaking, perspective taking, and problem solving that sets them apart from others within operational contexts. Our intent was to identify the strategic knowledge, skills and mental models they used to solve ill-structured, tactical problems. Perceptual skills and the ability to transform information into actions that enable tactical decision making under stress and time pressure were found to be critical markers of expertise.

We also were guided from the outset by an understanding that best practices can be captured and used to support decisionmaking and expertise development using CTA methods. These techniques for understanding expertise revealed that Human Tracking and Combat Profiling are complementary approaches for understanding the mindset of the quarry and proactively defeating him/her (see Figure 5).

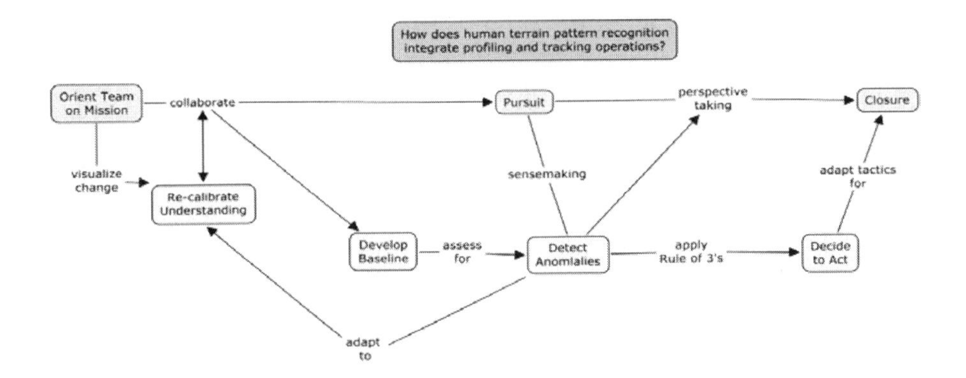

Figure 5. Linking tracking and profiling process stages merges the two disciplines.

Combat Hunter proficiency results in understanding the underlying explanations for patterns of life activities. Once the baseline has been calibrated, it can be adopted and used to make decisions and take actions. The can be updated over time based on a specific situation. Furthermore, the baseline provides a schema against which an informed observer can identify anomalies and make assessments about likely threats and next actions. When tracking and profiling are used in concert, they provide the Combat Hunter an advantage in the timing of actions and the accuracy of predictions. This combination is disruptive to the adversary and increases the Combat Hunter's ability to take actions left of bang – before the adversary has time to act.

Small units who learn to employ perceptual and cognitive skills have an improved ability to develop and share situational awareness. Shared awareness contributes to successful tactical performance. The ability to develop, maintain or update individual and team understanding of the adversary or quarry requires a combination of cognitive abilities, tactical skills and job knowledge. The end-state involves engaging with a particular adversary while managing risks and uncertainty.

However, Combat Hunter performance is not a universal solution to all problems. Knowledge of tactics, the ability to observe and relate information about a pattern of life across a team, the ability to adapt problem solving strategies based on perceptions, and the facility to act based on the perceptions are all enablers that contribute to mission success and survival.

Our approach for applying CTA methods to capture complex expertise, and then designing training from it, can be applied to other military training domains where efficient adaptability and resilience are the expected end-states.

ACKNOWLEDGMENTS

This work is supported in part by the Office of Naval Research. The views and conclusions contained in this document are those of the authors and should not be interpreted as representing the official policies, either expressed or implied, of the

Department of Defense or Office of Naval Research. The U.S. Government is authorized to reproduce and distribute reprints for Government purposes notwithstanding any copyright notation hereon. This work was supported, in part, by the Office of Naval Research project N00014-11-C-0193, Perceptual Training Systems and Tools (PercepTS).

REFERENCES

Crandall, B. W., Klein, G. & Hoffman, R. R. 2006. *Working minds: A practitioner's guide to cognitive task analysis,* Cambridge, MA, MIT Press.

Hoffman, R. & Militello, L. G. 2008. *Perspectives on cognitive task analysis: Historical origins and modern communities of practice,* New York, NY, Talor and Francis.

Klein, G., Moon, B. & Hofman, R. R. 2006. Making sense of sensemaking 2: A macrocognitive model. *Intelligent Systems, IEEE,* 21, 88-92.

Klein, G. A. & Hoffman, R. 1993. Seeing the invisible: Perceptual/cognitive aspects of expertise. *In:* Rabinowitz, M. (ed.) *Cognitive science foundations of instruction.* Mahwah, NJ: Lawrence Erlbaum & Associates.

Militello, L., Ross, W. A., Williams, G. & Scott-Donellan, D. Shifting Awareness Left of "Bang": Tracking and Profiling Expertise. 10th International Conference on Naturalistic Decision Making, 2011 Orlando, FL.

Militello, L. G. & Hutton, R. J. 1998. Applied cognitive task analysis (ACTA): a practitioner's toolkit for understanding cognitive task demands. *Ergonomics,* 41, 1618-41.

Ross, W. A., Militello, L. & Bencaz, N. 2010. Specification and Development of an Expert Model for "Combat Hunters". Orlando, FL: Cognitive Performance Group.

CHAPTER 3

Instructional Strategies for Cross-Cultural Perceptual Skills in Military Training

Katharine Lacefield, M.S., Ron Tarr, M.S., and Naomi Malone, M.S.

University of Central Florida
Orlando, Florida
Katharine.Lacefield@ucf.edu

ABSTRACT

Marines in combat operate in the context of dynamic environments where an acute awareness of sensory information is paramount for survival. Unlike many other skills that involve more concrete performance-based learning (e.g., properly firing a weapon), people often hold previously constructed conceptions, or mental models, about objects in the environment with which they interact. These models may make it difficult for people to understand new concepts without the inherent bias of their prior knowledge. Training perceptual skills creates a challenge unique to abstract knowledge acquisition for which appropriate instructional strategies may be particularly beneficial. Further complicating military training of perceptual skills are cultural dissimilarities in people and environments that Marines may encounter while in theater. Here, empirical and anecdotal data related to perceptual skills training for military populations, instructional strategies for perceptual training, and cross-cultural perceptual training are presented in discrete categories. To date, no published work combines research in these disciplines and avenues of training. The current paper aims to synthesize available research to determine optimal instructional strategies for perceptual training in cross-cultural domains designed for military use. Information regarding search methodology is provided. Additionally, recommendations are made to improve the current state of military perceptual training within a cross-cultural context.

Key words: instructional strategies, perceptual skills, cross-cultural, military, training strategies

1 INTRODUCTION

Marines operate in dangerous environments where an acute awareness of their surroundings and attention to sensory information are often critical for survival. Accordingly, perceptual training is a topic that should be an integral component of military training. However, unlike training for skills that involve more observable, performance-based learning (e.g., properly firing a weapon), the training of perceptual skills requires different learning strategies and approaches to instruction. Also, perceptual training is complicated by inherent cultural biases that can make it difficult for individuals to accurately perceive and assess new or foreign environments without understanding what is "normal" for that culture. In other words, extant research suggests that cultural characteristics may differentially affect perceptual skills in certain contexts (e.g., Ishii, 2011; Özkan, Lajunen, Chliaoutakis, Parker, & Summala, 2006). Therefore, instructional strategies that support the consideration of appropriate cultural nuances and how they affect perceptual skills may be necessary for successful learning. Training perceptual skills creates a unique challenge requiring instructional strategies for abstract knowledge acquisition that also account for cultural-sensitivity (of the context) and differential cultural backgrounds (of the observers). This complex approach to perceptual training is required to successfully prepare Marines for effective use of sensory information in dynamic, foreign operating environments.

Currently, discrete categories of empirical and anecdotal research exist that describe instructional strategies, perceptual skills training, and cross-cultural perceptual skills. However, no published work has combined the information in these categories to identify best practices for training perceptual skills while considering cultural differences and biases for military populations. In response, the current paper aims to provide an overview and synthesis of available research in each of these categories and to identify optimal instructional strategies to improve the current state of military perceptual training within a cross-cultural context.

2 OVERVIEW OF INSTRUCTIONAL STRATEGIES

Across educational domains, instructional strategies have been cited for their impact in augmenting learning (Gagné, 1997; Hirumi, 2002). Instructors who are able to effectively select and implement appropriate instructional strategies likely enhance their presentation of material and student learning (Gagné, 1997). Rather than using individual student learning style to improve knowledge delivery and acquisition, instructional strategies rely on four primary dimensions to optimize training (Vogel-Walcutt, Malone, Fiorella, 2011): (1) Knowledge type refers to the information being taught. Different knowledge types (e.g., declarative or procedural knowledge) benefit from different instructional strategies. For instance, whereas perceptual skills related to firing a weapon may be best taught using instrctional strategies geared and teaching basic skills, a different set of strategies likely are appropriate for a higher level perceptual skill set, such as accurately distinguishing combatants from allies (Gagné, 1997). (2) Expertise refers to the trainee's fund of knowledge prior to training. Levels of expertise range from novice to journeyman to expert, each with at least some variation in which instructional strategies best facilitate learning (Hoffman, 1996). (3) Group size refers to how trainees are grouped, ranging from individuals to large groups. As group size increases, various strategies using group dynamics become available for instructional use (Dick & Carey, 1996). (4) Environment refers to the setting, ranging from a classroom to a simulation, in which the information is being disseminated. Optimal strategy selection may vary based on whether the information or skill being taught is "live" or simulated (Dick & Carey, 1996).

With respect to military training of perceptual skills in cross-cultural contexts, instructional strategies may be implemented according to the four dimensions presented by Dick and Carey's (1996) model. One advantage of integrating instructional strategies into training is that, unlike the theory of culturally influenced "learning styles" that has been debunked in education literature (Jensen, 2000), instructional strategies have been robust in enhancing learning, regardless of a trainee's cultural background (Gagné, 1997; Hirumi, 2002). To date, there has been variation among researchers in their preferred mode of training perceptual skills and the degree to which instructional strategies have been integrated therein.

3 OVERVIEW OF PERCEPTUAL SKILLS TRAINING

Slight variations exist in how perception is currently defined. Each definition, however, suggests that sensory information is gleaned from a stimulus in the environment that can be used meaningfully by the perceiver (Grunwald, 2008; Moore, 1970; Raftopoulos, 2009). Perception initially is contingent upon the senses detecting information but relies on an additional process of cognitively processing the sensation. To date, a variety of research examining military training and perceptual skills has focused on training decision skills (Klein, 1997), situation awareness (Endsley, Robertson, & Porter, 1999), critical thinking skills (Freeman & Cohen, 1997a, 1997b; Cohen, Freeman, & Thompson, 1998), and stress-based training (Driskell et al., 2001).

Decision skills training (Klein, 1997) aims to facilitate the decision maker's domain expertise. Rather than attempt to compare possible decision outcomes in the short amount of time typically available to military personnel, decision making training aims to "size up" a situation with limited information to choose an appropriate course of action (e.g., Klein et al., 1986). Qualitative study results indicated that participants, who were comprised of noncommissioned military officers, would implement the strategies in their training regimens.

In an attempt to augment Klein et al.'s (1986) findings related to decision making, Freeman and Cohen (1997a, 1997b) and Cohen, Freeman, and Thompson (1998) examined appropriate instructional strategies in decision-making tasks. Training was implemented to enhance officers' critical thinking, skills to improve decision-making ability. Pre- and post-testing were conducted around training to help officers determine whether an approaching entity was a friend or enemy. Assessment was made regarding whether officers felt ready to act (i.e., had made a decision) or had determined that more information or time was necessary for them to make a final determination. Training occurred for 90-minutes in a classroom with a paper-based (Study 1) format or for 2-hour sessions with a PC-based simulator environment across two weeks (Study 2). Expert officers were compared with Navy post-graduates. In Study 1, post-testing occurred at the end of a long day of training, but changes in critical thinking skills approached significance. In Study 2, training had a significant effect across aspects of critical thinking, indicating that appropriate training can enhance critical thinking skills or metacognition, which can positively impact decision-making skills.

In addition to training decision making skills, Endsley, Robertson, and Porter (1999) trained 72 expert instructors with experience in aviation-related jobs across five domains: (1) improving trainers' ability to accurately perceive a given situation, (2) increasing the amount of feedback given, (3) teamwork improvement, (4) increasing verbal communication among team members during decision making, and (5) decreasing effect of distraction (e.g., noise) on situation awareness. Various instructional strategies, including formal instruction and group exercises, were used to meet the five stated goals. Subjective ratings indicated improvements in all domains addressed.

Driskell, Johnston, and Salas (2001) explored the impact of training on performance under stress. Training focused on three dimensions: (1) proving information to prepare the skills training, (2) acquiring and practicing new skills through rehearsal, and (3) practicing the new skills under simulated stress conditions to enhance their generalizability in combat. The study used a modified event-based approach to training by using one situation-specific stressor in a variety of tasks. Researchers examined the effect that training in conditions with one specific stressor would have on training in environments with novel stressors. To test this, researchers divided U.S. Navy technical school trainees into two groups who were exposed to a specific stressor (e.g., time pressure) and then to a novel stressor (e.g., auditory distraction). The group with training demonstrated a small reduction in stress when moving between stressors. Given the stressful nature of combat, even stress reduction consistent with a small effect size seems worthwhile if effective training is available.

Despite its inherent utility, little perceptual training currently is offered to most enlisted personnel. In the Marine Corps, for instance, only three enlisted infantry programs (i.e., marksmanship, IED defeat, and Combat Hunter training) explicitly integrate aspects of perceptual instruction. These programs include perceptual training in a nonintegrated, partial manner and without the aid of tailored instructional strategies (Schatz et al., 2010). Thus, our broad research goals are to better define a continuum of perceptual skills—specifically, those skills directly or indirectly influenced by culture—identify best practices and instructional strategies for engendering those competencies, and suggest instructional methodologies to support their instruction in military contexts. Consequently, research on cross-cultural perceptual skills needs to be explored in order to identify potential instructional strategies that can address cross-cultural issues.

4 REVIEW OF CROSS-CULTURAL PERCEPTUAL SKILLS

Current research suggests that cultural characteristics may affect perceptual skill acquisition (e.g., Ishii, 2011; Jeng, Costilow, Stangherlin, & Lin, 2011; and Özkan, Lajunen, Chliaoutakis, Parker, & Summala, 2006). For example, although the faces of other-race individuals typically are more slowly processed and more poorly remembered, training procedures have been developed that diminish differences between processing of other-race individuals (e.g., Chance, Lockwood-Turner, & Goldstein, 1982; Hughes, 2002). Considering that a significant number of American warfighters are in theater with culturally dissimilar individuals, augmented training in cross-cultural perceptual skills may be especially beneficial. Additionally, given that approximately 25 percent of the United States military identifies as belonging to a group other than Caucasian (Kane, 2005), understanding cultural nuances that may impact training strategies for perceptual skills may be crucial to increase the likelihood of the best outcome for all military personnel.

Some researchers have suggested that military personnel share a culture based on "the use of specific language, symbols, and hierarchy" therein (Greene, Buckman, Dandeker, & Greenberg, 2010, p. 958). Beyond these commonalities, additional cultural influences may affect warfighters in various ways: American warfighters interacting with the culture of the occupied nation, cultural conflict when American military personnel are of similar ethnic cultural background to the enemy, and ethnic cultural conflict within military culture. Each of these scenarios speaks to the need for an understanding of the role of culture in perceptual skills training for military personnel. For instance, cultural understanding between American warfighters in theater and the local civilians of that nation is necessary to "win hearts and minds" in combat settings to decrease hostility among non-combatant local civilians (McFate, 2005). When U.S. troops first entered Iraq, it is reported that they had little understanding of various aspects of Iraqi culture. In one situation, U.S. military personnel initiated a road-side check point in which they used the Western signal for "stop" (and outstretched arm, palm out) which was interpreted to mean "go" by Iraqi locals unaware of its meaning (McFate, 2005).

Overall, limited literature specific to cross-cultural perceptual skills training exists. A PsycINFO search was conducted using the primary keywords: *cross-cultural* and *perceptual skills*. Results were limited only to studies that: (1) focused on adult human subjects, (2) were written in the English language, and (3) were published between 2001 and 2011. This initial search yielded 36 results, of which 27 were discarded because they did not meet the research criteria for inclusion: 17 were discarded for irrelevance to the topic of interest (i.e., studies were related to "perceptions of" a cultural phenomenon rather than perceptual skills), eight were discarded for studying children and adolescents, and two were discarded for their focus on measure validation unrelated to perceptual skills. This left a total of eight articles specifically focused on perceptual skills in a cross-cultural context, and these tended to be comparative studies between two or more cultural groups. An additional two articles were found using the "snowball" method of finding relevant articles cited in the literature found in the original search (Elfenbein, 2006; Nixon &

Bull, 2005). Collectively, a review of recent literature examining perceptual skills in a cross-cultural context reveals a dearth of information pertaining to training differences in culturally dissimilar Marines as well as in culturally novel environments. Table 1 provides a summary of findings related to perceptual skills and perception, along with a description of how the study may be relevant to military personnel.

Table 1 Summary of Cross-Cultural Perceptual Skills Literature

Citation	Findings	Military Relevance
Özkan, Lajunen, Chliaoutakis, Parker, & Summala (2006)	Negative relationship between safety skill and number of driving accidents found for subjects from Greece and Iran, but positive relationship between perceptual skills and number of accidents found in Iran	Considerations may be made in driver perceptual training for military personnel of dissimilar cultural backgrounds
Ishii (2011)	Japanese and American participants showed mere exposure effect to animals in different backgrounds	Mere exposure paradigm can be considered for Marines in novel environments
Jeng, Costilow, Stangherlin, & Lin (2011)	Findings indicate that neural responses associated with individual harmonics dominate the pitch processing in the human brainstem, irrespective of whether the listener's native language is nontonal or tonal.	Training in tonal quality and verbal communication of the combatant's native language may be helpful for enemy recognition
Pascarella, Palmer, Moye, & Pierson (2001)	Students' involvement in diversity experiences during college had statistically significant positive effects on their scores on an objective, standardized measure of critical thinking skills.	Skills to promote cultural competency should be included in general military training
Hughes (2002)	The accuracy of 241 undergraduates in psychology in identifying a true coin from among a selection of similar fakes was low (41.9%), in line with findings from several other countries.	To increase facial recognition, military personnel should engage in continued exposure to warfighters and civilians while in theater
DeGutis, DeNicola, Zink, McGlinchey, & Milberg (2011)	Suggests that configural training with own-race faces may improve general configural/holistic attentional resources, which may help with identifying other-race faces	Facial recognition training may help U.S. troops to better identify and differentiate combatant and civilian faces to more accurately assess threat
Sheard (2009)	Scores on mental toughness questionnaire positively associated with successful sport performance.	Training exercises to increase mental toughness may help optimize combat performance

Spence, Yu, Feng, & Marshman (2009).	Training methods that develop basic spatial skills may be essential to achieve gender parity in both basic and complex spatial tasks.	If pre-assessment suggests disparities between male and female spatial skills, gender-specific training may be helpful
Elfenbein (2006)	Emotion perception skill in a cross-cultural context can be improved through training that includes diagnostic feedback that includes information additional to correct response.	Training in accurately perceiving emotions of people from other cultures may defuse potentially dangerous situations.
Nixon & Bull (2005)	Increasing cultural awareness in nonverbal communication styles leads to improvement in within - and cross-cultural perceptual accuracy. This study used a modified Interpersonal Perception Task presenting a variety of videotaped segments of people interacting in normal situations.	For military personnel to accurately perceive cross-cultural communications, they must be able to distinguish and attend to the cross-cultural cues that are important within the context of their specific environments.

5 RECOMMENDATIONS FOR MILITARY POPULATIONS

As our current findings demonstrate, a small number of studies have focused on perceptual skills in culturally diverse environments and situations. Conceivably, with regard to cross-cultural environments, situation awareness skills training has the potential to benefit most from added instruction regarding effectively reading and responding to cues specific to foreign surroundings. However, critical thinking and decision making can also benefit from perceptual skills training, to a lesser degree. Based on the existing research in perceptual skills training (and in particular, cross-cultural perceptual skills training), four instructional strategy recommendations are identified here that could have significant positive impact on training these skills in cross-cultural contexts.

5a Recommendation #1

To increase critical thinking perceptual skills in culturally dissimilar U.S. warfighters, diversity awareness training should be implemented prior to deployment. Instructional strategies for training diversity awareness can be implemented based on group size, knowledge level, and instructional format. Several specific instructional strategies were identified that may be useful to implement here. For example, increased feedback—an instructional strategy that can be effective across scenarios—was identified as one useful strategy for training (Endsley, Robertson, & Porter, 1999). Additionally, formal instruction, presentations, group exercises, discussion, active learning, inquiry, and discovery learning all were named as effective strategies in yielding significant training gains. In this case, the training conditions consisted of a large group of experts in a classroom with a live instructor to optimize training. Similar strategies employed in military classrooms have the potential to allow both experts and novices to learn more about cultural aspects of the civilians and combatants in the region to which they are to be deployed.

5b Recommendation #2

Research efforts in facial recognition by Hughes (2002) and DeGutis, DeNicola, Zink, McGlinchey, and Milberg (2011) suggest the potential for improvement in this perceptual skill with training. In combat, warfighters are tasked with making quick decisions based on the status (friend or foe) of people with whom they come into contact. Reactions can be further complicated when U.S. troops share ethnic or racial similarities with the combatant. Hughes' (2002) study provides evidence that the training used to help participants to identify novel stimuli other than faces (e.g., a coin) may generalize to improved facial recognition skills. It is recommended that, in an effort to reduce the potential for friendly-fire and other issues related to inaccurate facial recognition and decision making, instructional strategies based on the aforementioned dimensions are used to develop training for facial recognition skills for all military personnel deployed to foreign territory.

5c Recommendation #3

Before military personnel can develop cross-cultural perceptual skills necessary to accurately decode cues in their environment, they must know which cross-cultural cues require their attention. Instructional strategies that stimulate trainees' perceptions of cultural interaction scenes with no verbal explanations, followed by feedback on the correct response can significantly improve their cross-cultural perceptual accuracy (Nixon & Bull, 2009). Improvements in this domain may help to improve social cues and understanding of communication. One tool that can be modified for this purpose is the Interpersonal Perception Task (IPT: Constanzo & Archer, 1989), a strategy that presents short, videotaped segments of people interacting in their natural environment. After these segments, the participants normally respond to a series of questions, after which they are given the correct responses to those questions. This tool could also help warfighters improve their competencies in cultural understanding of the societal norms for the area in which they are in combat; thus, they may be better equipped to more effectively detect when something is incongruent with norms and potentially dangerous (e.g., a Middle Eastern woman traveling alone in a section of town that women do not usually go, or a Muslim woman who continues to make eye contact with men).

5d Recommendation # 4

As the US military transitions into a more global rather than regional focus (Dufour, 2007), recognizing the differences in the expression of emotion in diverse cultures becomes increasingly important. Further, an emphasis on "winning the peace" makes it important for military personnel to be able to accurately judge emotions to defuse potentially dangerous situations (Blaskovich & Hartel, 2008). Research suggests that practice alone can improve emotion recognition (Costanzo, 1992; deTurck, Harszlak, Bodhorn, & Texter, 1990; Grinspan, Hemphill, & Nowicki, 2003; Zuckerman, Koestner,& Alton, 1984). However there is evidence that adding diagnostic feedback to practice can further improve the accuracy of emotion recognition (Beck & Feldman, 1992; Grinspan, et al.). Additionally, a study specific to cultural emotional cues indicates that outcome feedback (in this case, providing information beyond the correct response) can significantly improve the accuracy of emotion recognition (Elfenbein, 2006).

6 Conclusions

As this review shows, there is a need for further empirical research to identify effective training strategies for cross-cultural perceptual skills that can be useful in military contexts. Current training applications (e.g. Tactical Iraqi, which includes only 30 gestures and no other perceptual skills) do not provide significant training in important perceptual skills such as facial and emotion recognition, body posture, gestures, eye contact, and oral (non-language) cues. To this end, the Perceptual Training Systems and Tools (PercepTS) team aims to synthesize "patterns of life" that may serve as the foundation for developing perceptual skills across cultural groups (see Schatz and Nicholson chapter, "Perceptual Training for Cross Cultural Decision Making (Session Overview)," for future goals and Wray's chapter, "Tailoring Culturally Situated Simulation for Perceptual Training," for further evidence supporting "patterns of life"). Identifying global, human commonalities that can be more uniquely determined based on specific combat locations may be a crucial step to making perceptual skills training across cultural contexts more easily generalized.

Additionally, although the current chapter elicited a dearth of literature on perceptual skills training in cross-cultural contexts, the four recommendations in the previous section point to the usefulness of two general strategies: First, practice-based, rather than lecture-based training immerses trainees in realistic situations, allowing them to train in a variety of different contexts. Second, trainees benefit most from training that is followed by feedback, preferably including additional diagnostic information rather than merely giving them the correct response.

ACKNOWLEDGEMENTS

This work is supported in part by the Office of Naval Research. The views and conclusions contained in this document are those of the authors and should not be interpreted as representing the official policies, either expressed or implied, of the Department of Defense or Office of Naval Research. The U.S. Government is authorized to reproduce and distribute reprints for Government purposes notwithstanding any copyright notation hereon. This work was supported, in part, by the Office of Naval Research project N00014-11-C-0193, Perceptual Training Systems and Tools (PercepTS).

REFERENCES

Abrami, P. C., Lou, Y., Chambers, B., Poulsen, C., & Spence, J. (2000). Why should we group students within-class for learning? *Educational Research and Evaluation, 6*(2), 158–179.

Beck, L., & Feldman, R. S. (1989). Enhancing children's decoding of facial expression. Journal of Nonverbal Behavior, 13, 269–278.

Blascovich, J. & Hartel, C.R. (2008). *Human behavior in military contexts*. Washington, D.C.: National Academies Press.

Chance, J. E., Lockwood-Turner, A, & Goldstein, A. G. (1982). Development of differential recognition for own- and other-race faces. *The Journal of Psychology: Interdisciplinary and Applied, 112*, 29-37.

Cohen, M. S., Freeman, J. T., & Thompson, B. (1998). Critical thinking skills in tactical decisionmaking: A model and a training strategy. In J. Canon-Bowers & E. Salas (Eds.), *Decision making under stress: Implications for training and simulation* (pp. 155–190). Washington, DC: APA.

Costanzo, M. (1992). Training students to decode verbal and nonverbal cues: Effects on confidence and performance. *Journal of Educational Psychology, 84*, 308–313.

DeGutis, J., DeNicola, C., Zink, T., McGlinchey, R., & Milberg, W. (2011). Training with own-race faces can improve processing of other-race faces: Evidence from developmental prosopagnosia. *Neuropsychologia, 49*(9), 2505-2513.

deTrurck, M. A., Harszlak, J. J., Bodhorn, D. J., & Texter, L. A. (1990). The effects of training social perceivers to detect deception from behavioral cues. *Communication Quarterly, 38*,189–199.

Dick, W., & Carey, L. (1996). *The systematic design of instruction* (4th ed.). New York, NY: Harper Collin.

Dreyfus, H. L., Dreyfus, S. E., & Athanasiou, T. (1986). *Mind over machine: The power of human intuition and expertise in the era of the computer*. New York: Free Press.

Driskell, J. E., Johnston, J. H., & Salas, E. (2001). Does stress training generalize to novel settings. *Human Factors, 43*, 99–110.

Dufour, J. (2007). The global deployment of US military personnel. *Global Research.* Retrieved from www.globalresearch.ca/PrintArticle.php?articleId=5564.

Elfenbein, H. (2006). Learning in emotion judgments: Training and the cross-cultural understanding of facial expressions. *Journal of Nonverbal Behavior, 30*(1), 21-36.

Endsley, M. R., Robertson, M. M., & Porter, K. (1999). Evaluation of a program for training skills related to team situation awareness in aviation maintenance. In *Proceedings of the 10th International Symposium on Aviation Psychology*. Columbus, OH: Ohio State University.

Freeman, J. T., & Cohen, M. S. (1997a). *Training for complex decision making: A test of Instruction based on the recognition/metacognition model* (Tech. Rep.). Naval AirWarfare Center Training Systems Division. Orlando, FL.

Freeman, J. T. & Cohen, M. S. (1997b). Training for complex decision-making: A test of Instruction based on the recognition/metacognition model. In *Proceedings of the 3rd International Command and Control Research and Technology Symposium 1996.* Monterey, CA.

Gagné, R., Briggs, L., & Wager, W. W. (1988). *Principles of instructional design* (3rd ed.). New York, NY: Holt, Rinehart and Winston, Inc.

Gagné, R. M. (1997). Mastery learning and instructional design. *Performance Improvement Quarterly*, *10*(1), 8-19. doi: 10.1111/j.1937-8327.1997.tb00027.x

Gagné, R. M., Briggs, L. J. (1979). *Principles of instructional design* (2nd ed.). New York, NY: Holt, Rinehart, & Winston.

Graham, C. R., Allen, S., & Ure, D. (2005). Benefits and challenges of blended learning environments. In M. Khosrow-Pour (Ed.), *Encyclopedia of information science and technology* (pp. 253–259). Hershey, PA: Idea Group Inc.

Greene, T., Buckman, J., Dandeker, C., & Greenberg, N. (2010). The impact of culture clash on deployed troops. *Military Medicine*, *175*(12), 958-963.

Grinspan, D., Hemphill, A., & Nowicki, S. (2003). Improving the ability of elementary schoolage children to identify emotion in facial expression. *Journal of Genetic Psychology*, *164*, 88–100.

Grunwald, M. (Cognition and Perception Ed.), (2008). *Human Haptic Perception.* Basel/Boston/Berlin: Birkhauser Verlag.

Hirumi, A. (2002). The design and sequencing of e-learning interactions: A grounded approach. *International Journal on E-Learning, 1*(1), 19-27.

Hoffman, D. L., & Novak, T. P. (1996). Marketing in hypermedia computer-mediated environments: Conceptual foundations. *Journal of Marketing*, *60*(3), 50-68. doi:10.2307/1251841

Hughes, B. M. (2002). Misremembering the appearance of common objects: Further cross-cultural confirmation. *Perceptual and Motor Skills*, *95*(3,Pt2), 1255-1258.

Ishii, K. (2011). Changes in background impair fluency-triggered positive affect: A cross-cultural test using a mere-exposure paradigm. *Perceptual and Motor Skills, 112*(2), 393- 400.

Jeng, F., Costilow, C. E., Stangherlin, D. P., & Lin, C. (2011). Relative power of harmonics in human frequency-following responses associated with voice pitch in American and Chinese adults. *Perceptual and Motor Skills*, *113*(1), 67-86.

Kane, T. (2005). Who bears the burden? Demographic characteristics of U.S. military recruits before and after 9/11. *Center for Data Analysis Report, 5-8.*

Klein, G. A., & Hoffman, R. R. (1992). Seeing the invisible: Perceptual-cognitive aspects of expertise. In M. Rabinowitz (Ed.), *Cognitive science foundations of instruction* (pp.203-226). Mahwah, NJ: Erlbaum.

Klein, G. A., Calderwood, R., & Clinton-Cirocco, A. (1986). Rapid decision making on the fireground. *Proceedings of the Human Factors and Ergonomics Society 30th Annual Meeting, 1*, 576–580.

Klein, G. A. (1997). Developing expertise and decision making. *Thinking and Reasoning, 3*, 337–352.

McFate, M. (2005). The military utility of understanding adversary culture. *Joint Forces Quarterly, 38*, 42-48.

Moore, M. R. (1970). The perceptual-motor domain and a proposed taxonomy of perception. *AV Communication Review 18*(4), 379-413.

Nixon, Y. & Bull, P. (2005). The effects of cultural awareness on nonverbal perceptual accuracy: British and Japanese training programmes. *Journal of Intercultural Communication, 9*, page numbers.

Özkan, T., Lajunen, T., Chliaoutakis, J. l., Parker, D., & Summala, H. (2006). Cross-cultural differences in driving skills: A comparison of six countries. *Accident Analysis and Prevention, 38*(5), 1011-1018.

Pascarella, E. T., Palmer, B., Moye, M., & Pierson, C. T. (2001). Do diversity experiences influence the development of critical thinking?. *Journal of College Student Development, 42*(3), 257-271.

Raftopoulos, A. (2009). *Cognition and perception.* Cambridge: MIT Press.

Schatz, S., Nicholson, S., Nicholson, D., Fautua, D. T., & Reitz, E. A. (2010). *CODIAC Program Guide.*

Sheard, M. (2009). A cross-national analysis of mental toughness and hardiness in elite university rugby league teams. *Perceptual and Motor Skills, 109*(1), 213-223.

Spence, I., Yu, J., Feng, J., & Marshman, J. (2009). Women match men when learning a spatial skill. *Journal of Experimental Psychology: Learning, Memory, and Cognition, 35*(4), 1097-1103. doi:10.1037/a0015641

Vogel-Walcutt, J.J., Malone, N., & Fiorella, L. (2011). Instruction Specific Model for Strategy Selection, AERA, Paper under review.

Zuckerman, M., Koestner, R., & Alton, A. O. (1984). Learning to detect deception. *Journal of Personality and Social Psychology, 46*, 519–528.

CHAPTER 4

Bridging the Gap between Humans and Systems: An HSI Case Study of Perceptual Training Systems and Tools

Denise Nicholson, Sae Schatz, Kathleen Bartlett

MESH Solutions, LLC – a DSCI Company
Orlando, FL
dnicholson@dsci.com, sschatz@dsci.com, kbartlett@dsci.com

ABSTRACT

Our team is currently addressing a range of perceptual–cognitive training challenges, including the development of effective, efficient, and sustainable Science and Technology (S&T) products that enhance Marine Corps perceptual instruction. Through theoretical inquiry, empirical investigation, and knowledge elicitation from key stakeholders associated with the Navy and Marine Corps, we have identified a set of deliverables that range from low-technology knowledge products to sophisticated culturally situated training environments, and we are actively developing these products through a Human–Systems Integration (HSI) approach. This approach includes the use of appropriate systems engineering processes, iteratively generating requirements, developing frequent "shared representation" artifacts that communicate assumptions, and threading an impact assessment plan throughout the development process. In this paper, we share lessons learned for research-based HSI development, in general, and discuss seven real-world HSI procedures used by the Perceptual Training Systems and Tools (PercepTS) project.

Keywords: Human–Systems Integration (HSI), system design, lessons learned, requirements development, perceptual training

1 HUMAN-CENTRIC RESEARCH AND DESIGN

The Human–Systems Integration (HSI) literature is ripe with examples of systems failure. Too often, developers focus on solving technology challenges and neglect human-centric issues, sometimes with devastating consequences. In 1981, the General Accounting Office reported that 50% of all military equipment failures were caused by human error; this followed a US Army report that found that many military human errors could be traced back to poor development processes that failed to sufficiently consider human performance concerns (Kerwin, Blanchard, Atzinger, and Topper, 1980). Driven by these revelations, the US military branches each began comprehensive HSI reforms in the early 1980s.

Today, HSI is a formalized development process, and an integrated US Department of Defense (DoD) HSI policy outlines recommendations for its employment (FY09 DoD HSI Management Plan). Research and development efforts that use HSI methods early and frequently throughout the acquisition process have higher rates of success, lower rates of downstream failure, and produce more cost-effective solutions across the system's life-cycle (Pew and Mavor, 2007). Additionally, HSI is not only "good business," the Office of the Secretary of Defense formally requires its use for US DoD projects (FY09 DoD HSI Management Plan).

As more Human, Social, Culture, and Behavior (HSCB) researchers develop cross-cultural trainers, social-science decision support systems, and other anthropologically informed technologies, we felt it timely to offer a primer on HSI for the Cross Cultural Decision Making (CCDM) conference community. This paper offers an introductory overview of HSI. It also presents a brief case study of HSI employment, using our current Perceptual Training Systems and Tools (PercepTS) project as an example, and broad-spectrum lessons learned for research-based HSI development efforts are described throughout.

1.1 Introduction to Human–Systems Integration

HSI is a philosophy and set of processes that focus on systems-level human performance concerns throughout research, development, and acquisition. Its purpose is to mitigate the risk of downstream system failure. More formally, the DoD Handbook on human engineering process offers this definition:

> HSI is a comprehensive management and technical strategy to ensure that human performance, the burden design imposes on manpower, personnel, and training (MPT), and safety and health aspects are considered throughout system design and development. HSI assists with the total system approach by focusing attention on the human part of the total system equation and by ensuring that human-related considerations are integrated into the system acquisition process (MIL-HDBK-46855A, Section 5.1.2, 1999, p. 19).

Good HSI practice promotes several core principles (see Table 1 for a summary). First, HSI continuously emphasizes human performance concerns during the research, design, and development stages of the system acquisition process. In other words, "the human in acquisition programs is given equal treatment to hardware and software" (FY09 DoD HSI Management Plan, p. 3).

Table 1 Core HSI Tenets

1. Emphasize human performance issues early in the design process
2. Emphasize system-level outcomes (across the HSI domains)
3. Focus on life-cycle (not just immediate) costs/benefits
4. Realistically facilitate multidisciplinary design processes

Second, HSI attempts to optimize overall system-level performance. This means that researchers and developers evaluate outcomes, such as costs and benefits, at the comprehensive system level and not simply at the individual component levels. That is, HSI practitioners strive to achieve the greatest *total* efficiency and effectiveness for a whole system, within given resources; this may mean that individual components, by themselves, cannot fully achieve optimization. HSI practitioners recognize several domains of interest across which they attempt to optimize, including manpower, personnel, training, safety and occupational health, human factors engineering, habitability, and survivability (see Table 2).

Table 2 Summary of HSI Domains (FY09 DoD HSI Management Plan)*

Manpower	People necessary to operate, maintain, and support a system
Personnel	Knowledge, skills, and attitudes the system's operators, maintainers, and support personnel must possess
Training	Training required to support a system's operators, maintainers, and other support personnel
Safety/Health	Considerations for acute or chronic illnesses, disabilities, or death for system operators, maintainers, and other support personnel
Human Factors	Considerations for potential injury or errors caused by the physical, cognitive, or affective aspects of human–machine interfaces
Habitability	Physical conditions, living conditions, and, if appropriate, personnel services (e.g., mess hall) affecting system performance
Survivability	For systems that might encounter physical threats, the likelihood of injury for systems' operators, maintainers, and support personnel

* *While various agencies formally recognize slightly different HSI categories, these categories provided by the DoD's* Human Systems Integration Management Plan *(2009) are commonly used.*

Third, HSI practitioners take a long view of their systems' performance, attempting to maximize benefits—while controlling costs and mitigating risks—

across a system's entire life-cycle. Many different life-cycle models exist, but in general, they all include initial development phases, followed by implementation, operation, and eventual retirement stages. Bahill and Dean (2009) surveyed a variety of systems engineers who, by consensus, offer these seven typical system life-cycle phases: (1) Discovering system requirements; (2) Investigating alternatives; (3) Full-scale engineering design; (4) Implementation; (5) Integration and testing; (6) Operation, maintenance and evaluation; (7) Retirement, disposal and replacement.

Finally, in order to achieve the major goals thus far described in this section, HSI practitioners must facilitate a multidisciplinary design process. This often entails "translation" between specialists in different disciplines (e.g., between sociologists and computer scientists) and the negotiation of requirements among interested parties (e.g., brokering compromises between training specialists and manpower analysts). In practice, this means that HSI practitioners spend considerable time eliciting inputs from various stakeholders, documenting assumptions, clarifying friction points, and developing "shared representations" that transform these requirements and analyses into meaningful, unambiguous formats. Some examples of these shared representations are listed in Table 3.

Table 3 Example Shared Representations to Facilitate HSI Development

Personas	Scenarios	Network Diagrams
Architectural Diagrams	Storyboards	Work Domain Breakdowns
Abstraction Hierarchies	Concept Maps	Process Diagrams
Interface Mock-ups	UML Diagrams	Prototypes

1.2 HSI Risk Mitigation

The ultimate purpose of HSI is to mitigate the risk of system failure. Systems can fail for a wide variety of reasons. Heuristically, these reasons fall into two large categories: technological issues (e.g., flaws in the system design) and logistical issues (e.g., budgetary and schedule issues). Failure may occur during any phase of the system life-cycle. For example, a project may be cancelled during the engineering design phase because of cost and schedule overruns. Or a system may be terminated during the operation phase because it suffers from high operating costs and becomes too expensive to maintain. According Pew and Mavor (2007), three of the most common causes of system failure include:

- Underutilization or disuse due to difficult, inefficient, or dangerous design
- Serious compromises in system performance from human error
- High operational and maintenance (O&M) costs and challenges

By adhering to the four key tenets of HSI philosophy, discussed in the previous section, system designers can reduce the likelihood of these kinds of system failures. In addition to following sound HSI philosophy, practitioners can use a variety of well-defined HSI processes to further mitigate the risk of life-cycle system failure. These HSI processes include, for example:

- Formally identifying goals and writing requirements
- Architecting solutions to meet the requirements
- Conducting cost–benefit risk mitigation analyses
- Performing "system scoping" to maximize return-on-investment (ROI)
- Actively facilitating multidisciplinary design, "translating" between teams

1.3 Areas of Study Related to HSI

HSI is related to several other disciplines. Most notably, HSI closely resembles systems engineering, and formally, an effective systems engineering process should already include HSI processes. In real-world practice, however, systems engineers often inadvertently neglect human concerns. That is, "there has been a continuing concern that, in each phase of development, the human element is not sufficiently considered along with hardware and software elements" (Pew and Mavor, 2007, p. 9). It was for this reason that HSI developed into a unique specialty area.

Similarly, from a definitional perspective, "HSI is synonymous with the traditional definition of *human factors* in its broadest sense" (Malone, Savage-Knepshield, and Avery, 2007). Again, in real-world practice, researchers tend to conceptualize human factors and ergonomics (HF/E) from a more limited perspective. Generally, practitioners define HF/E narrowly, emphasizing immediate operator issues and neglecting broader life-cycle considerations, the range of applicable domains, and system-level emergent properties.

Ultimately, HSI fills practical gaps in both systems engineering and HF/E. It draws upon systems engineering perspectives to underscore broad-spectrum system concerns and life-cycle management issues, and it pulls methodologies and best practices from HF/E, in order to give emphasis to the human in the system.

3 PERCEPTS HSI USE CASE

The Schatz and Nicholson chapter in this volume, entitled "Perceptual Training for Cross Cultural Decision Making (Session Overview)" offers a detailed description of the Perceptual Training Systems and Tools (PercepTS) project. PercepTS is a Science and Technology (S&T) research initiative, sponsored by the US Office of Naval Research. Its purpose is to deliver training methodologies and technologies that will enhance military personnel's perceptual skills, including their abilities to detect, interpret, synthetize, and anticipate complex behavioral cues at the micro-, meso-, and macro-cultural levels.

The PercepTS team is employing a development approach that emphasizes the four HSI tenets described above and uses a range of HSI processes, such as iteratively generating requirements, coordinating closely with stakeholders, and threading an impact assessment plan into the development process. As of the writing of this paper, the PercepTS team has completed about seventeen months' of effort and plans to complete roughly four additional years' of research and development. The following sections briefly describe some of the HSI activities the PercepTS team completed in the first year of the project. This information is intended to provide an example of real-world HSI use as well as offer lessons learned from a personal (i.e., subjective) perspective.

3.1 Baseline the Status Quo

Much of the work during the project's first year involved "baselining," i.e., theoretical investigations of the current academic body of knowledge and empirical data collection of existing relevant systems. These analyses helped determine the status quo in military perceptual training, articulated the project's theoretical foundations, and informed future system requirements. Naturally, reviewing status quo gaps and opportunities before attempting to address them is commonsense. The PercepTS program went beyond the obvious rudimentary approach, however, to conduct extensive baseline inquiries, including a large-scale literature review, development of a research-based taxonomy of perpetual skills, and detailed empirical analyses of the most relevant existing military perceptual training.

These baselining efforts were conducted *in lieu of* initial product development in the project's first year. This may seem inefficient; however, we believe that the product development process will be more successful—and ultimately more economical—as a result of the significant baselining investment. Already, the in-depth, first-hand knowledge gained from baselining has had considerable benefits, greatly influencing the resulting system design, facilitating enhanced interactions with stakeholders, and accelerating the development of shared mental models within the PercepTS team. Further, in future years, the quantitative and qualitative data collected during the baselining phase will be compared against outcome data from the PercepTS system, enabling detailed assessment of the prototype's capabilities and demonstrating its (anticipated) gains over the status quo.

Lesson-Learned: Investment in first-hand theoretical and empirical baselining pays off. When time and resources allow, include comprehensive baselining as part of an effective HSI research and development approach.

3.2 Collect a Wide Range of Stakeholder Opinions

While not all opinions are accurate or relevant, collection of a wide range of diverse inputs can help scope the contextual problem space, illuminate stakeholders' subcultures, uncover potential system design flaws, and reveal additional targets-of-opportunity for system requirements. For the PercepTS project, we have actively sought reactions and recommendations from a wide variety of stakeholders, end-users, and experts. We have solicited input from individuals at different levels of expertise (i.e., not just experts but novices, as well), with different degrees of influence, holding different military ranks, and across a variety of occupational fields. We have talked with individuals from the primary and secondary end-user populations, and solicited opinions from decision-makers and other power-brokers who may interact with the system. We have also sought open-ended advice from subject matter experts in the operational and academic communities. When feasible, we try to bring shared representations (e.g., design drawings or prototypes) to these discussions. Most people can better articulate their thoughts if they can respond to some artifact. As we collect new responses, we integrate these recommendations into future iterations of the artifact, as appropriate.

Lesson-Learned: Be creative in identifying stakeholders; consider diverse groups of individuals who may have an interest in each of the HSI domain areas, including end users, decision makers, system operators, maintenance personnel, and trainers. Make sure to "speak the right language;" that is, have enough first-hand knowledge of the problem space (e.g., through baselining) to effectively communicate at a deep and detailed level with the stakeholders. When feasible, bring shared representations to which the stakeholders can react.

3.3 Explicitly Align Solutions with Gaps/Opportunities

As a result of the baselining and knowledge elicitation efforts, our team developed greater insight into the problem space. This allowed us to articulate much more targeted recommended solutions. Great care was taken to explicitly articulate how the gaps and opportunities align with each specific recommendation, and how these recommendations connect to our proposed solutions. For the PercepTS effort, 10 unique gaps/opportunities were ultimately identified from the baselining phase. Based upon that research, along with stakeholders' input and considerations for the constraints of the problem space, we devised short-, mid-, and long-term courses of action to address each gap/opportunity. This yielded a table with thirty cells. For each cell, we estimated the level of effort (i.e., cost) and prospective gains (i.e., benefits), which allowed our team, as well as other decision-makers, to readily identify the approaches with the highest ROI and to begin a phased approach to addressing the array of recommendations.

Lesson-Learned: Create clear, easy-to-read descriptions that align specific gaps and opportunities to proposed solutions. Include rough estimates of costs and benefits for each proposed approach. Clearly articulate the rationale for these decisions. Consider creating brief tables, executive summaries, or other succinct artifacts of this content to which stakeholders can rapidly respond.

3.4 Dream Big, Then Scope Down

Early in the design process (i.e., immediately following completion of the baselining data collection, approximately one year into the project), the PercepTS team began forming a vision for the ultimate system end-state. This "big dream" of the final deliverable cannot be realistically fielded; however, the brainstorming exercise proved valuable. The process strengthened team members' shared mental models, and we were able to discuss system goals from a purely technical perspective, without worrying about constraints. Importantly, though, the team's effort did not end once the ultimate vision was developed. Although unfettered brainstorming stimulated our creativity and helped focus discussions on technical issues, the eventual system must function under real-world constraints. Hence, after the team articulated a detailed "perfect version" of the final system, we adjusted this vision to meet real-world practicalities, such as development schedules, acquisition budgets, manpower and personnel limitations, and throughput realities. Issues involving each of the HSI domains, across the proposed system's breadth and eventual life-cycle, were considered. This two-pronged approach encouraged a great deal of creativity while also ensuring that realistic constraints were considered.

Lesson-Learned: When initially developing a design concept, permit unfettered creativity. Consider what the "perfect" version of the system might look like, without real-world constraints. However, once this vision is solidified, then review it with a pragmatic eye, and re-scope it as necessary.

3.5 Build Many Shared Representations

One of the most important pieces of advice is to develop "shared representations" frequently, iteratively, and in many different forms. As described above, a shared representation is a tangible artifact that helps explicitly reveal the designers' assumptions in a succinct, clearly communicated, unambiguous manner. For the PercepTS project, we have already developed several of these artifacts and used them to facilitate intra-team communication, to elicit knowledge and feedback from stakeholders, and as mechanisms to encourage creative design. Some of these artifacts, such as the concept art for the final system (see Schatz and Nicholson chapter in this volume) have been more enduring, while other shared representations have been quickly created and then quickly discarded as part of an iterative design process. Useful PercepTS shared representations have included the baselining recommendations table (described in 3.3), summary overview reports, one-page project overviews, concept art of the final products, use-case scenarios describing end-users interactions with deliverables in a narrative form, storyboards of training plans, and conceptual architectural diagrams of software solutions.

Lesson-Learned: Develop artifacts to explain design assumptions; create these shared representations frequently. Create many different forms, since each type will reveal unique properties. Use these artifacts to solicit feedback from teammates and stakeholders.

3.6 Devise an Iterative Development Plan

The PercepTS HSI development effort employs an iterative, spiral process model that includes early and frequent verification and validation testing, concurrent development of requirements and prototypes, ongoing risk-mitigation evaluation, and iterative phases with built-in decision-gates, milestones, and anchor points. Plainly stated, now that the baselining effort is complete, the PercepTS team is iteratively defining system requirements, while simultaneously creating partial prototypes and assessing the quality of previously developed prototypes through a variety of testing and evaluation methods.

Lesson-Learned: Do not try to make a perfect design or product the first time. Work iteratively, initially developing many low-cost design prototypes, and solicit extensive feedback from all possible stakeholders. Continue to collect feedback and, if appropriate, adjust the project plan throughout the entire development process.

3.7 Plan for Ongoing Test and Evaluation

Ongoing testing of prototypes is critical for successful system development. Frequently, developers save the bulk of system assessment until the final phases of the project, and, too often, only a small set of criteria are assessed. In our experience, however, regular testing and evaluation across a broad set of factors best supports system development.

As described above, the PercepTS project initially collected qualitative and quantitative data from the existing status quo. These data provide a comparison unit against which future iterations of the PercepTS system can be measured. Formal annual evaluations of PercepTS prototypes are planned. These evaluations will involve empirical data collection with end-user populations (e.g., enlisted Marines) in realistic settings (i.e., not a laboratory). Semi-regular evaluations are also scheduled, as necessary. For instance, laboratory experiments have been conducted, and will continue to be executed in the future, as empirical questions present themselves. In addition to ad-hoc experimental trials, formal verification and usability testing are planned as needed. Finally, informal evaluations are also conducted regularly, as opportunity allows. These evaluations may include user testing of prototypes or the solicitation of structured (e.g., survey) or unstructured (e.g., comments) feedback from stakeholders.

During each of these assessment opportunities, our team plans to collect data on one or more categories of relevant criteria: learning effectiveness, learning efficiency, reactions, and technology capabilities. Each of Kirkpatrick's (1967) levels of training assessment relate to "learning effectiveness." Learning efficiency includes metrics such as the amount of instructors required to facilitate a training event. Reactions include subjective feedback, and technology issues encompass the range of hardware/software factors, including usability and technical performance.

Lesson-Learned: Define comprehensive categories of assessment criteria from the beginning, and evaluate prototypes against these criteria frequently, using both formal and informal methods. Ensure that the assessment criteria relate to relevant life-cycle factors and that they comprehensively address the HSI domains, as appropriate.

4 CONCLUSION

Good HSI practice advises practitioners to consider human performance issues early in the design process, explore system-level outcomes, focus on life-cycle costs and benefits, and realistically facilitate multidisciplinary design processes. The academic literature lists a variety of procedures that can help meet these goals. For instance, formal requirement generation and documentation processes can be readily found, structured system design approaches exist, and numerous of cost–benefit algorithms can be uncovered. However, it is important to consider the practicability of the HSI approaches and to tailor them to each project's specific needs.

This paper provides a personal account of some HSI processes used to aid development of a US Marine Corps perceptual trainer. The team is conducting other HSI actions, in addition to those listed here; for example, we are currently completing the first iteration of formal requirements authoring, a technology trade-off evaluation, and an instructional system design analysis. The seven items listed above were included for their unique lessons-learned and likely relevancy to other HSCB-related research and development initiatives.

ACKNOWLEDGMENTS

This work is supported in part by the Office of Naval Research. The views and conclusions contained in this document are those of the authors and should not be interpreted as representing the official policies, either expressed or implied, of the Department of Defense or Office of Naval Research. The U.S. Government is authorized to reproduce and distribute reprints for Government purposes notwithstanding any copyright notation hereon. This work was supported, in part, by the Office of Naval Research project N00014-11-C-0193, Perceptual Training Systems and Tools (PercepTS).

REFERENCES

Bahill, T. A. and F. F. Dean. 2009. What Is Systems Engineering? A Consensus of Senior Systems Engineers. Accessed September 9, 2009, http://www.sie.arizona.edu/sysengr/whatis/whatis.html

Department of Defense. 1996. *Human Engineering Program Process and Procedures* (MIL-HDBK-46855A). Washington, D.C.: Government Printing Office

Kerwin, W. T., Blanchard, G. S., Atzinger, E. M., & Topper, P. E. (1980). *Man/machine interface – a growing crisis* (DTIC Accession No. ADB071081). Aberdeen Proving Ground, MD: U.S. Army Material Systems Analysis Activity.

Kirkpatrick, D. L. 1967. Evaluation of training. In *Training and Development Handbook*, eds. R. Craig and I. Mittel. New York: McGraw Hill.

Malone, T., P. Savage-Knepshield, and L. Avery. 2007. Human-Systems Integration: Human factors in a systems context. *Human Factors and Ergonomics Society Bulletin* 50: 1–3.

Pew, R. W., A. S. Mavor, A. S., and the Committee on Human-System Design Support for Changing Technology. 2007. *Human-system integration in the system development process: a new look.* Washington, D.C.: National Academies Press.

Office of the Secretary of Defense. 2009. *FY09 Department of Defense Human Systems Integration Management Plan,* Version 1.0. Washington, DC: ODUSD(A&T), ODUSD(S&T) Director of Biological Systems.

CHAPTER 5

Tailoring Culturally-Situated Simulation for Perceptual Training

Robert E. Wray, III

Soar Technology, Inc.
Ann Arbor, MI 48015
wray@soartech.com

ABSTRACT

Simulation allows learners to practice skills simulating physical and even interpersonal interactions. Such practice is critically important for ill-defined domains such as cross-cultural observation and interaction. Pedagogical experience manipulation (Lane and Johnson, 2008) is an emerging complement to traditional educational technologies. It is applied during practice to scaffold, to challenge, and to engage a learner using primarily intrinsic (within the simulation) events and outcome biases. Experience manipulation, in effect, is designed to amplify the learning signal within practice environments. We describe an implemented software architecture that can tailor practice on-the-fly to reinforce pedagogical goals. It employs probabilistic estimates of learner proficiency, agent-based situation interpretation and decision-making, and constraint-based expert models. We outline the architectural components of dynamic tailoring and existing tailoring strategies and discuss how we are reusing and extending the tailoring capability for a new application focused on cross-cultural perceptual training.

Keywords: training, pedagogical experience manipulation, dynamic tailoring

1 INTRODUCTION

Individualized, simulation-based training has the potential to revolutionize adult learning. Simulation technology can allow learners to practice skills simulating physical and even interpersonal interactions. Such practice is critically important for domains such as cross-cultural observation and interaction, in which few fixed or definite "rules" govern acceptable or appropriate interpretation of behavior. Such ill-defined domains also present challenges for traditional educational technologies, such as precision in diagnosing student errors, which then makes extrinsic or explicit tutoring of the learner subject to sometimes distracting and extraneous interruptions.

Pedagogical experience manipulation (Lane and Johnson, 2008) is an emerging complement to traditional educational technologies. Experience manipulation applied within practice is used to scaffold, to challenge, and to engage a learner using intrinsic (within the simulation) events and outcome biases. Experience manipulation, in effect, is designed to amplify the learning signal in practice environments, customizing learning content and reducing extraneous "noise" for each individual learner. Experience manipulation also does not typically require precise diagnosis of individual student deficits. Intrinsic manipulations, especially those represented via virtual human actors, are less evident to the learner and typically less distracting when inappropriate (Lane and Wray, 2012).

This paper describes a software architecture that can modify practice on-the-fly to deliver pedagogical experience manipulation, reinforcing the pedagogical content of practice. This dynamic tailoring capability includes: 1) estimating learner proficiency using uncertainty models, 2) agent-based situation interpretation and decision-making, and 3) expert modeling. The primary goal of the architecture is to provide a general-purpose tool for delivering tailored experience in simulation-based practice environments. Below we outline current progress and limitations with respect to this goal. A secondary goal, only briefly discussed herein, is to provide a tool for evaluation and comparison of alternative tailoring strategies (Wray et al., 2009). We describe several tailoring strategies used for a conversational, intercultural learning application including biasing of character affect, conversational elaboration of pedagogical content, and interruptions from turn-taking in conversation.

A newer application of the architecture is addressing perceptual training. Rather than a face-to-face interaction, in this domain learners will be observing culturally-situated behavior from a distance. Their goal is to learn to recognize typical and specific "patterns of life" and to be able to classify the tactical relevance/importance of deviations from typical patterns. We discuss how the existing tailoring strategies from the architecture can be refined and adapted for this domain and offer suggestions for new tailoring strategies afforded by more remote interactions. This analysis suggests that significant reuse of the dynamic tailoring capability is feasible, supporting the ambition and potential value of a general architecture to support dynamic tailoring for simulation-based practice.

2 THE DYNAMIC TAILORING ARCHITECTURE

A general-purpose software system typically distinguishes an architecture, those core functions provided to support the operation of the system, and the content with which the architecture acts to produce some result. An architecture is designed to require minimal changes from one application to the next, requiring primarily changes in content as the architecture is re-applied to other application domains.

Our ambition in developing a dynamic tailoring architecture is to build a general-purpose software architecture for tailoring that can be used in a variety of simulation-based training systems and domains. As described below, we have created a functional implementation of this architecture and have demonstrated it in a few domains. At this stage in its development, the architecture might be more aptly characterized as a framework because, as we explore initial applications of the architecture and identify new requirements for it, the architecture itself is being expanded to include additional functionality. As an example, the Pedagogical Manager component evolved out of the observed need to encapsulate the consideration of the implications of the instructional context ("What are appropriate pedagogical strategies given the learner's current situation and progress?") from interpretation of the student's actions (Monitor) and the particular ways in which the system can act to change simulated experience (Experience Manager).

Another important architectural goal is the representation of application- and domain-specific content as (declarative) data. This goal is important for two reasons: 1) it ensures content is readily separable from one application of the architecture to the next and 2) it gives users of the architecture the ability to define and to encode content data without having to understand the specific details of the implementation of the architecture. This approach also facilitates creation of content authoring tools. Long-term, such tools will shift the responsibility of content development to training-content developers. As a consequence of this goal, the implementation reflects a choice to prefer simpler context-representation schemes that enable a much wider audience for content authoring.

Figure 1 illustrates the current architecture. There are three core functional components (boxes) and four primary representational components (database icons). The representational components define the form in which application- and domain-specific content must be represented. The following subsections outline the role of each functional and representational component.

2.1 Functional Components of Dynamic Tailoring

Monitor: The Monitor observes learner actions, interprets those actions in the context of the learning situation (using the expert model), assesses the learner's behavior in terms of active learning objectives, and then classifies the observed behavior using the behavior ontology. The Monitor, Pedagogical Manager, and Experience Manager are all implemented using Soar as an agent architecture (Wray and Jones, 2005).

Pedagogical Manager: The Pedagogical Manager's role is to maintain an estimate of proficiency for each learning objective, to mediate between extrinsic stimuli (such as explicit instruction and feedback provided by an ITS) and the intrinsic adaptations of tailoring, and to choose between alternative instructional strategies. For example, a "challenge" tailoring strategy can be used to when a learner has demonstrated high levels of competence, but it may also be triggered as "preparation for future learning" (Bransford and Schwartz, 1999). Longer-term, the Pedagogical Manager will also mediate choices between affective and domain-content tailoring strategies.

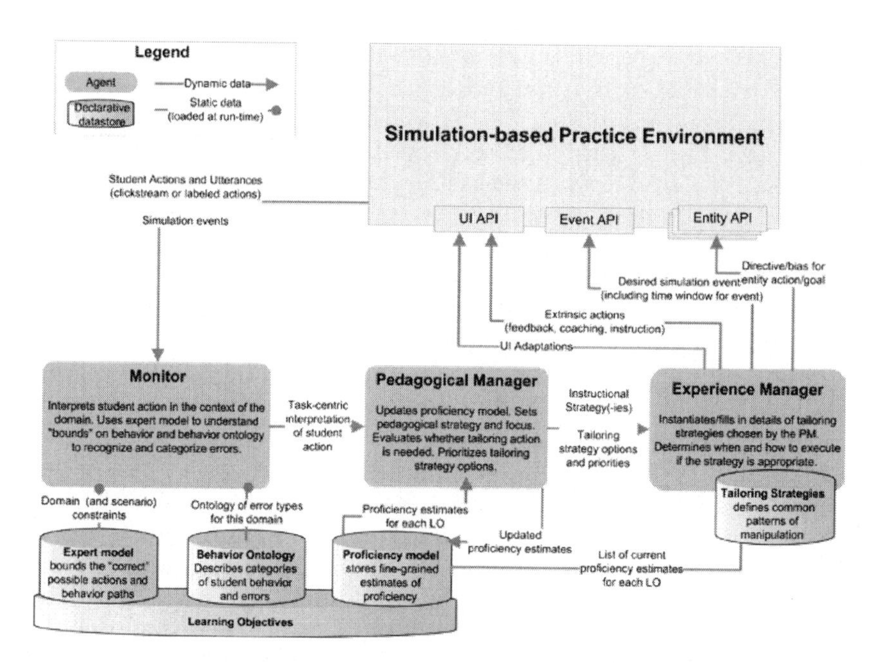

Figure 1 Overview of the Dynamic Tailoring Architecture.

Experience Manager: The Experience Manager build on design and data structures from Magerko's (2007) Interactive Drama Architecture, which tailors user experience for games and learning. Our architecture extends the tailoring strategies available. The Experience Manager chooses and instantiates tailoring strategies based on general recommendations from the Pedagogical Manager. For example, the Pedagogical Manager may recommend a tailoring strategy that is intrinsic, meant to scaffold, and focuses on one or more enumerated learning objectives. The Experience Manager then evaluates tailoring strategy options to determine which strategies can be used to satisfy the request.

2.2 Representational Components of Dynamic Tailoring

Expert Model: The expert model describes behavioral constraints and preferences. It is defined in a declarative model representation loosely derived from constraint-based tutoring (Mitrovic and Ohlsson, 1999). The Monitor then uses this description to detect conflicts between learner actions and specified constraints.

Behavior Ontology: The expert model allows the Monitor to recognize behavioral constraint violations, but does not, by itself, enable diagnosis of the cause of the error. Fine-grained diagnosis of the learner's action, as in model-tracing expert models (e.g., Anderson et al., 1995) is often costly and difficult for users to author. As a middle ground, we are developing a behavioral ontology that is used to identify the type of error that a constraint-violation represents. For example, some specific constraint can be violated for different reasons (e.g., wrong step order vs. incomplete application context) and the choice and instantiation of the tailoring strategy may be able to tailor learner feedback according to the error type, avoiding focus on just the error itself, which is a poor feedback strategy (Shute, 2008). We are today using a bottom-up catalog of error types but, long-term, we will be incorporating more complete and empirically-grounded catalogs of learner behavior (e.g., Van Lehn, 1990). Applying formal semantics for the behavior ontology will enable consistent and reusable error categorization across domains.

Proficiency Model: The ability to tailor experience to modulate the learning situation is predicted on an estimate of the learner's skill and knowledge. The architecture's proficiency model is derived from the classic work on SHERLOCK (Katz et al., 1998). For each learning objective, the learner's proficiency is modeled via a "fuzzy variable" with four states (limited, partial, non-automated, automated) which map loosely to progressive learning/mastery of a concept/skill. After each learner action and interpretation by the Monitor, the Pedagogical Manager updates the fuzzy variables, which serve as nominal estimates of proficiency for the applicable learning objectives. This approach provides an operational semantics of the fuzzy variable which tailoring then uses to evaluate possible strategy choices.

Learning Objectives: The expert model, behavior ontology, and proficiency model all are organized by a representation of the knowledge, skills, and abilities for which the learner must demonstrate proficiency. In the current implementation, we use a simple, hierarchical (parent-child, with multiple inheritance) representation of the learning objectives. As the behavior ontology matures, we expect the learning objectives representation will also adopt a richer semantic representation, enabling the functional components to reason about the implications of student behavior and tailoring options based on the type and properties of specific learning objectives.

3 TAILORING STRATEGIES

A tailoring strategy is defined by its purpose(s) and its effect(s). The dynamic tailoring system can be used for distinct purposes: scaffolding/fading, challenging/assessing, engaging (as in adjusting the level of challenge to maintain interest, Vygotsky, 1978), and individualization. Its tailoring strategies can act in four different categories: 1) outcome manipulation (modifying or modulating the effects of actions, 2) character direction (suggesting character reactions, utterances, gestures, etc.), 3) simulation/event manipulation: (introducing or suppressing events that the native simulation would not have produced any pre-emptive event introduced to support learning), 4) Gameplay manipulation (modifying how the simulation is experienced). Examples of gameplay manipulation include *choice manipulation* (modifications of the actions available to the leaner) and *narrative manipulation* (blocking routes of exploration until experiential objectives are met).

We use the term "strategy" to describe tailoring mechanisms because a single tailoring strategy can be used to support multiple purposes, may seek to produce effects of different and multiple types, and may choose alternate effects based on the specific learning objectives identified as relevant by the Pedagogical Manager. Most of the current tailoring strategies focus on a specific type of effect and can be used for different purposes. Table 1 summarizes these strategies.

The strategies in Table 1 are primarily reactive (taken in response to an observed learner action) but plan-based tailoring strategies can likely be readily integrated as well. Similarly, because the tailoring system was originally developed in conjunction with an Intelligent Tutoring System, we have focused primarily on intrinsic tailoring of the simulation experience. However, especially for delivery of simple feedback, we are extending the Experience Manager to be able to deliver timely extrinsic feedback in addition to intrinsic experience manipulation.

Table 1 Examples of Implemented Tailoring Strategies

Name (effect)	Description
Tailored selection (Outcome manipulation)	Many simulation systems employ probabilistic models to parameterize the outcomes of events. Tailored selection influences the outcomes probabilistic selection. For example, when a novice learner is learning to recognize cause-and-effect relationships, it may be useful to have the tailoring system bias selection toward an obvious presentation of the effect.
Character response tailoring (Character)	The tailoring system can choose or bias the choice of character utterances. We focus on biasing on the consistency of a character's response to a learner action and the pedagogical content of the response.
Pedagogical elaboration (Character)	Pedagogical elaboration appends character utterances with additional information relative to current learning objective(s). The goal of this scaffolding strategy is to provide feedback akin to extrinsic mediation, but delivered intrinsically, via the characters, rather than via an ITS.
Character affect tailoring (Character)	Tailoring can bias the "mood" of a character and its emotional response to a situation. For example, biasing affect can provide a simple way to make gestures more or less subtle without needing direct control of individual gestures.
Initiative taking (Event/ Character)	In a turn-based conversation, it is possible to surprise a learner by having the character make a statement "out of turn" when the learner is inattentive. The specific utterances of the character are chosen based on the affective goal and the learner's level of proficiency in the current stage of the meeting. A more general form is to introduce some surprising event, which could also be environmental (a ringing phone). This strategy is predicated on being able to estimate attention.
Action-choice highlighting (Gameplay)	This is a scaffolding and fading strategy. The tailoring system highlights options with the intent of making it easier for the novice learner to recognize actions that are appropriate for the current situation.

These tailoring strategies are implemented today as rules in the Experience Manager. Via Soar knowledge encapsulations (problem spaces and operators), the rules associated with specific tailoring strategies are separable from one another and reusable from one application to another. However, in this representation, they are not readily authorable by non-programmers. We are investigating the use of declarative schemas, as demonstrated in other applications (Taylor and Sims, 2009), to expose strategies for authoring. These schemas are designed to be authorable and are then imported into Soar where domain-general agent knowledge is used to interpret external events and act based on the descriptions in the schemas.

4 TAILORING STRATEGIES FOR PERCEPTUAL TRAINING

We are currently investigating a novel application of the dynamic tailoring architecture to the challenge of perceptual training. As outlined in the Schatz and Nicholson chapter in this volume, we are focusing specifically on the domain of combat profiling and an immersive practice environment, the "Virtual Ville," that will allow practice of combat profiling skills. Combat profiling is a process by which an observer develops a baseline of normal activity, the "pattern of life" for a particular location, recognizes and categories anomalies from the baseline, and then evaluates those anomalies to generate potential actions. The Virtual Ville will be an immersive, whole-task practice environment that helps trainees developing combat profiling skills from a remote observation location, such as an observation post established outside a village or town.

The process of learning combat profiling interacts with cross-cultural interpretation and decision making and these factors must be integrated in the design of the Virtual Ville. For example, practice will necessarily be embedded within a specific cultural context (specific in the sense that there are some cultural patterns; the culture giving rise to the pattern can be artificial or constructed). As a consequence, it is important that learning scenario within the Virtual Ville provide cultural contexts that can be manipulated to emphasize principles and skills independent of a specific culture.

We plan to use dynamic tailoring to provide more varied and learner-targeted experiences in this domain. Based on observations of live training and interactions with other researchers working on the Virtual Ville, we have identified three classes of desirable pedagogical manipulations that a tailoring system should support:

1. **Scenario Complexity**: A primary independent variable is the pedagogical complexity of the scenario. Complexity can include the physical size of the virtual village as well as the number of virtual characters in the village, the observability of key events in the scenario (non-observable, indirectly observable, partially observable, and fully observable), the number of key events that have to be observed to culminate in a prediction, and the number of distracter events. Rather than a fixed level of complexity defined at scenario start, tailoring offers the ability to match complexity to the observed capabilities of the trainee team.

Table 2 Mapping Desired Manipulations to Existing Tailoring Strategies

Name (effect)	Pedagogical Category	Example(s)
Tailored selection (Outcome manipulation)	Signal-to-noise (group) Signal-to-noise (population)	• Bias number of phone calls contemporaneous with HVI call • "Benign" POL disruption ("holiday")
Response tailoring (Character)	Complexity Signal-to-noise Instructional Intervention	• Reaction/disruptions of POL as a response to an insurgent action ("clear the market") • Use virtual team to call attention to or distract attention from key events • Use virtual team to encourage greater reporting frequency and reporting detail
Ped Elaboration (Character)	Instructional intervention	• Use virtual team to prompt use of domain language
Character affect tailoring (Character)	Signal-to-noise (individual)	• Subtlety/explicitness of kinesic and proxemic cues of HVIs • Subtlety/explicitness of non-HVI kinesic cues ("suspicious behavior") • Subtlety/explicitness of distracter actor's kinesic cues
Event triggers (Event/ Character)	Signal-to-noise Scenario Complexity	• Events used to trigger or distract attention (chickens and goats) • Events that result in reducing or increasing local population density ("clear market")
Narrative manipulations (Character/event)	Scenario Complexity	• Choose specific location of HVI rendezvous (observable, partially observable, etc.) • Narrative permutations (size of insurgent group)
Action-choice highlighting (Gameplay)	Signal-to-noise	• Deliver highlights or pointers to key areas and key actors?

2. **Signal-to-Noise**: Another key manipulation in this environment is the relative "overtness" of behavioral cues. For example, the gestures and stances of an individual (kinesics) can be varied from being highly exaggerated (a clearly nervous or agitated state) to highly subtle (little physical indications of internal affect). Tailoring will likely require modification of signal-to-noise of behavioral cues at an individual (micro), group (meso), and population. For example, at the population level, tailoring could be used to increase or decrease adherence of individuals to specific

patterns of life active for a scenario. Varying signal-to-noise at the population level will both add nuance to developing baseline patterns and provide baseline anomalies that should be ignored.

3. **Instructional Interventions**: Tailoring can deliver learner feedback and instructional interventions more intrinsically. One component of the Virtual Ville will be a "virtual observation post" that is "manned" by virtual characters (and the instructor). Pedagogical roles of the virtual OP will be eliciting observations from trainees ("What's the white truck doing?"), encouraging use of domain concepts ("Is it kinesics that leads you to that conclusion?"), and delivering feedback ("Are you watching your sector? – What's X doing?"). An important but secondary role of the virtual OP will be fill observational gaps that allow the trainees to proceed successfully in the scenario. Tailoring can also be used to enable more explicit and extrinsic cues. For example, a team that was continually missing key events might be presented with highlighting or other indicators within the rendered environment to help bring their attention to learning-relevant cues.

Table 2 outlines how specific instances of the desired manipulations for the Virtual Ville map to the tailoring strategies introduced in Table 1. The potential reuse of the existing tailoring strategies in a new and reasonably different domain suggests the potential feasibility and value of using a common, reusable software infrastructure for dynamic tailoring. Although mapping the desired manipulations into the existing strategies does require design and implementation effort, that effort should be markedly reduced in comparison to the effort that would have been required to develop a specialized tailoring system for the Virtual Ville.

5 CONCLUSIONS

This paper has presented an implemented dynamic tailoring architecture, an outline of the tailoring strategies afforded in the current implementation, and a preliminary analysis of the feasibility of the reuse and refinement of the existing tailoring strategies and architecture to support a new application, an immersive "Virtual Ville" developed to support practice and training of combat profiling. One of the primary advantages of a dynamic tailoring architecture, in comparison to a customized tailoring solution for this domain is this reuse of the software architecture, which should offer significant return on investment over a solution developed solely for use in a single domain.

ACKNOWLEDGMENTS

This work is supported in part by the Office of Naval Research. The views and conclusions contained in this document are those of the authors and should not be interpreted as representing the official policies, either expressed or implied, of the Department of Defense or Office of Naval Research. The U.S. Government is authorized to reproduce and distribute reprints for Government purposes notwithstanding any copyright notation hereon. This work was supported, in part, by the Office of Naval Research project N00014-11-C-0193, Perceptual Training Systems and Tools (PercepTS). We would also like to thank Sae Schatz and J.T. Folsom-Kovarik for collaborative discussions that contributed to the ideas in this paper and H Chad Lane and Brian Stensrud for earlier contributions to the development of the dynamic tailoring architecture.

REFERENCES

ANDERSON, J. A., CORBETT, A. T., KOEDINGER, K. & PELLETIER, R. 1995. Cognitive Tutors: Lessons Learned. *The Journal of the Learning Sciences,* 4, 167-207.

BRANSFORD, J. D. & SCHWARTZ, D. L. 1999. Rethinking Transfer: A Simple Proposal With Multiple Implications. *In:* IRAN-NEJAD, A. & PEARSON., P. D. (eds.) *Review of Research in Education.* American Educational Research Association: Washington.

KATZ, S., LESGOLD, A., HUGHES, E., PETERS, D., EGGAN, G., GORDIN, M. & GREENBERG, L. 1998. Sherlock 2: An intelligent tutoring system built on the LRDC framework. *In:* BLOOM, C. P. & LOFTIN, R. B. (eds.) *Facilitating the development and use of interactive learning environments.* Mahwah, NJ: Erlbaum.

LANE, H. C. & JOHNSON, W. L. 2008. Intelligent Tutoring and Pedagogical Experience Manipulation in Virtual Learning Environments. *In:* COHN, J., NICHOLSON, D. & SCHMORROW, D. (eds.) *The PSI Handbook of Virtual Environments for Training and Education.* Westport, CT: Praeger Security International.

LANE, H. C. & WRAY, R. E. 2012. Individualized Cultural and Social Skills Learning with Virtual Humans. *In:* DURLACH, P. J. & LESGOLD, A. M. (eds.) *Adaptive Technologies for Training and Education.* New York: Cambridge University Press.

MAGERKO, B. 2007. Evaluating Preemptive Story Direction in the Interactive Drama Architecture. *Journal of Game Development,* 3.

MITROVIC, A. & OHLSSON, S. 1999. Evaluation of a constraint-based tutor for a database language. *International Journal of Artificial Intelligence in Education,* 10, 238-250.

SHUTE, V. J. 2008. Focus on Formative Feedback. *Review of Educational Research,* 78, 153-189.

TAYLOR, G. & SIMS, E. 2009. Developing Believable Interactive Cultural Characters for Cross-Cultural Training. *HCI International.*

VAN LEHN, K. 1990. *Mind Bugs,* Cambridge, MA, MIT Press.

VYGOTSKY, L. S. 1978. *Mind and society: The development of higher psychological processes,* Cambridge, MA, Harvard University Press.

WRAY, R. E. & JONES, R. M. 2005. An Introduction to Soar as an Agent Architecture. *In:* SUN, R. (ed.) *Cognition and Multi-agent Interaction: From Cognitive Modeling to Social Simulation.* Cambridge, UK: Cambridge University Press.

WRAY, R. E., LANE, H. C., STENSRUD, B., CORE, M., HAMEL, L. & FORBELL, E. 2009. Pedagogical experience manipulation for cultural learning. *Workshop on Culturally-Aware Tutoring Systems at the AI in Education Conference* Brighton, England.

CHAPTER 6

Do You Hear Me Now? Verbal Cues to Enhance Perception

Kathleen M. Bartlett, Sae Schatz, Denise Nicholson
MESH Solutions, LLC – a DSCI company
Orlando, Florida
kbartlett@dsci.com, sschatz@dsci.com, dnicholson@dsci.com

ABSTRACT

Communication works best when the message is clearly, efficiently delivered to an audience that perceives (and acts on) the intended outcome of the message. However, this communication goal is not always achieved. While the denotations (dictionary definitions) of language activate our conscious response, the connotations (emotional impact) of words shape the reaction of our unconscious mind. Recent research (e.g., Brooks, 2011) suggests that unconscious thought often directs human action and influences human perception. Thus, the emotional impact of language influences audience perception of the message.

Words represent a cognitive tool-kit with which speakers, writers, and audiences shape (and share) their individual perceptions of reality. To more accurately and effectively influence audience perception of messages, senders must recognize that the words (and actions) they employ do not have universal connotations. Thus, the intended meaning of a message may be far different from the perceived meaning. Decisions made based on inaccurate (or misunderstood) messages may be disastrous, and the inability of any group to communicate causes dysfunction.

To address these linguistic challenges and achieve more effective communication, particularly among US Marine Corps (USMC) personnel, this paper proposes strategies for the use of verbal cues intended to improve audience perception of messages. Strategic word choice (diction), in addition to spoken and written delivery techniques, can improve communication outcomes and encourage integrated perception.

Keywords: communication filters, persuasive language, verbal cues, perception

1 COMMUNICATION FOR ENHANCED PERCEPTION

Communication is key to good interpersonal relationships (Vaynerchuk, 2011), and use of effective communication strategies represents a critical teamwork skill. For military teams, the ability to communicate can mean the difference between life and death. US Marine Corps (USMC) small unit leaders, for example, must be certain that their instructions are heard, understood, and remembered—even under duress. This is true on the battlefield, and it also applies to the range Marines' activities, such as during surveillance, reconnaissance, and other intelligence gathering duties. As part of these activities, Marines must precisely, efficiently, effectively articulate (and decide how to communicate) the information they gather in a way that can be correctly heard and accurately interpreted by the appropriate receivers (e.g., intelligence analysts).

Effective communication strategies may assist in perceptual training: tactical use of verbal cues may enhance interaction among team members and improve outcomes of exchanges with external individuals. Combat Hunter, a USMC program of instruction (POI) prepares Marines for these types of operational challenges via training in situation awareness, sensemaking, mental simulation, and dynamic decision-making (Schatz et al., 2010). Combat Hunter trainees are encouraged to "read" their environment (Schatz, Taylor, Nicholson, et al., 2011), but they need good communication skills to effectively express and accurately convey their observations. In support of Combat Hunter, the PercepTS project is developing a suite of instructional methodologies and technologies that improve the training of advanced perceptual skills and situation awareness (Schatz and Nicholson, 2011; see also the session overview chapter by Schatz and Nicholson in this volume); inclusion of strategies for effective communication may enhance that effort.

Understanding the influence of communication filters and the connotative properties of diction, for example, might help small unit leaders and their teams better predict how different audiences will receive and interpret messages across any information environment. For instance, UK and Afghan national audiences will perceive differently the announcement of UK withdrawal of combat troops from Afghanistan by 2014 (Butcher, 2012). This difference in perception may result in divergent decisions among coalition teams; effective communication strategies can mediate those differing responses.

Increased awareness of verbal cues that persuasively influence communication outcomes can help mitigate (or prevent) verbal perceptual conflicts. A study of persuasive language used by hostage negotiators (Gilbert, 2000), for example, reveals strategies that can impact audience perception of message. Advertisers, too, have mastered the art of communicating their message via verbal (and visual, auditory, and other perceptual) cues intended to increase recall and generate predictable responses (Dunn, 1965). Strategic diction, in addition to audience-specific delivery techniques, can also improve communication outcomes and encourage integrated perception, leading to greater retention, deeper understanding, and an enhanced ability to act, and to persuade others to act.

1.2 Communication Process

We convey our thoughts, feelings, and vitally important messages using an imperfect system of signs and symbols; diction is slippery, and the confusion created by multiple meanings of words is compounded by diverse factors influencing interpretation of those signs. We use signs (semiotics, the relation between symbols and the way people use them, Radford, 2003) because there is something to be said (Eco, 1999). A stimulus or event creates the need to communicate. In response to stimuli we send verbal and/or non-verbal messages. If a stimulus creates an urgent need to hurry, "Move!" might be shouted to a companion while beginning to run. If the companion is a person who cannot run, then the message might be "Get out of the way!"

Messages should be audience-centric; the best communicators make deliberate choices to tailor content and delivery of the message to the intended audience. As Thompson notes in *Verbal Judo*, "If you know yourself but don't know your opponent, you are lucky to win 50% of the time. If you know yourself and your opponent, you can win 100% of the time" (1994). Effective communication requires critical analysis and awareness of intent; who is the receiver, and what does the message need to accomplish?

The communication process, including audience analysis and selection of cues to enhance message impact, might be depicted as the following equation (where the = sign means *determine*; syntax means the arrangement of words):

message + audience = style + tone = diction + syntax + other cues

For a written message, the style of the message may be dictated by the format. Written messages require additional steps:

message + audience + organizational plan = style + tone = diction + syntax + editing + other cues

Other cues to enhance audience perception of messages might involve non-verbal actions, such as body language or facial expressions, while written communication might include visual images, graphical representations, or even animated illustrations for digital text. Since critical verbal cues can get lost in too much perceptual clutter, however, other cues should serve to reinforce (not overpower) the message.

2 REACHING THE AUDIENCE

Human beings possess a limited information processing capability (Fiske and Taylor, 1991). The human mind can take in 11 million pieces of information at any given moment, but the most generous estimate is that people can consciously attend to about of forty of these (Wilson, 2002). In order for a message to have a long-lasting impact, it needs to evoke an emotionally charged reaction (Vaynerchuk, 2011).

Listeners are inundated by messages, and attention is fleeting. People fail to pay attention to messages that they do not care about. Thus, it behooves the sender of a message to make the receiver care, this means (among other things) that sender of a message must care about the content, as well (Vaynerchuk, 2011). The persuasive communicator defines the message through the creation of context and personal commitment. Using verbal cues that demand immediate attention, a persuasive message emotionally hooks the audience, despite pervasive, sometimes intrusive, communication filters.

2.1 Communication Filters

If everyone had the same perception of events, communicating would be easier; we could assume that others would understand our message and intent. Instead, each person has a unique perception of reality, based on his or her individual experience and many other variables. Each variable acts as a filter that shapes a person's unique impressions of reality (Ober, 2009).

Communication filters influence perceptions of messages and shape the outcomes of verbal and non-verbal exchanges. Filters include the following:

Age	Personality	Mood
Gender	Experience	Distractions
Culture	Emotions	Knowledge
Expectations	Beliefs	Physical State
Attitude	Semantics	Socioeconomic Status

Filters can determine what is heard, how it is processed, and whether it is recalled, understood, and put into practice. Communication outcomes are shaped not only by the message, but also by the receiver's personal attributes.

Therefore, when conducting pre-communication audience analysis, we must ask "who are they, what are their filters, what do they know, what do they need to know, and how will they learn, listen, or respond to the message?" When possible, speakers should monitor audience response and seek feedback that affirms understanding and attention. Feedback may include expressions of interest, questions, apathy, resistance, or boredom, and an effective communicator adjusts delivery in response to audience feedback. When messages are sent to a remote audience, a sender can seek feedback via a request for the receiver to quickly summarize what he or she heard, in order to confirm shared perceptions.

2.2 Communication and the Unconscious Mind

Some researchers (Wilson, 2002; Kahneman, 2011) suggest that the unconscious mind shapes much of our perception and drives many of our actions. Use of loaded language (highly connotative diction) may engage and activate the unconscious mind. Our conscious minds may not always recognize the effects of loaded language (*Don't think of a cool glass of water*), but nonetheless the unconscious often responds (*Are you thirsty?*). Loaded language is recalled because it:

- Attracts attention (Danger);
- Evokes a sensory response (Fourth of July);
- Carries a predictable emotional message (innocent children); or
- Conveys strong positive or negative connotations (famous or infamous historical figures).

Loaded questions may infer a predetermined conclusion or leading assumption to influence audience perception:

- "Do you feel proud of that performance?"
- "Is that the best you can do?"

When language evokes emotional responses, the message is more often heard and recalled. Choosing the right word may seem like a simple task, but nuances of meaning and variations in interpretation may result in confusion. The words we choose should not be random; they signify how we view the world and how we want others to view the world (Fenstermacher et al., 2012). Or, as the colloquial Marine Corps saying goes, "Words have meaning." Using words with a predictable connotative emotional impact can shape audience response, increase recall, and reinforce message. See the Appendix that follows this chapter for examples of highly connotative language.

2.3 Non-verbal Perceptual Cueing

A person's thought process is very closely tied to his or her physiology. We can read the signals embedded in the body language of our audience; for example, slumped shoulders, downcast eyes, drooping head, lack of animation, and other cues signify a loss of attention. The non-verbal cues of a speaker can reinforce or contradict the intended message and thereby impact audience response. Think of the iconic body language of some highly animated preachers: pulpit pounding, leaning forward, and gesturing to the heavens. Animated behaviors generate emotional involvement among listeners. These types of cues are often called paralanguage: communication cues sent via body language, gestures, facial expressions, or gaze that reinforce message and increase audience attention. An act as simple as changing the proximity between sender and receiver can influence communication outcomes.

3 PERSUASION

Persuasive communication involves the psychology of compliance: the ability to get one party to agree to a set of conditions or desires of another party; persuasion exerts power over individual perceptions and actions (Gilbert, 2000). Cialdini calls predictable persuasive patterns "click, whirr" behaviors: "Click and the appropriate tape is activated; whirr and out roils a standard sequence of behaviors" (Cialdini, 2001). The key to persuasive communication is determining what exactly triggers the "click, whirr" response for a specific audience and message.

In a famous experiment measuring the effect of language on behavior, social psychologist Ellen Langer asked 120 participants to cut in line, in front of fellow, unsuspecting students at the library photocopier. In one case, the line-jumping student said, "Excuse me, I have five pages. May I use the Xerox machine?" This case received a compliant response 60 percent of the time. In the second case, the student asked, "Excuse me, I have five pages. May I use the Xerox machine because I am in a rush?" This instance had a 94 percent compliant rate. In the third case, where the student asked, "Excuse me, I have five pages. May I use the Xerox machine, because I have to make copies?" nearly 93 percent complied with the request; no new information was provided in the third case versus the first case; the difference was the inclusion of the word "because" (Langer et al., 1978). Cialdini notes that the word "because" triggered a "click/whirr" response (Cialdini, 2001). Training and practice in use of appropriate words to trigger desired audience responses can enhance communication outcomes.

4 CONCLUSION

Effective communication among teams, like USMC squads and platoons, is a critical for efficiently and effectively achieving collective goals. Effective communication (1) minimizes the risk of misunderstandings; (2) clearly conveys the desired message in a memorable way; (3) maximizes learning in training, and (4) reduces time requirements via efficient use of powerful, precise diction. By strategically choosing the content and delivery of verbal cues best suited to their audience, USMC small unit leaders may be able to stimulate thinking, improve recall, enhance understanding, persuade other to act, and build cohesion among group members. Training small unit leaders to use high-impact verbal cues may enable their messages to be better heard and remembered.

ACKNOWLEDGMENTS

This work is supported in part by the Office of Naval Research. The views and conclusions contained in this document are those of the authors and should not be interpreted as representing the official policies, either expressed or implied, of the Department of Defense or Office of Naval Research. The U.S. Government is authorized to reproduce and distribute reprints for Government purposes notwithstanding any copyright notation hereon. This work was supported, in part, by the Office of Naval Research project N00014-11-C-0193, Perceptual Training Systems and Tools (PercepTS).

REFERENCES

Brooks, D. 2011. *The social animal: The hidden sources of love, character, and achievement.* New York: Random House.

Butcher, F. 2012. How language and narratives impact upon cross-cultural decision making. *Proceedings of the Cross Cultural Decision Making Conference*, San Francisco, CA

Cialdini, R. B. 2001. *Influence science and practice*, Fourth Edition. Boston: Allyn and Bacon.

Dunn, S. W. 1965. *Study of the influence of certain cultural and content variables on the effectiveness of persuasive communications in the international field.* Madison: Wisconsin Univ.

Eco, U. 1999. *Kant and the Platypus: Essays on language and cognition.* Translated by Alastair McEwen. Orlando, FL: Harcourt.

Fenstermacher, L., L. Kuznar, and M. Yager. 2012. Analysis of discourse and discursive practices for indications and warnings. *Proceedings of the Cross Cultural Decision Making Conference*, San Francisco, CA.

Gilbert, H. T. 2000. *Persuasion detection in conversation.* Master's Thesis, Naval Post Graduate School, Monterey, CA.

Kahneman, D. 2011. *Thinking Fast and Slow.* New York: Farrar, Straus, & Giroux.

Langer, E. J., A. Blank, and B. Chanowitz. 1978. The mindlessness of ostensibly thoughtful action: The role of "placebic" information in interpersonal interaction. *Journal of Personality and Social Psychology*, Vol. 36, No. 6.

Ober, S. 2009. *Contemporary business communication*, 7th Ed. Boston: Houghton Mifflin Company.

Radford, L. 2003. Gestures, speech, and the sprouting of signs: A semiotic-cultural approach to students' types of generalization. *Mathematical Thinking and Learning* Vol. 5 No. 1.

Schatz, S. and D. Nicholson. 2011. *PercepTS Training and Tools: 2011 Program Summary* (COR_11_1209_01). Orlando, FL: MESH Solutions, LLC.

Schatz, S., E. Reitz, D. Nicholson, & D. Fautua. *2010 Expanding Combat Hunter: The science and metrics of Border Hunter.* In Proceedings of the Interservice/Industry Training, Simulation and Education Conference (I/ITSEC). Orlando, FL.

Schatz, S., A. Taylor, D. Nicholson, et al. 2011. *PercepTS Combat Hunter Baselining Technical Report*, November. Orlando, FL: MESH Solutions, LLC.

Tversky, A. and D. Kahneman, 1974. Judgment under uncertainty: Heuristics and biases. *Science 185.*

Vaynerchuk, G. 2011. *The Thank You Economy.* New York: Harper Collins.

Wilson, T. D. 2002. *Strangers to Ourselves: Discovering the Adaptive Subconscious.* Cambridge, MA: Harvard University Press.

APPENDIX: HIGHLY CONNOTATIVE LANGUAGE

Loaded Language

Positive connotations	Negative connotations
Democracy	Dictatorship
Freedom	Captivity
Hero	Coward
Success	Failure
Strong-willed	Pig-headed
Diligent	Lazy
Winner	Loser
Strong	Weak

Power Words and Phrases

You know best	Your last chance
Play the game	You can't be serious
Think about it	Worst nightmare
Love of my life	Living in denial
Master of the universe	Say your prayers
Right hand of God	Beyond help
Want to fight	End of the road
To the death	In your dreams
Fight for your rights	Over the edge

Words or phrases that weaken position

I am trying…In my opinion…I am not sure…It goes without saying…
Needless to say…There is/There are/There were (passive voice)

Control Words

Choice, Power, In-charge, Decision, Confidence/Confident

Trigger Words

Exclusive	Urgent	Because
Discover	Earn money	Now
At last	Explore	Must
The only	Secrets of	Free
Sex	First	Deadline
New	When you	Imagine
Quiz	Person's name	Today

CHAPTER 7

Recognizing Patterns of Anomie That Set the Conditions for Insurgency

Karen Guttieri, Curtis Blais, Rob Shearer and Sam Buttrey

Naval Postgraduate School
Monterey, CA, USA
guttieri@nps.edu, rlsheare@nps.edu, sebuttre@nps.edu, clblais@nps.edu

Jack Jackson, Tom Deveans

US Army TRADOC Analysis Center, Monterey
Monterey, CA, USA
lajackso@nps.edu,tmdevean@nps.edu

Niklaus Eggenberger-Argote, Adrien Gschwend

Swiss Academy for Development
Biel/Bienne, Switzerland
eggenberger@SAD.ch, gschwend@SAD.ch

ABSTRACT

A proactive approach to the problem of insurgency requires analysis to begin before violence starts – to develop a working model of the origins of insurgency. This project seeks to identify potential interrelations between anomie— the loss of compelling norms that enable populations to meaningfully interpret social change — and the support of and participation in non-state armed groups. The project is conducting micro and macro level research and analysis to contribute to the recognition of indicators that support stakeholders in the timely identification and mitigation of unstable social structures. Micro-level research comprises qualitative and quantitative surveys conducted in the Niger Delta, while the macro-level research utilizes economic and social data from fifty-five nation states in Africa. Insights gained from the micro-level analysis will guide final model building efforts, in contrast to the more common data mining approaches that tend to over-fit models with spurious correlations. The project's multi-level approach is a methodological advancement and its outputs include new empirical grist for fellow scholars and field practitioners. This paper provides a summary of the project and describes the micro-level and macro-level research findings to date from the work.

Keywords: anomie, insurgency, field research

1 INTRODUCTION

Anomie— the loss of compelling norms that enable populations to meaningfully interpret social change — provides a plausible framework for explaining support to and participation in non-state armed groups. This project takes a multi-level approach. Macro-level research utilizes economic and social data from fifty-five nation states in Africa, and micro-level research comprises qualitative and quantitative surveys conducted in the Niger Delta. The macro level analysis investigates linkages between structural factors (regime type, economic situation, etc.), social change (globalization, deteriorating environment, modernization, etc.), anomie, and insurgency. The survey questionnaire investigates micro-level factors, based upon the Swiss Academy for Development (SAD) Anomie Scale that includes core elements of discontent, distrust, pessimism and individual anomie. At the macro-level, we observe patterns suggesting a relationship between anomie proxies and susceptibility to insurgency. Micro-level analysis is in progress to offer insight into the role of anomie, as a step toward identifying potentially beneficial interventions. This chapter outlines the approach and findings to date.

2 ANOMIE AND INSURGENCY

The sociological concept of anomie is based on the theories of Emile Durkheim and Robert Merton, emphasizing the role of the social system in setting the norms that regulate human drives. Investigating the consequences of rapid social change, such as the transition from feudal to industrial society, Durkheim (1997, 1933) observed that individuals can lose their reference points for determining their status and role. A society can only bear change if its population can apply a meaningful interpretation to social reality. If this is not possible, social change leads to crisis-laden insecurity and instability: Traditional values, norms and reference points erode, resulting in a loss of individual social orientation (anomie). Anomie describes a disoriented state of mind, caused by uncoordinated, sudden, non-anticipated social change or a lack of (blocked) social change. Under conditions of normlessness, or anomie, drives are no longer regulated and the valid regulations and norms of behavior lose their function.

Merton (1938, 1959) elaborated on Durkheim's concept of anomie, focusing on the strain produced by a social structure itself. Merton (1938, 1959) claims that anomie emerges when the social structure prescribes specific goals to the majority of its members but does not provide the legitimate means to achieve these goals to a portion of them. Anomie results from the discrepancy between culturally defined goals and means. Deviant behavior is caused by social pressure. Therefore, society is structurally anomic insofar as legitimate means to achieve generally valid goals are not distributed equally among all societal groups (Legge, Davidov, and Schmidt, 2008). Cloward (1959) follows Merton's theory, arguing that a person's position in a certain society determines not only his/her access to legitimate means but also his/her access to illegitimate means. Anomie has found wide application in the literature on criminology. Steven F. Messner and Richard Rosenfeld's *Crime and the American Dream* (2008) argues that a society that over-values material goods, and under-values the morality of means used to acquire them, more or less invites criminal behavior.

Recent approaches to anomie link the theories of Durkheim and Merton to the context of globalization. Lanyon et al. (1995) suggest that it is above all the modern processes of globalization, mass migration and urbanization as outputs of our times that lead to an increase in the complexity of social life and can therefore cause social anomie. The process of globalization leads to accelerated social change, increasing complexity of social life due to increasing competition between different value systems, and more widespread social inequality. Passas (2005), for example, sees a greater amount of illegitimate means being used in poorer countries while they try to adapt to new global circumstances. These theories are in accordance with assertions that a state of "normlessness" (i.e., the breakdown of values and the disruption of the moral order) very rarely exists (Lanyon et al., 1995). Rather, anomie results in the state of flux of competing core and peripheral values of a society. For purposes of this research, we propose the following working definition of anomie, based on a synthesis of the prevailing theories:

> *Anomie is a condition of social structures whose regulative and integrative social forces are weak...On the individual level, social anomie is reflected by the psychological state of anomia, associated with great difficulties in individual adaptation resulting in a loss of general social orientation...feelings of insecurity and marginalization, and the questioning of the legitimacy of core values. Previously valid behavioral norms as well as personal compentencies disintegrate.*

Ricigliano (as cited in Dudouet, 2009) identifies resistance/liberation movements as "attempts…to challenge or reform the balance and structure of political and economic power, to avenge past injustices and/or to defend or control resources, territory or institutions for the benefit of a particular ethnic or social group" (p 5). Insurgency in many respects represents a deviance. If, following McAdam, Tarrow and Tilly (2003), *contained politics* operate with well-established means, insurgency might be considered *transgressive* politics that is engaging newly self-identified actors and innovative action. If researchers intend to produce insights that will enable a more proactive approach to conflict management, a conceptual shift in the problem framing and analysis is needed. The emergence of non-state armed groups needs to be understood as a dynamic process rather than based on static causes. Yet much of our current research on insurgency provide insufficient granularity necessary to understand social dynamics. More importantly, research tends to focus on situations that are already violent; it still lacks for a working model of the origins of insurgency. Today, there is broad consensus that the underlying causes are multidimensional. However, most of the studies either look at macro- or micro-level factors, in isolation from the other. Approaches such as Collier and Hoeffler's "loose molecule" hypothesis (1998), reduce the causality of conflicts to the question of material incentives ("greed") and neglect the demand-side such as specific grievances within the population. Research on social forces is in short supply compared to focus on political leaders and organizations as agents, and on economic growth, topography, or demography as structural factors. Societies can convey exhaustion of war, demand peace, and promote social and political space for negotiations, and thus affect the duration of conflicts. On the other hand, communities can also inhibit social negotiations; for example when they are the carriers of nationalist narratives, racism, etc. mobilizing sentiments that are counterproductive to peace processes (Heiberg, O'Leary, and Tirman, 2007).

Smith (2004) proposes the study of background causes (basic elements of social and political structure), the mobilization strategy (political behavior, the causes for which people fight), triggers (events, actions by significant actors), and catalysts (that affect the intensity of the conflict). Bjorgo (2005) distinguishes structural causes, facilitator (accelerator) causes, motivational causes, and triggering causes. On a motivational level, Guichaoua (2007) distinguishes three crucial considerations for potential members of non-state armed groups: (1) economic considerations (or "greed"); (2) a feeling of danger which rouses the "desire of protection against fuzzily identified risks (criminality, unknown future, menace from other ethnic groups, etc.)" (p 27); and (3) social proximity to militia insiders.

Based on different descriptions of insurgency, we identify the insurgency as an organized movement with a political effort (in contrast to groups with a private agenda, such as drug cartels, for example) and a specific aim that may use violent or non-violent means, (e.g., political mobilization or propaganda), to challenge or weaken the existing government or established leadership for control of all or a portion of its territory, or force political concessions. Insurgencies require the active or tacit support of some portion of the population involved.

3 STUDY APPROACH

A conceptual framework for studying patterns of anomie that set the conditions for insurgency must synthesize anomie theory and existing knowledge of predictors of the participation in non-state armed groups. Meanwhile, predictors related to anomie only represent one aspect of a bigger and more comprehensive picture. The conceptual framework in Figure 1 aims to reflect this variety of causes at different levels of society as well as their interactions.

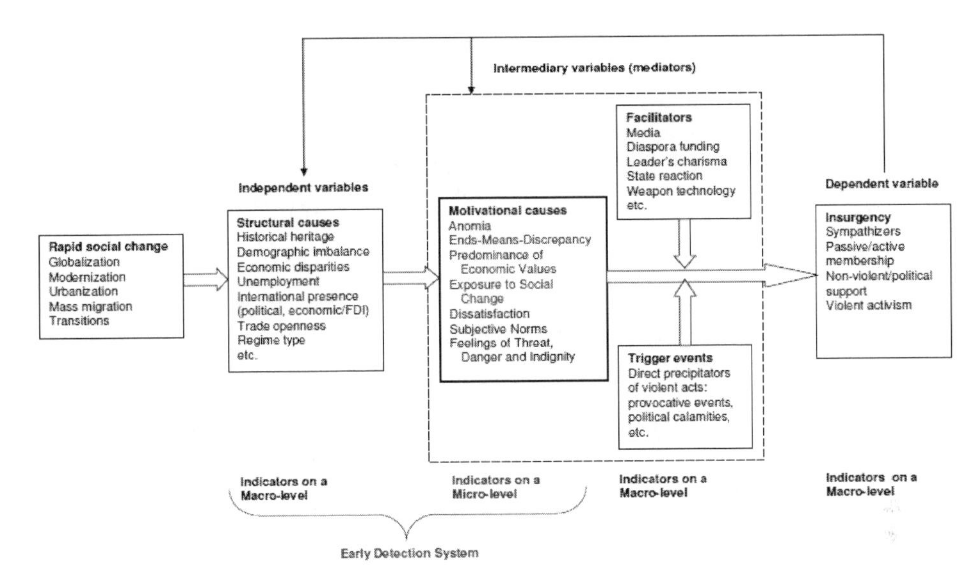

Figure 1 Framework based upon SAD Anomie Construct

The framework forms the basis for developing research questions and hypotheses to be verified or disproved through desk research (in the case of macro-level indicators) and empirical field research (in the case of micro-level indicators). Intermediary variables such as motivational causes, facilitators, and trigger events mediate between macro-level structural causes (independent variables) and insurgency (dependent variable). Structural causes and motivational causes are considered part of an indicator system aiming at early detection of conditions promoting likelihood of insurgency.

The primary objective of the research on the micro-level is to analyze to what extent anomie on an individual level is a motivational factor that can explain the support of and participation in non-state armed groups, and what factors may lead to anomie. Taking into account the previously discussed theories of anomie, we will examine three potential motivational causes of insurgency: (1) (exposure to) social change and *anomia*, or anomie at the individual level (Durkheim); (2) ends-means discrepancy (Merton); and (3) the prevalence of economic factors (Messner and others).

4 HYPOTHESES

Based on the current research in the fields of anomie and insurgency we aim to investigate the following assumptions: 1) Social anomie is reflected and thus can be measured on the individual level; 2) Anomie is a condition of social structures whose regulative and integrative social forces are weak. The weakness of these structures has either 2a) resulted from rapid structural change whereby the processes which reinforce social integration decline in salience and force, or 2b) it is structurally inherent, caused by the discrepancy between culturally defined goals and accessible legitimate means, or 2c) it resulted from the predominance of economic values over other competing institutional values; 3) We expect the participation in a non-state armed group to show a catalyst function: People who participate in a non-state armed group should therefore show a lower level of dissatisfaction and/or anomia than people who "merely" support non-state armed groups without participating in them. Given the conceptual framework described above, the following hypotheses are based on the theory of Durkheim:

> H1.a: People who heavily experience social change (regarding pace, scope, and evaluation) show higher levels of anomia than people who experience social change weakly.
>
> H1.b: People with a high level of anomia show a higher level in the support of non-state armed groups than people with a low level of anomia.

The following hypotheses are based on the theory of Merton:

> H2.a: People with a low socio-economic status show a higher level of ends-means-discrepancy than people with a high socio-economic status.
>
> H2.b: People with a highly vulnerable socio-demographic status show a higher level of ends-means-discrepancy than people with a less vulnerable socio-demographic status.
>
> H2.c: People with a high level of ends-means-discrepancy show a higher level in the support of non-state armed groups than people with a low level of ends-means-discrepancy.

The following hypotheses are based on alternative explanatory paths:

> H3.a. People with a close social affiliation to militia or people supporting non-state armed groups (descriptive norm) perceive a stronger injunctive norm towards non-state armed groups (i.e. they think that relevant people would like them to support non-state armed groups) than people with no such affiliations.
>
> H3.b. People with a injunctive norm supporting non-state armed groups and their behaviour show a higher level of support of non-state armed groups than people with no such injunctive norm.

5 MACRO-LEVEL STUDY

The experts working with SAD to adapt the Anomie Scale identified four potential economic sources of anomic strain: cost of living, employment opportunities, quality of housing, and access to communication. The experts also identified seven potential social source of anomic strain: access to health services, access to infrastructure, educational opportunities, quality of the natural environment, security of life & property, inter-ethnic relations, and traditions.

Working from these variables, we identified macro level proxy indicators to test their hypothesis that anomic strain at the nation state level explains the presence of insurgency within a nation state. We collected data for these macro level indicators from the World Development Indicators database, for fifty-four nation states in Africa, from 1980 to 2005, and conducted a series of hypothesis tests, testing the statistical significance of each macro proxy indicator in explaining the presence of an insurgency. We then developed several logistic regression models to explain the presence of an insurgency in terms of a collection of the macro proxy indicators.

The primary source of data is the 2008 World Development Indicators database. This database contains over 800 indicators by nation state and year for over 200 nations. A significant challenge has been dealing with missing observations (replaced with the regional mean for the given year, for our purposes). We also collected insurgency data from the 2008 RAND Counterinsurgency Study identifying 87 insurgencies that have occurred since the end of World War II.

We fit logistic regression models to the data, examining economic, political, and social indicators against the occurrence of insurgency. We created a preliminary measure for macro-level anomie as the weighted sum of economic, political, and social goals-means differences. Four inferences emerged from this exploratory analysis: (1) non-insurgent states experience all levels of anomie; (2) the likelihood of insurgency increases with anomie; (3) insurgency requires a minimum level of anomie; and (4) a high level of anomie is not a sufficient condition for insurgency. These inferences support several of our assumptions and suggest that our hypothesized relationship between anomie and insurgency is plausible. We have classified patterns of anomie that led to insurgency using the k-nearest neighbor algorithm. The algorithm recognized a high percentage of nation states that experienced insurgency, but had a high false positive rate. We infer from these results that our assumption that high levels of anomie are a necessary, but not sufficient, condition for insurgency is correct.

In a second step, we obtained a ranking of the most impactful macro predictors based upon the magnitude of the coefficients. Since we had normalized the macro predictor scores, the coefficients represent the change in the log odds of an insurgency for an increase in each predictor by one standard deviation. The logistic regression model ranked the macro predictors in the following order of impact: 1 - public spending on education, 2 – mortality rate under five, 3 – political rights, and 4 – pollution.

As a third step, we propose a methodological triangulation, assessing the convergence of results from both the macro logistic regression and the micro survey. The triangulation will compare which predictors the desk and field approaches found the most impactful on the likelihood of an insurgency.

6 MICRO-LEVEL STUDY

The SAD Anomie Scale has been applied and adapted to research in numerous cultural contexts. We sought a location for field research characterized by existing or high likelihood of social unrest, yet still reasonably safe enough for field research teams to conduct interviews. Among a number of options, we selected the Niger Delta region, with further scoping of the data collection to 10 districts in the city of Port Harcourt, where local faculty and students from Port Harcourt University could be enlisted and trained to perform the data collection.

To adapt the scales used in the questionnaire to local context, we conducted interviews with experts knowledgeable about the situation in the greater Port Harcourt area. The experts were primarily selected by the Center for Ethnic Conflict Studies (CENTECS), our local research partner in Port Harcourt. The experts included persons from the academic field as well as civil society. The interviews helped frame the questions and how responses would be interpreted. Two principal areas were identified for determination: social change and relative deprivation.

Social change can be defined as a permanent and rapid transformation of a given social organization and its internal dynamics. The perception of change has two aspects: scope and pace. Whether social change actually affects the psychological well-being, however, depends on how people appraise it. For each potential area of social change, we asked three questions: (1) what level of social change in a certain area is perceived; (2) how slow of fast is this change taking place (pace); and (3) their evaluation of the change. Interaction with the experts identified the following indicators: level of living costs for housing, food, clothing, etc.; access to infrastructure like roads, electricity, pipe-born water, etc.; level of corruption; level of political thuggery; job opportunities; level of security of life and property; natural environment; right to political participation; education opportunities; health situation; communication systems; performance of the federal government; performance of the state government; performance of the local government; traditions; influence of militant groups; and inter-ethnic relations. Findings were reflected in survey questions such as:

- In your experience, to what extent has [the level of living costs, etc.] changed in Port Harcourt and your community in the last five years?
- In your opinion, how fast or slow is this change?
- How positive or negative do you see this change?

With respect to relative deprivation, Merton claims that anomie results from the discrepancy between goals and means, often operationalized as relative deprivation. Relative deprivation results from the recognition of a discrepancy between the actual level of satisfaction of one's needs and the expected (or desired) level of satisfaction of one's needed. An individual can compare oneself to other persons at the same time (synchronic reference group) or to one's own situation at another time (diachronic reference group). Unfavorable comparisons (the cognitive component of relative deprivation) can generate feelings of deprivation (the affective component of relative deprivation). Although relative deprivation can occur on the individual or collective/group level, the concept of collective relative deprivations finds support in diverse studies on insurgency which highlight the influence of feelings of threat, danger, and indignity on a collective level as an alternative explanatory path in the model.

Interactions with the experts identified the following indicators: employment opportunities; educational opportunities; qualify of housing; access to infrastructure;

marital situation (only asked for individual relative deprivation); security of life and property; access to health services; quality of the environment on lives in; opportunities for political participation; opportunities to benefit from governmental projects; and right of self-determination. Resulting questions for the survey included:

- Compared to other persons, are you better or worse off regarding [employment opportunities, etc.]? How do you feel about this? [individual relative deprivation]
- Compared to now, how do you think your life situation will be in five years regarding [employment opportunities, etc.]? How do you feel about this? [individual relative deprivation]
- Compared to other ethnic groups, is your ethnic group better or worse off regarding [employment opportunities, etc.]? How do you feel about this? [collective relative deprivation]
- Compared to now, how do you think your life situation will be in five years regarding [employment opportunities, etc.]? How do you feel about this? [collective relative deprivation]

The questionnaire was translated into Pidgin English as the lingua franca of the research area based on forward- and back-translation. The survey was pre-tested and shortened in response to feed-back.

Figure 2 Photographs showing some of the local conditions for conducting the field research. Right, streets of Ogbakiri.

Our research design required a minimum of 700-750 valid questionnaires. Taking into account a maximum of 25% of incomplete questionnaires, we aimed at a total of 1000 collected questionnaires. The sampling procedure included cluster sampling in which the greater Port Harcourt-area (GPH) was grouped into geographical clusters which cover various social strata in the city, hot-spots of insurgency and supposed safe-havens; and ensure quantitative/qualitative representativeness of GPH. Socio-economic (rich, middle and poor) and socio-demographic categories (rural vs. urban) as well as areas with high and low intensity of insurgency were being covered. 10 such clusters were sampled: Okrika, Gborokiri, Township, Diobu, Rumuokoro, Igwuruta, G.R.A. Phase II, Ogborghoro, Ogbakiri, and Rumuokurishi. Systematic sampling - a random technique relying on the probability principle – selected the streets and buildings that are representative of the clusters. Finally, a simple random technique determined the person to be interviewed. The supervision, monitoring and quality control comprised interview observations and individual supervision, group supervisions, checking of questionnaires, data entry, data cleaning, and data analysis, and last but not least seeking help if necessary.

In accordance with the research protocol approved by the Institutional Review Boards of the Naval Postgraduate School and the Swiss Academy for Development, survey participation was confidential and based upon informed community and individual consent. Approximately 1000 questionnaires needed to be entered. For quality purposes data entry was done by means of double entry. This means all questionnaires were entered twice by two different data entry clerks. The logical data cleaning efforts prepared the data for preliminary analysis.

Factorial and principle components analysis were used to analyze the dimensionality of the variables (one-dimensional or multi-dimensional). The principle component analysis confirms a pattern of multi-dimensionality. Structural Equation Modeling (SEM) can specify and estimate models of linear relationships among variables in a cross-sectional multisample design (MacCallum & Austin, 2000) and evaluate how well the collected data corresponds with the data implied by the model. Several tests are currently in progress, first of alternative mediation models such as absolute and relative depravation to compare their fit, using "classical" regression mediation models. A second step will combine these predictors in one model. Theoretically, it is more sound to treat the alternative predictors as multiple predictors of insurgency and include them in a path model (using the scale scores) or even better a structural equation model (treating the single items as manifest variables). We will properly specify the measure model and structural model with exogenous and endogenous factors and their expected covariances based on theoretical reflections.

7 CONCLUSION

Anomie theory provides a potential theoretical approach to the interrelationships between such processes and the development of insurgent movements against a (legal or illegal, legitimate or illegitimate) authority, including the psychological level. However, an examination of the explanatory power of anomy theory in this field has yet to be concluded. If the macro model is correct, is it reasonable to obtain survey results similar to those obtained by the survey? We propose to answer this question by calculating a p-value for the survey results. If we assume that the macro model is correct (reasonable given its theoretical underpinnings and low classification error rate) we can use this to determine the probability of obtaining the survey's ranking. This will inform the update to analysis of this chapter. The results of the study will be of interest to those who are confronted with such potentially conflict-ridden situations. The results should assist them to detect risky social developments and to implement preventive measures accordingly.

ACKNOWLEDGMENTS

The authors would like to acknowledge the Office of Naval Research for support to basic research in this project, CENTECS, and the participants in the field study.

REFERENCES

Bjorgo, T. 2005. Root Causes of Terrorism: Myths, reality and ways forward. Routledge: London.

Cloward, R. A. 1959. Illegitimate means, anomie, and deviant behavior. American Sociological Review. Vol 24(2):164-176.

Collier, P., and A. Hoeffler 1998. On economic causes of civil war. Oxford Economic Papers. 50(4): 563-573.

Dudouet, V. 2009. From War to Politics: Resistance/Liberation Movements in Transition. Berghof Research Center: Berlin.

Durkheim, E. (1997; 1933). The Division of Labor in Society. New York, Free Press.

Fischer, M., Niklaus Eggenberger, Robert Shearer, Karen Guttieri, Curtis Blais, Leroy Jackson and Richard Brown (2009). Anomie and Insurgency Research Protocol Naval Postgraduate School, Swiss Academy for Development, U.S. Army Training and Doctrine Command-Monterey.

Guichaoua, Y. 2007. "Who joins ethnic militias? A survey of the Oodua people's congress in southwestern Nigeria." Crise Working Paper No. 44. Accessed February 29, 2012. http://www.crise.ox.ac.uk/pubs/workingpaper44.pdf.

Heiberg, M., B. O'Leary, and J. Tirman 2007. Terror, Insurgency, and the State: Ending Protracted Conflicts. University of Pennsylvania Press, Philadelphia.

Lanyon, A., L. Han-Lin, Q. Wang, and J. Western 1995. Anomie: A Revival. Unpublished seminar paper.

Legge, S., E. Davidov, and P. Schmidt 2008. Social structural effects on the level and development of the individual experience of anomie in the German population. International Journal of Conflict and Violence. Vol 2 (2): 248-267.

MacCallum, R. C. & Austin, J. T. (2000). Applications of Structural Equation Modeling in Psychological Research. In: Annual Review of Psychology, 51, 201-226

McAdam, D., Sidney Tarrow and Charles Tilly (2003). Dynamics of Contention. Cambridge, Cambridge University Press.

Merton, R. K. 1938. Social structure and anomie. American Sociological Review. Vol 3 (5): 672-682.

Merton, R. K. 1959. Social conformity, deviation, and opportunity structure: a comment on the contributions of Dubin and Cloward. American Sociological Review. Vol 24 (2): 177-189.

Messner, S. F., H. Thome, and R Rosenfeld 2008. Institutions, anomie, and violent crime: clarifying and elaborating institutional-anomie theory. International Journal of Conflict and Violence. Vol 2(2): 163-181.

Passas, N. 2005. Global Anomie Theory. The Essential Criminology Reader. Westview: Boulder, Colorado. 174-182.

RAND 2008. War by Other Means: Building Complete and Balanced Capabilities. The RAND Corporation.

Smith, D. 2004. Trends and causes of armed conflict. In Berghof Handbook for Conflict Transformation, eds. Fischer, M., H. J. Giessmann, and B. Schmelzle. Berghof Research Centre for Constructive Conflict Management: Berlin.

Section II

Tactical Culture Training: Narrative, Personality and Decision Making

CHAPTER 8

Assessing Changes in Decision Making as a Result of Training

David A. Kobus, Erica P. Viklund

Pacific Science & Engineering Group
San Diego, CA, USA
davidkobus@pacific-science.com / ericaviklund@pacific-science.com

ABSTRACT

Warfighters are required to make a multitude of decisions every day, often under stress and with incomplete information. Many of these decisions involve sociocultural considerations that can significantly impact mission effectiveness. While immersive training for decision making has intuitive appeal and been embraced by the US military, few empirical measures have been identified to objectively assess the effectiveness of this type of training. This chapter describes the methods and procedures employed during Spiral 2 of the Future Immersive Training Environment (FITE) Joint Capabilities Technology Demonstration (JCTD) to develop an instrument for assessing training-related changes in decision making skills of infantrymen. Results indicated that more "expert-like" post training decisions occurred for many of the decision theme areas after five training scenarios. Changes in decision making were related to the reinforcement received for a particular decision theme during after-action discussions. Further work is needed to establish reliability and validity of using this type of measure to assess training effectiveness.

Keywords: Decision making, Situational Judgment Test, Training

1 OPERATIONAL PROBLEM

Infantrymen of all ranks and experience levels are required to make decisions that may have an effect not only on the immediate tactical situation, but also on larger operational and strategic issues. The information processing demands placed on modern Warfighters are enormous, and decision makers in the field are usually not afforded the time to combine sequential, procedural, and comparative methods to solve problems. Crucial to enhancing performance under such demanding conditions is training that quickly advances Warfighters beyond declarative knowledge to a level of mastery that supports expert-level decision making. Further, pre- and post-training assessments are required to determine the effectiveness of such training. This chapter briefly outlines a method to assess changes in decision making as a direct result of immersive training using situational judgment tests (Kobus, Kobus, Ostertag, Kelly &Palmer, 2010b).

The Future Immersive Training Environment (FITE) was a Joint Forces Command sponsored Joint Capabilities Technology Demonstration (JCTD) with the explicit mission of providing "military trainers with sufficient enablers to train close combat tasks in a realistic, fully immersive training environment that creates and reinforces complex (tactical and human dimension) decision making skills." This effort was conducted in two Spirals. In Spiral 1, the focus was on individually worn virtual reality (IWVR) systems and is reported elsewhere (Kobus, Palmer, Kobus, & Ostertag, 2010a). During Spiral 2 the training environment was expanded to blend both virtual and real events together in a mixed-reality environment (Camp Pendleton's Infantry Immersion Trainer (IIT)). The specific focus was to demonstrate the capabilities of a facility based mixed-reality training system to create conditions that would allow a trained infantry squad to experience complex, realistic environments and situations. The Operational Demonstration (OD) consisted of five training scenarios designed to provide multiple operationally relevant human decision making opportunities in situations similar to what may be experienced in an irregular warfare environment. A key objective was the development of a methodology to assess the impact of training upon decision making within the immersive mixed-reality environment.

1.2 Assessment of Tactical Decision Making: The Situational Judgment Test (SJT)

Situational Judgment Tests (SJTs) have high face validity and are often used to assess decision making, tacit knowledge, and practical "know-how" (Weekley & Ployhart, 2006). These tests usually start with a participant reading, viewing or listening to a brief scenario, after which they rate or rank a set of courses of action (COAs) based on either their perceived effectiveness or how likely the participant would be to do the action. SJTs allow the assessment of decision making behavior for complex situations by presenting the participant with incomplete or ambiguous cues regarding a specific "real-world" event. The freedom to rate COAs as equally effective/ineffective acknowledges the fact that in complex situations, such as those involving sociocultural considerations, there is often not a single "right" COA. Developed correctly, the SJT methodology provides a unique and viable measurement tool for assessing changes in the decision making process as a result of training. This method allows for the assessment of each squad member's decision making independently, and does not interfere with training or with the after action review (AAR) process.

2 METHOD

2.1 Situational Judgment Test (SJT) Development

The first step in employing the SJT methodology in the context of the FITE JCTD was to identify specific decision making events within the five training scenarios to be used in the OD. Common decision making categories or "themes" were then identified across scenarios. A final set of 27 themes was identified for inclusion in the SJT. The decision themes are listed in Table 1.

Table 1 Decision Themes identified across the five OD scenarios.

Decision Themes		
1. Identify Individuals for questioning	2. Recognizing/dealing with suspicious behavior	3. Establishing security
4. Intra-squad communication	5. Responding to request of village elder	6. Establishing cordon
7. Entry/search homes / buildings	8. Prioritizing multiple ongoing events	9. Communicating with adjacent unit / QRF
10. Interaction with villagers	11. Detecting anomalies in baseline	12. Assessing crowd behavior
13. Conduct security inspection	14. Tactical questioning	15. Handling casualties
16. Patrol route selection	17. Detaining a villager / HVI	18. Controlling an unruly crowd
19. Interacting with foreign females	20. Communication with HQ	21. Cultural awareness / sensitivity
22. Finding suspicious/illegal materials	23. Escalation of force	24. Integrating interpreter / ANA into mission
25. Dealing with an uncooperative / confrontational villager	26. Respond to contact	27. Collecting / assessing / disseminating Intel

Two vignettes (decision dilemmas) with possible COAs were developed for each decision theme. Vignettes, and the decisions required, were similar to the situations encountered during the training scenarios, but differed in detail. Each vignette was developed to provide sufficient context and amplifying information to support evaluation of each of the COAs while avoiding an emphasis on specific tactics and unit standard operating procedures (SOPs). In an attempt to limit the time requirement to complete the task, the number of COAs was restricted to five for each vignette. Two vignettes for each decision theme, along with their COAs were then refined and finalized through an iterative review process. USMC and FITE subject matter experts (SMEs) validated the vignettes and COAs to ensure that the vignettes were realistic and provided the information necessary to support rating of each of the COAs, and that the COAs were appropriate to the dilemma depicted in the vignette. Once finalized, one vignette for each decision theme was assigned to SJT Set 1, and the other was assigned to Set 2 (random assignment).

2.2 SJT Administration Software Development

One of the primary concerns of using the SJT method was the extra burden upon trainees for reading and maintaining concentration over a long series of vignettes. To minimize the burden, a computerized version of the SJT was developed to provide the trainee the option to read and/or listen to a recorded version of the vignette (male voice) while viewing the screen. Volume control and the ability to pause or replay audio were available options. Figure 1 shows an example screen shot of the trainee interface.

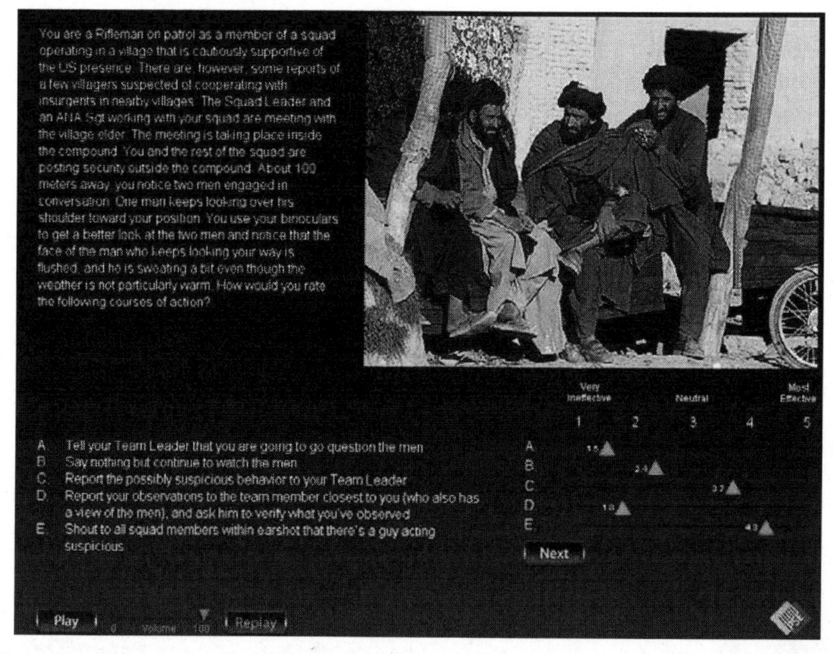

Figure 1 Example screen shot of SJT presentation and data collection software showing dilemma, COAs and slide rating bar for each COA.

After listening to or reading a given vignette, trainees read and assessed the effectiveness of each COA, using the mouse to slide a triangle marker to a rating next to the COA. Ratings ranged from 1 (Very Ineffective) to 5 (Most Effective). The trainee was required to provide a rating for each COA. All markers were initially placed on the far left of the sliding scale (position 1). The marker was highlighted (intensity increased on triangle) after it was selected by the trainee and displayed the rated value of the COA as they moved across the scale. When the trainee finished rating each of the COAs, he could then select the "NEXT" button to continue to the next vignette. Trainees were informed that each COA should be rated independently; therefore COAs could be rated equally if their effectiveness was perceived to be equal.

2.3 SME Database

A variety of Subject Matter Experts (SMEs) were used throughout the development process to assess and validate test items. A final set of 11 SMEs was identified by the FITE-JCTD manager for development of the SME database (SME-DB). All SMEs were current or former military with at least one combat tour. Each of the SMEs had served multiple tours as a small unit leader and was deemed an expert in small unit training by their respective services. They completed the computerized version of both SJT Set 1 and 2 during separate sessions and provided ratings for all COAs for each vignette. The ratings for each COA were averaged across SMEs, providing an average expert rating for each item and forming the SME-DB.

2.4 Participants

Twenty-five USMC infantrymen from the two squads participating in the FITE OD completed the SJTs. One trainee was dropped from all further analyses because he did not complete all of the SJT items. Three additional trainees were dropped due to their completing one or both versions of the SJT too quickly to provide valid responses (< 10 minutes). For all analyses, Squad 1 had 11 trainees with a mean age of 21.8 years, and Squad 2 had 10 trainees, with a mean age of 21.2 years. Squad 1 trainees' rates ranged from E-2 to E-5 with the greatest number of trainees (5) being E-4. Five of the trainees had some experience training in immersive environments, with three having gone through Mojave Viper training and two previously receiving training in the IIT. Three members had previous combat experience. Squad 2 trainees' rates ranged from E-2 to E-5 with the greatest number of trainees (6) being E-3. One trainee had received training at Mojave Viper, and another had previous training in the IIT. Only one squad member had previous combat experience.

2.5 Procedure

Trainees completed one SJT set prior to participating in any of the training scenarios to serve as a baseline measure. Half of the trainees completed Set 1 items as the pre-test, and half completed SJT Set 2 items. After completion of the final scenario and AAR, trainees completed an SJT post-test of the opposite item set. Squad number, position in squad, and team number were also recorded to assist in further analysis after the post-test. Trainees were also asked if they felt their decision-making had changed due to the training they had received.

Members of the research team observed each scenario to verify which decision themes the squads actually encountered during training. Observations were based upon what could be viewed from cameras at the IIT instructor workstation, and audio that was monitored via radio or lapel microphones. These observations were an attempt to validate the decision making training experience that each squad received.

After each of the training scenarios, Company-Level Intelligence Cell (CLIC) debriefs, and detailed AARs were conducted. Members of the research team observed the debriefs / AARs to determine which decision themes were identified and discussed. These data were used to identify which decision themes were being reinforced with each squad.

3 RESULTS

Each squad participated in the same five standardized scenarios, but differences in how the squads approached the scenarios meant that they did not necessarily experience the same decision themes within a given scenario. All but one of the decision themes was experienced by both squads at least once across the training scenarios. The absence of one of the decision themes ("controlling an unruly crowd") resulted from an administrative change in the execution of the scenarios. These data reflect of what the squad actually experienced, providing verification of the opportunities that each squad had to practice in making decisions related to each decision theme. Seventy-two percent (72%) of trainees indicated the training positively affected their operational decision making.

3.2 SJT Results: Comparison of Trainee to SME Responses

Each trainee's effectiveness ratings for each COA in each vignette were compared to the SME database (SME-DB) by calculation of Root Mean Square Error (RMSE) differences. RMSE is a measure of the differences between values predicted by a model or an estimator and the values actually observed. Average differences were then calculated for each decision theme for each trainee, separately for the SJT pre-test and post-test. The number of individuals showing improvement in their decision making (as defined by their post-test responses more closely resembling the SME-DB than their pre-test responses did) was then tabulated for each decision theme, and is displayed as the percentage of all trainees in Figure 2. Results show that changes in decision making varied greatly among the different decision themes. In nine of the 27 decision themes, more than half of the trainees displayed more expert-like decision making on the post-test. Decision themes showing the greatest improvement (2, 3, and 15) were specifically related to individual squad activities such as intra-squad communication, entering or clearing a building, and detaining personnel. Analysis of scenario and debrief / AAR observation data indicated that these results are consistent with the hypothesis that the decision themes showing greatest improvement were also the ones that were repeated during scenarios and reinforced during debriefs. Results may also reflect differences in training requirements among decision theme areas.

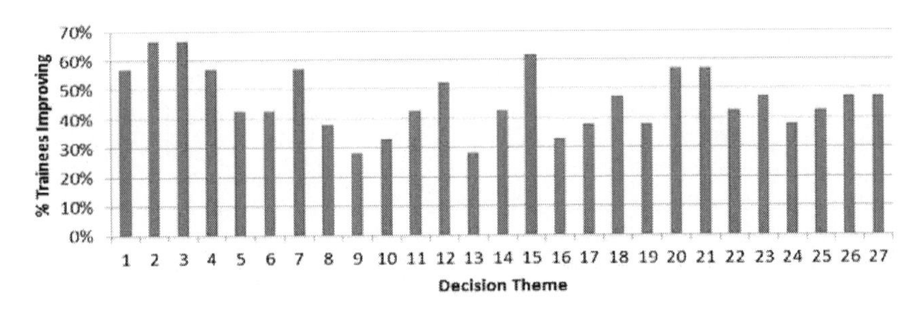

Figure 1 Percentage of trainees demonstrating more expert-like decision making.

In addition, there was a great deal of variation among individuals in how many decision themes demonstrated improvement. Percent of decision themes showing change toward more expert decision making was lowest for squad leaders (26%) and greatest for inexperienced riflemen (85%). This result may be reflective of the expert-like decision making exhibited by squad leaders prior to training. This notion is supported by squad leaders' relatively expert-like pre-test responses.

Table 2 Decision themes on which decision making became more expert-like (positive alpha value), as assessed separately for each squad.

Squad 1	Squad 2
Intra-squad communication	Intra-squad communication
Interacting with foreign females	Interacting with foreign females
Tactical questioning	Tactical questioning
Detaining a villager/HVI	Detaining a villager/HVI
Establish cordon	Establish cordon
Cultural awareness/sensitivity	Cultural awareness/sensitivity
Recognizing/dealing with suspicious behavior	Interaction with villagers
Respond to contact	Conduct security inspection
Handling Causalities	Patrol route selection
Entry/search of homes/buildings	Communicating with adjacent unit/QRF
	Assessing crowd behavior
	Integrating interpreter/ANA into mission
	Identify individuals for questioning

Differences between squads led to further analyses of the squad results independently. Due to the smaller sample size, a different measure (Krippendorff's alpha) was used to assess the level of agreement between squads and SME-DB. Krippendorff's alpha (α) was computed for each trainee for each decision theme. Values fall between 0 and 1, with 1 representing perfect agreement between individuals and SMEs, and 0 representing no relationship. This measure signifies whether individual trainees responded more or less like SMEs across the set of COAs associated with each vignette. Difference scores were then used to assess pre- to post-training changes. A positive value reflected improvement (responses more similar to SMEs), and a negative value would indicate less expert-like decision making. Squad 1 demonstrated improvement (positive alpha) on 10 of the decision themes, whereas Squad 2 demonstrated improvement on 13 of the decision themes. Specific decision themes are listed in Table 2.

A paired *t-test* was computed for each decision theme for each squad to assess whether the observed changes between the pre- and post-test were statistically significant. Squads demonstrated statistically significant improvement on several of the decision themes, as shown in Table 3. If a decision theme listed in Table 2 for a squad does not appear for that squad in Table 3, that means that while there was pre- to post- test improvement in decision making for that theme, the change was not statistically significant.

Table 3 List of decision themes showing statistically significant improvement between pre- and post-test.

Squad 1	Squad 2
Entry/search of homes/buildings	Intra-squad communication
Recognizing/dealing with suspicious behavior	Tactical questioning
Respond to contact	Detaining a villager/HVI
	Cultural awareness/sensitivity
	Integrating interpreter/ANA into mission

Interestingly, post-training decision making for both squads regarding communication with higher HQ *was further* from the mean of the SME-DB – indicating less expert-like decision making after training for this theme. One possible reason for this was that during the AARs a great deal of emphasis (reinforcement) was placed upon the importance of communicating with higher HQ. This emphasis may have caused the trainees to rate "report to higher HQ" as a highly effective choice regardless of the dilemma; a response given less weight in the SME-DB. Another possible explanation was that "higher HQ" was interpreted differently by the trainees and by the SMEs. For example, squad members may have interpreted "higher HQ" as their squad leader, whereas the SMEs may have interpreted it to mean Platoon Commander, Company Commander, etc.

4 CONCLUSIONS

A reliable, valid assessment tool is critical for evaluation of training environments and measurement of training-related changes in decision making. While further research is needed to validate and maximize the value of SJTs for meeting these requirements, results from this demonstration suggest that the SJT methodology may provide a useful assessment tool for assessing changes in decision making as a result of training. The FITE scenarios were not designed to address sociocultural decision making explicitly, but the scenarios, vignettes, and COAs can be tailored to a variety of training applications, including those with an emphasis on sociocultural awareness and decision making. Further, different scenario presentation formats (e.g., video-based scenarios) with differing levels of fidelity can be explored in order to identify those which are best suited to sociocultural decision making training.

A fully developed SJT approach has the potential to provide a method for assessing sociocultural competency and effects of sociocultural training by means of: 1) a baseline assessment of sociocultural competency for specific regions of the world (individual level of competency), 2) assessment of the training effectiveness of any sociocultural training venue (by quantifying changes in sociocultural competency between pre- and post-training levels); and 3) tailored training derived from assessment of individual competency with diagnostic results identifying where additional training is needed. The end result is a highly useable, scientifically based product that meets both sociocultural training and evaluation needs.

ACKNOWLEDGMENTS

The authors would like to acknowledge the funding and support of Mr Clarke Lethin and Mr. Jay Reist, Technical and Operational managers respectively of the FITE-JCTD project and Dr. Roy Stripling from the Office of Naval Research. We would also like to thank Mr. Jason Kobus, Mr. Matthew Kelly, Mr. Jared Ostertag, Mr. Geoffrey Williams and Mr. Bill Ross for their assistance in software development, data collection and statistical analyses.

REFERENCES

Bandura, A., Ross, D., & Ross, S.A. (1961). Transmission of aggressions through imitation of aggressive models. *Journal of Abnormal and Social Psychology, 63*, 575–582

Kobus, D. A., Palmer, E. D., Kobus, J., & Ostertag, J. (2010a). *Physiological and Behavioral Assessment of Presence: FITE JCTD Spiral 1 Operational Demonstration (Tech Report 10-10)*. Pacific Science & Engineering Group, San Diego, CA.

Kobus, D. A., Kobus, J., & Ostertag, J., Kelly, M., & Palmer, E. D., (2010b). *Assessing changes in tactical decision making: FITE-JCTD Spiral 2 Operational Demonstration (Tech Report 10-13)*. Pacific Science & Engineering Group, San Diego, CA.

Weekley, J.A. & Ployhart, R.E. (Eds.). (2006). *Situational Judgment Tests: Theory, Measurement, and Application*. Mahwah, NJ: Lawrence Erlbaum Associates, Inc.

CHAPTER 9

Methods for Capturing Cultural Lessons Learned and Training Cross-Cultural Skills

Michael J. McCloskey, Kyle J. Behymer

361 Interactive, LLC
Springboro, OH
mike@361interactive.com

ABSTRACT

In the past, US military forces undertook unilateral actions against adversaries, quickly assumed leadership in multinational operations, and worked in coalitions with well-marked responsibilities requiring little direct collaboration. Times have changed. To achieve mission success in contemporary operating environments, our Soldiers must now regularly assess and interact with individuals from other cultures while undertaking combat and counterinsurgency operations, disaster relief, stability-support operations, and foreign forces training as a means of ensuring stability in various regions of the world. At all levels of command, they must consider the immediate and long-term impact of planned actions on foreign nationals. Further, the requirement to deploy forces to a large range of possible settings leaves military trainers with an enormous task of understanding and developing *generalizable* cross-cultural skills in their Warfighters; both for current operations and the foreseeable future. To address this demand, we are leveraging the expertise of Army Special Forces and Civil Affairs cultural trainers and Soldiers to develop *CultureGear*, a cognitively-authentic, research-based tool that will be integrated into existing JFKSWCS cultural training curricula. *CultureGear* is based on an operational model of cross-cultural competence that leverages decision-centered observation/interpretation scenarios, perspective taking exercises, and competence assessment instruments. Cognitive task analysis methodologies have been leveraged to identify the critical skills and aptitudes required to assess complex cultural situations, accurately assume the perspectives of foreign nationals, and recognize the impact of one's own biases in situation assessments.

Users of *CultureGear* will be able to select and participate in cultural scenarios based on a variety of search parameters, including region, rank, conflict/situation, core skills involved, and others. The final tool will also allow trainers and selected deployed personnel to develop and upload their own cultural scenarios, incorporating imagery and lessons-learned obtained in theatre.

Keywords: cross-cultural competence, cultural training, cognitive skills, cultural awareness, observation skills, competence assessment

1 UNDERSTANDING CROSS-CULTURAL COMPETENCE

For Soldiers and Marines in the contemporary operational environment, performance beyond typical military technical and tactical skills is increasingly required. For example, Army small unit leaders and Special Forces Soldiers must often interact with, engage, and persuade local populations in foreign countries to achieve mission objectives. To address these new requirements, training approaches that are grounded in both military expertise and social science research concepts are needed (Holmes-Eber, Scanlon, and Hamlen, 2009). Despite the growing awareness within the military of the importance of understanding how cultural variables can impact mission performance and the proliferation of cross-cultural training tools, many challenges remain for the research community to develop effective training. Most significantly, despite a wealth of experience available from Soldiers and Marines with multiple deployments, cultural information on current operations is limited and difficult to find (Holmes-Eber, Scanlon, and Hamlen, 2009). Few mechanisms exist for easily converting lessons learned from previous deployments into training scenarios. Additionally, although cross-cultural competence is universally recognized as important for military operations, the research community has struggled to agree on a consistent definition of cross-cultural competence (3C) (see Abbe, Gulick, and Herman, 2007, for a discussion of this issue).

Before developing effective 3C training tools, an operational definition of 3C that captures the field requirements of these deployed military personnel in novel cultural environments must first be identified. To this end, the research team capitalized on proven social science research techniques to capture the components of operational 3C associated with effective performance in the field. Cognitive Task Analysis Interviews were conducted with over 200 Soldiers and Marines who had experienced significant amounts of cross-cultural interaction during their deployments. The Critical Decision method (CDM) is a particular cognitive interview technique for learning from specific, non-routine events which challenge a person's expertise (Klein, Calderwood, and MacGregor, 1989). Using recollection of a specific incident as its starting point, CDM employs a semi-structured interview format with specific, focused probes to elicit particular types of information from the interviewee. Critical incidents were elicited wherein Soldiers recalled the challenging aspects of cross-cultural encounters or assessments. Although recall of specific events cannot be assumed to be perfectly reliable, the method has been highly successful in eliciting perceptual cues, information sources, sequencing of decisions and judgments made, and details of judgment and decision strategies that are generally not captured with traditional reporting methods (Crandall, Klein and Hoffman, 2006). In addition, team ranking task interviews were conducted during which Soldiers evaluated and described their peers based on perceived cultural ability, as well as simulation interviews during which Soldiers responded to cognitively authentic deployment scenarios. These data were compiled and analyzed with a focus on identifying patterns and frequencies in the presence and/or mention of specific KSAAs (knowledge, skills, abilities, and aptitudes). This yielded a set of 29 cognitive, affective, and behavioral KSAAs, which were later comprised to form an overall 3C construct wherein different levels of competence were described in terms of the KSAAs that were present (or absent), as well as the levels of development for KSAAs within each component. The 29 KSAAs were consolidated though a series of analyses into a final construct that included five 3C components (see Figure 1), and five stages of development. A detailed description of the interview collection techniques, data analyses conducted, and the overall 3C model can be found in McCloskey et al., 2010.

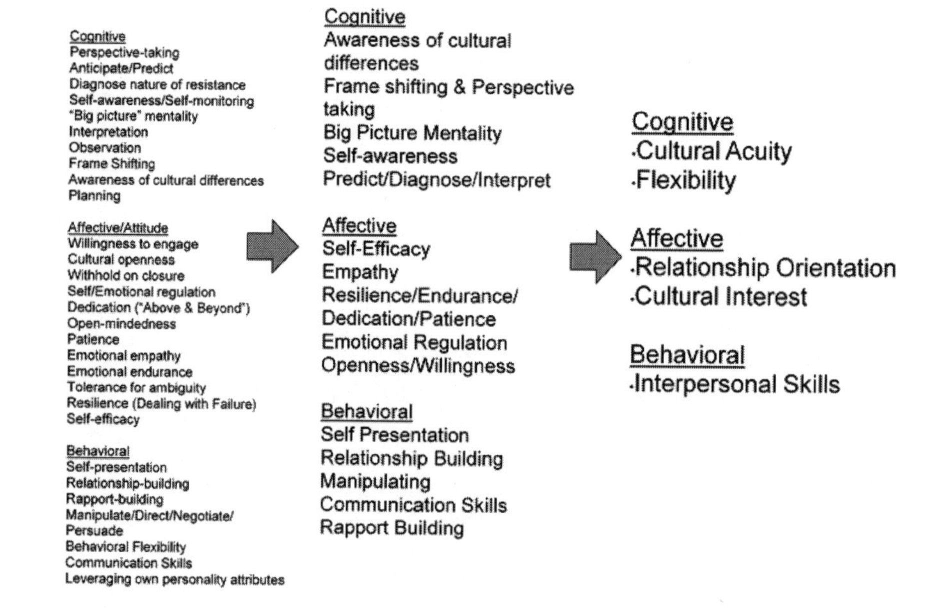

Figure 1: Knowledge, skills, abilities, and aptitudes found to influence general cross-cultural performance in the field.

2 TRAINING COGNITIVE CROSS-CULTURAL SKILLS

While the aforementioned research focused on the development of a computer-based 3C *assessment* tool, we have since leveraged many findings from the research and applied them to the development of CultureGear, a general 3C *training* tool that is currently being tailored to needs of the John F. Kennedy Special Warfare Center & School (JFKSWCS). CultureGear focuses on training the general cognitive skills that we identified as being critical to the mission success of small unit leaders operating in novel cultural environments. Four training modules (discussed in-depth below) have been developed for the following skills: cultural/self-awareness, cultural observation and assessment, perspective taking, and mindful preparation.

Additionally, our research supported the assertion that when it comes to general cross-cultural competence, not everyone will be starting from the same point. Soldiers will have different general skills and attitudes that they bring to the table upfront. To address this, culture-general training needs to include an assessment of each trainee's starting point. To this end, we are incorporating an assessment module within CultureGear. This module is based on our model of cross-cultural competence that was developed for a related, Army-sponsored effort. In addition to receiving customized feedback on their individual strengths and weaknesses, the user will be guided to specific modules of the tool where they can address their weaknesses and learn how to best leverage their cultural strengths. Complete details of this assessment module can be found in McCloskey et. al. (in publication).

2.1 Cultural/Self-Awareness

One of most common cognitive abilities of successful Soldiers in unfamiliar cultural territory is to be consciously aware of their own cultural biases and the impact on their assessments – they are very aware of how their own perceptions of people and situations are altered by their own cultural views. To promote this self-awareness in trainers, CultureGear contains an awareness module wherein trainees provide their levels of agreement on a variety of statements that reflect each of the PMESII-PT dimensions (Political, Military, Economic, Social, Infrastructure, Information, Physical Environment and Time). After providing their reactions, trainees are then shown how their perceptions of "normal" or "common" behaviors and attitudes can differ greatly from those of individuals in other cultures. Trainees can then compare their perceptions and beliefs to general tendencies of other specific cultures of their choosing as shown in Figure 2. The intent here is to provide some initial exposure to junior Warfighters to the idea that cultures do differ in meaningful, significant ways, and that everyone has their own cultural biases that can impact their assessments.

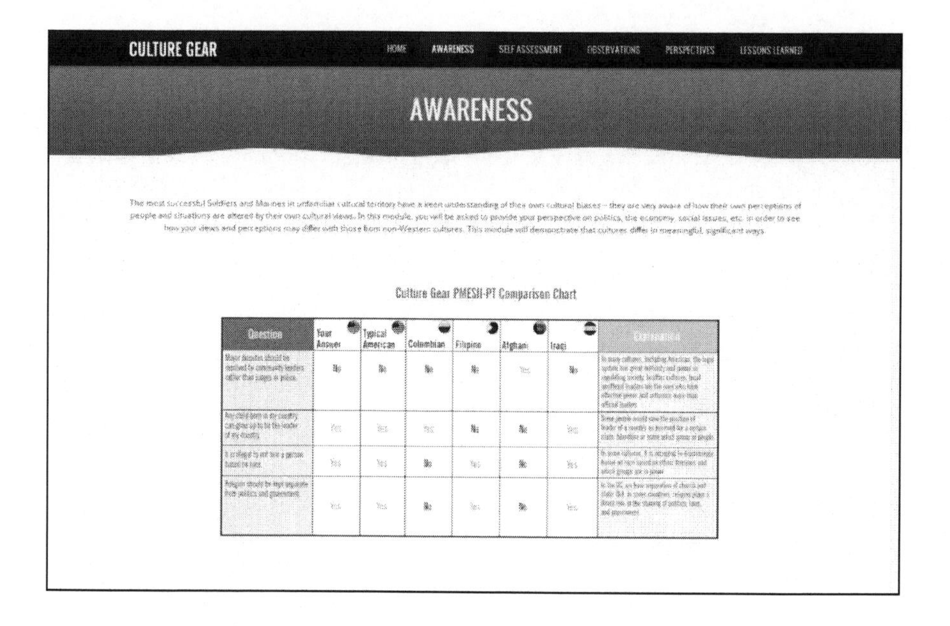

Figure 2: CultureGear cultural perception/belief comparison screen, leveraging PMESII-PT framework.

2.2 Cultural Observation and Assessment

Warfighters who operated most effectively within different cultures also possessed keen observational and assessment skills (McCloskey, et al. 2009). They had the ability to quickly and accurately assess cross-cultural interpersonal dynamics and situations and choose appropriate courses of action based on their assessments. This critical assessment skill, which we refer to as *Cultural Acuity*, is a key focus of CultureGear. To promote Cultural Acuity, we have developed training scenarios wherein Soldiers can practice building their cultural acuity/assessment skills. For the observation scenarios, Soldiers are provided with multimedia representations of different scenes (imagery, video, etc.). For each representation, the trainee is prompted to identify critical elements within the scene that would impact their cultural assessments of the area, and then provide rationales for each. After then making actual assessments based on their observations, trainees are shown how their observations and assessments differed from those of more culturally-experienced Warfighters (Figure 3).

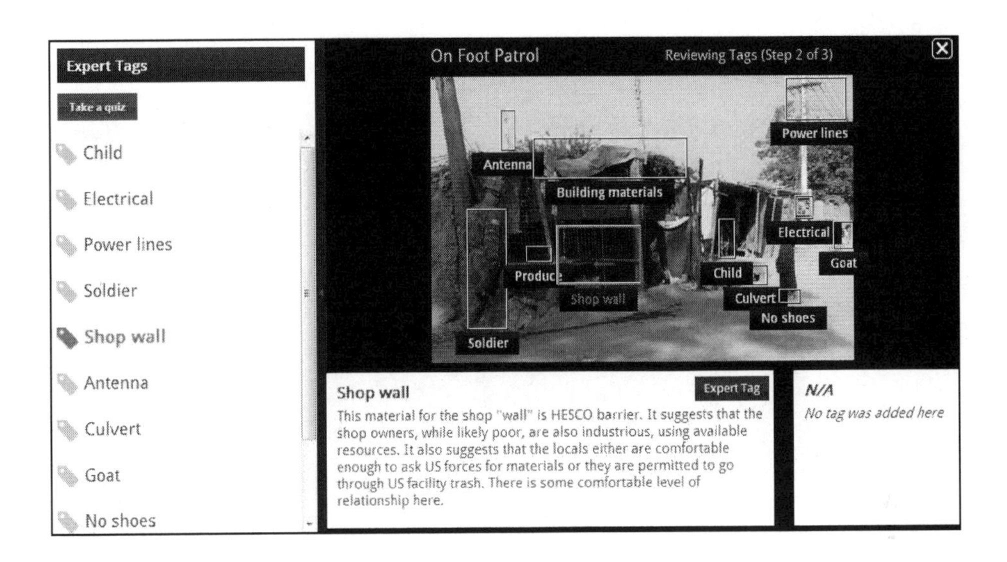

Figure 3: Scene Analysis Screen from CultureGear wherein a trainee compares their assessments to those of culturally-experienced Warfighters.

Initial expert comments were elicited through interviews with Soldiers recently returned from deployments who had missions that required extensive cross-culture interaction (training foreign national counterparts, for example). We have found that allowing inexperienced Warfighters to first make their own assessments, and then compare them to the assessments of more experienced Warfighters is a powerful learning tool (McCloskey and Behymer, 2010; McCloskey, 2009; McCloskey, 2009; McCloskey and Baxter, 2008). This side-by-side assessment comparison is a concept that is continually integrated throughout the training experience.

An additional critical functionality is being added to the observation scenario module based on feedback from the user community. Trainers and small unit leaders often have their own stories and even supporting imagery. In addition to seeing the value of the preexisting scenarios, tool reviewers expressed the desire to upload their own imagery, video, and text-based lessons-learned, and incorporate them into the training experience. As a result, we have developed functionalities that give trainers and small unit leaders the capability to upload their own media and annotate them with their own lessons learned, incorporating their uploaded information into an interactive scenario that they create. A quiz creation functionality is also being developed that will allow trainers/uploaders to provide opportunities for learners to not just observe and interpret, but also make assessments and select courses of action based on their assessments. This is all tied to a Google-Earth interface that automatically geo-tags the uploaded imagery/information and places it on an interactive map (Figure 4). Users can also tag the relevant PMESII-PT elements in the imagery that they upload. This promotes a greater degree of ownership in the training and also has great potential for creating a "corporate memory" of imagery/scenarios to facilitate continuously current and relevant training.

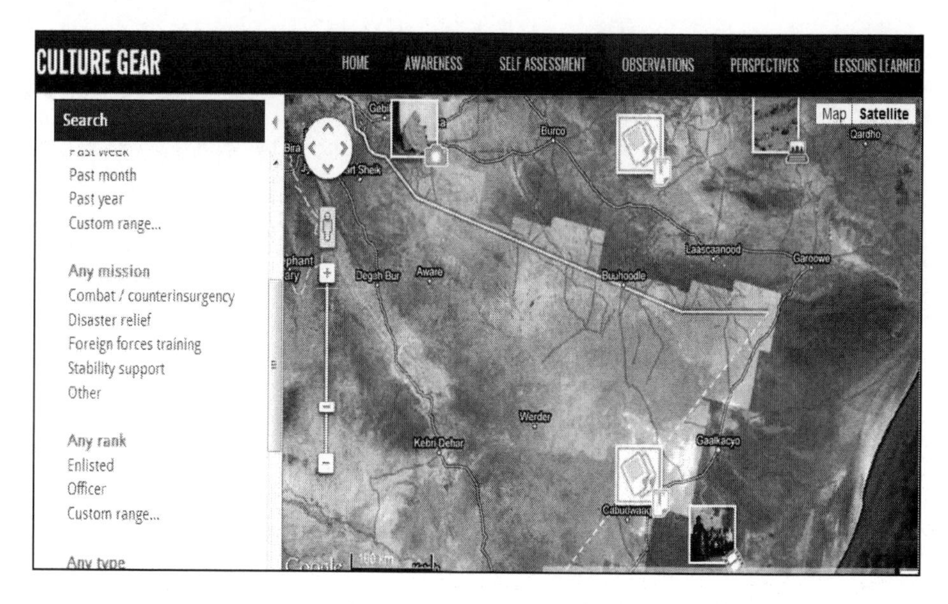

Figure 4: Google Earth Interface showing selection criteria, different media types and navigation options.

2.3 Perspective Taking

Perspective taking is defined as the ability to see things from another's point of view (Davis, 1983). Rentsch et al. (2007) hypothesized that understanding the thoughts and rationales of those from other cultures and to then use this understanding to guide mission-related actions could improve mission performance. Within this module, trainees are presented with real-world imagery of deployed Soldiers interacting with a local populace. They are then asked to assess how the Soldiers are likely being perceived by the locals based on their observations. Perspective taking training has met with great success utilizing this type of scenario-based approach, where trainees are asked to assume the roles of adversaries or collaborators (Redmond, 1995). In this module Soldiers and Marines can practice perspective taking by applying the knowledge they obtain and the critical skills they learn in other areas of the tool. Much like actual field situations, the scenarios are crafted so that they challenge the trainee to identify observable cues, recall or consult information about PMESII-PT elements, and practice putting themselves in the place of allies, adversaries or host nation individuals.

2.4 Mindful Preparation

Through our interviews we learned how skilled Soldiers and Marines prepare for deployments or overseas assignments which will involve significant cultural interaction. Many learn the customs, talk with those who have been deployed there previously, or research the region's history, geography, and economy. However, only the best go a step further and actively compare this newly found knowledge with the same type of information about the United States. Only by making these comparisons can Soldiers and Marines fully appreciate the meaning behind the numbers. By making comparisons to what they are used to in the U.S., Soldiers and Marines can form expectancies of how the region's people will view the world, the U.S., and themselves.

In the preparation module (shown in Figure 5), trainees are given a visually-based tool to compare the United States with other countries on specific PMESII-PT variables. The underlying data is imported from a wide variety of publicly available data sources such as the CIA factbook which are frequently updated. Trainees are also able to see how specific variables have changed over time, which can add to their understanding of the region.

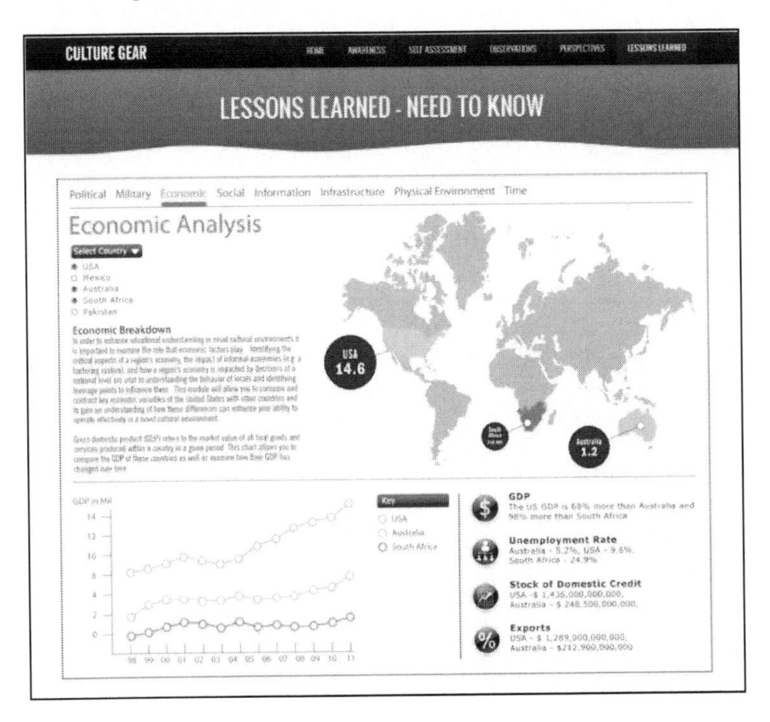

Figure 5: CultureGear preparation module

3 CONCLUSIONS

In order for *CultureGear* to be viewed as a needed asset versus another burden on our small unit leader's fully packed schedule, we have incorporated the feedback of trainers and end-users at all stages of development. The final stage in this effort will be to evaluate *CultureGear* to ensure that the tool truly improves the skills we hypothesize that it does. Our measures of success will be marked improvements in performance as rated by leaders and peers, and perhaps most importantly, acceptance by the user community.

ACKNOWLEDGMENTS

The authors would like to acknowledge Dr. Ivy Estabrooke and Mr. Gary Kollmorgen of the Office of Naval Research (ONR) Human Social Cultural and Behavioral Sciences (HSCB) program for their continued support of, and interest in, this effort. The authors would further like to acknowledge Dr. David Brand, Ms. Lorae Roukema, and LTC David Walton of the US Army JFKSWCS for their vision and enduring support of our research and development of CultureGear. The authors would finally like to acknowledge all our Armed Forces for continually placing themselves in harm's way to keep our country safe and strong.

REFERENCES

Abbe, A., L. M. V. Gulick, and J. L. Herman. 2007. *Cross-cultural competence in Army leaders: A conceptual and empirical foundation. (*Study Report 2008-1). Arlington, VA: U.S. Army Research Institute for the Behavioral and Social Sciences.

Crandall, B., G. A. Klein, and R. R. Hoffman. 2006. *Working minds: A practitioner's guide to cognitive task analysis.* London, England: The MIT Press.

Davis, M. H. 1983. Measuring individual differences in empathy: Evidence for a multidimensional approach. *Journal of Personality and Social Psychology* 44: 113-126.

Klein, G. A., R. Calderwood, and D. MacGregor. 1989. Critical decision method for eliciting knowledge. IEEE Transactions on Systems, *Man and Cybernetics* 19(3): 462-472.

Holmes-Eber, P., P. M. Scanlon, and A. L. Hamlen. 2009. *Applications in operational culture: Perspectives from the field.* Quantico, Virginia: Marine Corps University.

McCloskey, M. J., K. J. Behymer, E. L. Papautsky, and A. L. Grandjean, in publication. *Measuring learning and development in cross-cultural competence.* (Technical Report). Alexandria, VA: U. S. Army Research Institute for the Behavioral and Social Sciences.

McCloskey, M. J., K. J. Behymer, E. L. Papautsky, K. G. Ross, and A. Abbe. 2010. *A developmental model of cross-cultural competence at the tactical level.* (Technical Report 1278) Alexandria, VA: U. S. Army Research Institute for the Behavioral and Social Sciences.

McCloskey, M. J. 2009. *CultureGear: A Cognitive Approach to Promoting Cross-Cultural Perspective Taking Skills.* Defense Equal Opportunity Management Institute Biennial EO, Diversity, and Culture Research Symposium, Patrick AFB, FL.

McCloskey, M. J. 2009. *Cognitive Interfaces to Support Cultural Competence Assessment.* Proceedings of the Human, Social, Cultural and Behavioral Summit, Office of Naval Research, August 1-3, 2009.

McCloskey, M. J., A. L. Grandjean, K. J. Behymer, and K. G. Ross. 2009. *Assessing the development of cross-cultural competence in Soldiers.* (Technical Report). Alexandria, VA: U. S. Army Research Institute for the Behavioral and Social Sciences.

McCloskey, M. J. and K. J. Behymer. 2010. *Modeling and Assessing Cross-Cultural Competence in Operational Environments*, Proceedings of the 1st International Conference on Cross-Cultural Decision Making, CRC Press, Taylor & Francis, Ltd.

McCloskey, M. J. and H. C. Baxter, H.C. 2008. *Promoting cross-cultural perspective taking skills in operational environments.* Proceedings of the Interservice/Industry Training, Simulation, and Education Conference, Orlando, FL.

Redmond, M. V. 1995. A Multidimensional Theory and Measure of Social Decentering. Journal of Research in Personality 29(1): 35-58.

Rentsch, J., A. Gunderson, G. Goodwin, and A. Abbe. 2007. Conceptualizing multicultural perspective taking skills. (Technical Report 1216). Alexandria, VA: U. S. Army Research Institute for the Behavioral and Social Sciences).

CHAPTER 10

Credibility Assessment and Inference for Fusion of Hard and Soft Information

R.C.Núñez[*], *T.L.Wickramarathne*[*], *K.Premaratne*[*], *M.N.Murthi*[*], *S.Kübler*[‡], *M.Scheutz*[◊], *M.A.Pravia*[+]

[*]University of Miami
Coral Gables, FL, USA
{nunez, t.wickramarathne}@umiami.edu, {kamal, mmurthi}@miami.edu

[‡]Indiana University
Bloomington, IN, USA
skuebler@indiana.edu

[◊]Tufts University
Medfort, MA, USA
mscheutz@cs.tufts.edu

[+]BAE Systems
Burlington, MA, USA
marco.pravia@baesystems.com

ABSTRACT

Effectively combining multiple (and complementary) sources of information is becoming one of the most promising paths for increased accuracy and more detailed analysis in numerous applications. Neuroscience, business analytics, military intelligence, and sociology are among the areas that could significantly benefit from properly processing diverse data sources. However, traditional methods for combining multiple sources of information are based on slow or impractical methods that rely either on vast amounts of manual processing or on suboptimal representations of data. We introduce an analytical framework that allows automatic and efficient processing of both hard (e.g., physics-based sensors) and soft (e.g., human-generated) information, leading to enhanced decision-making in multi-source environments. This framework combines Natural Language Processing (NLP) methods for extracting information from soft data sources and the Dempster-Shafer (DS) Theory of Evidence as the common language for data representation and inference. The steps in the NLP module consist of part-of-speech tagging, dependency parsing, coreference resolution, and a conversion to semantics based on first order logic representations. Compared to other methods for handling uncertainties, DS theory provides an environment that is better suited for capturing data models and imperfections that are common in soft data. We take advantage of the fact that computational complexity typically associated with DS-based methods is continually decreasing with both the availability of better processing systems, as well as with improved processing algorithms such as conditional approach to evidence updating/fusion. With an adequate environment for numerical modeling and processing, two additional elements become especially relevant, namely: (1) assessing source credibility, and (2) extracting meaning from available data. Regarding (1), it is clear that the lack of source credibility estimation (especially with human-generated information) could direct even the most powerful inference methods to the wrong conclusions. To address this issue we present consensus algorithms that mutually constrain the data provided by each of the sources to assess their individual credibility. This process can be reinforced to get improved results by incorporating (possibly partial) information from physical sensors to validate soft data. At the end of a credibility estimation process, every piece of information can be properly scaled prior to any inference process. Then, meaning extraction (i.e., (2)) becomes possible by applying the desired inference method. Special consideration must be taken to ensure that the selected inference method preserves the quality and accuracy of the original data as much as possible, as well as the relations among different sources of information and among data. To accomplish this, we propose using first-order logic (FOL) in the DS theoretic framework. Under this approach, soft information (in the form of natural language) is analyzed syntactically and for discourse structure, and consequently converted into FOL statements representing the semantics. Processing of these statements through an "uncertain logic" DS methodology renders bodies of evidence (BoE) that, combined with experts' opinions stored in knowledge bases, can be fused to provide accurate solutions to a wide variety of queries. Examples of queries include finding or refining groups of suspects in a crime scene, validating credibility of witnesses, and categorizing data in the web. When hard-sensor data is also incorporated in the inference process, challenging applications such as multi-source detection, tracking, and intent detection, could also be addressed with the proposed solution.

Keywords: Evidence Fusion, Source Credibility, Credibility Estimation, Consensus, Belief Theory, Theory of Evidence, First Order Logic, Meaning Extraction, Natural Language Processing, Dependency Parsing, Coreference Resolution

1 INTRODUCTION

The development of new sensing and data acquisition technologies is occurring at such a fast pace that it is triggering a need for more sophisticated meaning extraction and inference methods. These inference schemes need to take advantage of the increased amounts of information, producing more accurate and complete solutions to varied problems. This information, in general, can be classified as "hard" or "soft". "Hard information" refers to information generated by physics-based sources, and "soft information" refers to information generated from human-based sources, including human reports, text and audio communications, and open sources such as newspapers, radio/TV broadcasts, and web sites.

Solutions for meaning extraction and inference have typically targeted either hard information (e.g., sensor networks) or soft information (e.g., data mining). However, simultaneously using both hard and soft information is still mostly a human-intensive task, with very little research addressing this hard/soft information fusion application (Pravia, et. al., 2008).

Aimed at addressing this issue, we introduce a general model for automated analysis of hard and soft information. As an application of this technology, consider the following scenario.[1] A team of experts is trying to assess the credibility of witnesses of a crime scene. The messages provided by the witnesses, as they were documented, are:

Witness 1 (W1): "The suspect was driving a black SUV";
Witness 2 (W2): "The suspect was driving a white sedan";
Witness 3 (W3): "The suspect was driving a white vehicle".

Each of the witnesses was asked to rate, from 0 to 100, how certain he/she was on the information they provided. They answered 80%, 90%, and 95%, for W1, W2, and W3, respectively. In addition to the information provided by witnesses, the team of experts has access to video surveillance (VS) reports that identified the suspect's vehicle as a light-colored sedan. This type of report has been characterized as being 98% accurate.

Having this information, is it possible to estimate the credibility of the witnesses? Is it possible to refine the crime-scene scenario? Although this simple scenario can be easily solved by humans, our work aims at defining a framework that allows automatic analysis of this type of events, especially when hundreds of thousands of pieces of information are available, all of them potentially providing valuable information for the human experts.

A general framework for solving this type of problem is shown in Figure 1. In this framework, soft data is converted into first-order logic (FOL) constructs by a Natural Language Processing (NLP) module. FOL is preferred for the semantic representation because it preserves a higher amount of information compared to other methods (e.g., RDF Graphs). A combination of semantic representation

---

[1] We introduce this scenario as a running example that will allow us to easily describe each step of our hard and soft information fusion process.

methods with higher and lower levels of detail could also be used as means of reducing complexity. These logic constructs are quantified and mathematically modeled (e.g., via probability or belief functions), and their credibility is assessed. An alternative for assessing credibility is based on finding consensus among information sources. Distance to consensus can be used as a measure of credibility. With such a measure, the data could be properly weighted for further processing in meaning extraction and inference.

In the remainder, we introduce the methods that we have designed for each of the components in our hard/soft information fusion framework, and, as an illustration of the techniques, we apply them for processing the scenario described above.

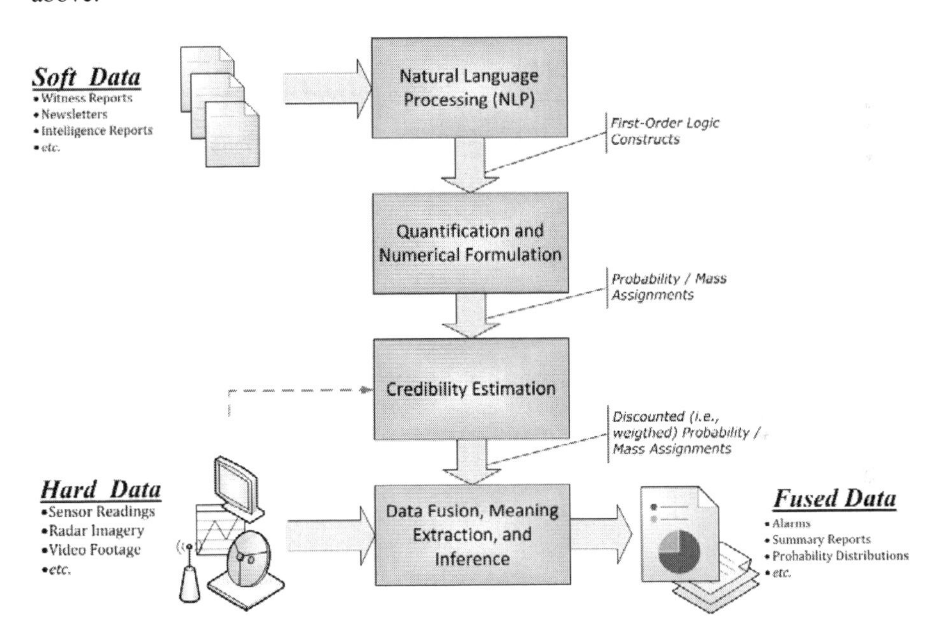

Figure 1 Fusion of hard and soft information.

2 BUILDING NUMERICAL MODELS FROM TEXTUAL DATA

2.1 Natural Language Processing

The goal of the NLP module is to accept plain text, such as witness statements, interrogation protocols, texts from the WWW, or any other textual source, and produce an analysis that allows further processing in the Dempster-Shafer (DS) framework. Overall, the NLP analysis faces two problems: (1) efficiency concerns and (2) out-of-domain data. With regard to efficiency concerns, syntactic parsing and coreference resolution are computationally very intensive steps. Additionally, a traditional NLP architecture is built as a pipeline, in which texts are processed sentence by sentence, i.e. one sentence is analyzed by the first module, then sent to the second, etc. We approach this problem by using a fully incremental architecture, in which words are processed as they come in, and partial results are directly passed to the next module without waiting for the full sentence to be analyzed. This necessitates changes in the individual modules, which cannot rely on having access to context beyond the word currently processed. Problem (2) refers to the fact that all NLP modules are based on supervised machine learning, and thus need to be trained. The only available training set is often a part of the Wall Street Journal, which has been annotated on different levels (cf. e.g., Marcus et al., 1993). For this reason, we need to develop methods that allow us to adapt the learned model to the domain of texts that need to be analyzed (Kübler et al., 2010, Kübler and Baucom, 2011).

The first step in the NLP module is part-of-speech (POS) tagging, which assigns words classes to words in a sentence. Thus witness statement W1 would be assigned the following parts of speech: The/DT suspect/NN was/VBD driving/VBG a/DT black/JJ SUV/NNP. 'The' and 'a' are articles, 'suspect' is a noun, 'was' is a verb in past tense, 'driving' a present participle, 'black' is an adjective, and 'SUV' is a name. We use an n-best, anytime implementation of a Markov model tagger.
The POS tagged sentences serve as input for a dependency parser. Dependency parsing performs a syntactic analysis. We use a dependency parser, which assigns dependency relations to pairs of words in a sentence. The dependency analysis for W1 is shown in Figure 2. The analysis shows that 'suspect' is the subject of 'was', 'driving' is a verbal complement of 'was', 'SUV' is the direct object of 'driving', and the two articles modify the nouns. As parser, we use MINK (Cantrell, 2009), a fully incremental implementation of MaltParser (Nivre et al., 2007).

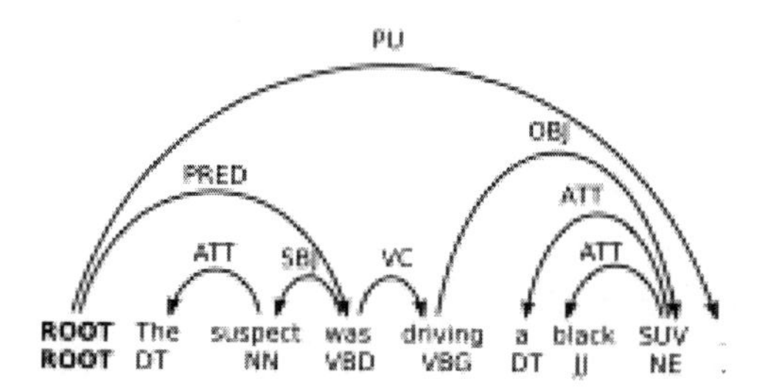

Figure 2 The dependency analysis of W1

After the syntactic analysis, we perform coreference resolution. In this step, we determine which expressions in the document refer to the same entity. That is, if we had a fourth witness statement:

> **Witness 4 (W4)**: "I saw the suspect driving a black SUV; he parked the vehicle right in front of the shop",

then coreference resolution will tell us that 'the suspect' and 'he' refer to the same person and that 'a black SUV' and 'the vehicle' refer to the same car. For coreference resolution, we use UBIU, a robust, multilingual system using memory-based learning for classifying pairs of potentially coreferent mentions (Zhekova and Kübler, 2010, Zhekova and Kübler, 2011). For the future, we are planning on extending the system to cross-document coreference resolution, which will enable us to cross-link mentions of persons across different documents, thus giving us sets of sentences that can be used for DS inference.

After coreference resolution, we convert the sentence into a semantic representation based on first order logic statements. Thus, W1 in our running example (see Section 1 above), would be represented as □x, y: WasDriving(x, y) ^ Suspect(x) ^ Black(y) ^ Suv(y), stating that there is an entity x who is a suspect, and an entity y which is an SUV, and that x was driving y. The conversion is lexicon-based (Kübler et al., 2011).

2.2 Basic Probability Assignments for FOL Constructs

With a set of FOL sentences available, the next problem that must be tackled is that of quantifying this information in such a way that uncertainty in the information is properly modeled, and that the inference engine can use it.

Although FOL has traditionally been one of the preferred modeling frameworks for resolution and inference, FOL is not designed for handling problems with uncertainties. Methods such as uncertain reasoning and probabilistic logic extend FOL for solving these problems (Genesereth and Nilsson, 1987). These methods improve the scope of problems that can be solved based on FOL, but they do not address the cases of incomplete data, and they cannot be used for easily describing information uncertainty by lower and upper probability bounds. We propose modeling data and making inference using Dempster-Shafer (DS) theory (Shafer, 1976) for addressing these issues. DS theory has been successfully used in applications such as rule mining (Hewawasam, et. al., 2007) and target identification (Ristic and Smets, 2005).

Current approaches for building DS models for logic operators provide models for particular uncertain operators in propositional logic. For example, Benavoli, et. al., (2008) introduce a method for modeling uncertain implication rules using DS models. We are enhancing model-building strategies by defining a method that incorporates logic quantifiers (hence, FOL models), and preserves fundamental logic properties (e.g., associativity, commutativity, distributivity, and idempotency) for a set of basic logic operators (i.e., `not`, `and`, `or`, and implication rules).

Based on our FOL models for DS, the witness sentences in our example can be converted into probability (or mass) assignments. Let us assume that the output of the NLP stage for our running example is the following:

W1: `∃x,y: WasDriving(x, y) ^ Suspect(x) ^ Black(y) ^ Suv(y);`

W2: `∃x,y: WasDriving(x,y) ^ Suspect(x) ^ White(y) ^ Sedan(y);`

W3: `∃x,y: WasDriving(x, y) ^ Suspect(x) ^ White(y) ^ Vehicle(y);`

where $x \in \Theta_{suspects}$, and $y \in \Theta_{vehicles}$. $\Theta_{suspects}$ and $\Theta_{vehicles}$ are called the frame of discernment (FoDs), i.e., the sets that define the groups of elementary events related to the problem.

For a decision process, we typically need to incorporate some domain knowledge. In this case we assume:

$\Theta_{\text{vehicles}} = \{ \Theta_{\text{color}} \times \Theta_{\text{type}} \}$;
$\Theta_{\text{color}} = \{$ white, silver, red, brown, black $\}$;
$\Theta_{\text{light}} = \{$ white, silver $\} \subset \Theta_{\text{color}}$;
$\Theta_{\text{dark}} = \{$ brown, black $\} \subset \Theta_{\text{color}}$;
$\Theta_{\text{type}} = \{$ sedan, jeep, SUV, truck $\}$.

DS theory models are defined by mass assignments. A mass assignment (or basic belief assignment) is a mapping $m_\Theta(\cdot)$: $2 \rightarrow [0, 1]$, such that $\sum_{A \subseteq \Theta} m_\Theta(A) = 1$ and $m_\Theta(\emptyset) = 0$. The mass assignment measures the support assigned to proposition $A \subseteq \Theta$. The triple $\{\Theta, \Im, m(\cdot)\}$, with $\Im$ being the set of all elements for which m(A) > 0, is called the body of evidence (BoE).

The mass assignments corresponding to witness statements and video surveillance reports are:

W1: $m_1(\{\text{black}\} \times \{\text{SUV}\}) = 0.80$; $m_1(\Theta_{\text{vehicles}}) = 0.20$;
W2: $m_2(\{\text{white}\} \times \{\text{sedan}\}) = 0.90$; $m_2(\Theta_{\text{vehicles}}) = 0.10$;
W3: $m_3(\{\text{white}\} \times \Theta_{\text{type}}) = 0.95$; $m_3(\Theta_{\text{vehicles}}) = 0.05$;
VS: $m_4(\{\text{white, silver}\} \times \{\text{sedan}\}) = 0.98$; $m_4(\Theta_{\text{vehicles}}) = 0.02$.

These mass assignments are obtained by using DS fusion based on the conditional update equation (CUE) (Premaratne, et. al., 2009). The fusion operators must be properly tuned for handling logic operations. The mass assignments can then be used for credibility assessment and inference, as is described next.

3 ESTIMATING CREDIBILITY OF SOURCES

It is very important, especially when dealing with multiple sources of information, to account for the credibility (i.e., trustworthiness) of the sources. In the DS framework, it is possible to account for this credibility by a procedure called discounting.

The issue becomes then, being able to estimate this credibility measure (in some applications such as judge or jury trials, the whole problem is precisely assessing the credibility of witnesses). When an adequate number of sources are considered, it is not unreasonable to assume that the truth is reflected in the majority opinion. If this majority opinion can be established via some rational aggregation procedure, the very aggregate, often referred to as a consensus, can in turn be used for credibility estimation.

We propose a consensus-based technique for credibility estimation of evidence in the absence of the ground truth. The credibility of a BoE E (i.e., a particular piece of evidence) can be defined as (Wickramarathne, et. al., 2012):

$$\text{Cr}(E) = \left(1 - \text{dist}(E, E^t)^\lambda\right)^{1/\lambda},$$

with E^t denoting a BoE that contains the ground truth, and $\lambda \in \Re^+$. This definition requires the computation of consensus BoEs. These BoEs can be computed using an iterative procedure based on the CUE. A detailed description of this credibility estimation technique as well as of the BoE update procedure can be found in (Wickramarathne, et. al., 2012).

Based on this methodology, the consensus BoE in our example is given by:

$$m_{\text{consensus}}(\ \{\text{white}\} \times \{\text{sedan}\}\ ) = 1.00,$$

and the distance to consensus as well as the credibility of the sources become:

dist(WS1, Consensus) = 0.9154; Cr (WS1) = 0.0846
dist(WS2, Consensus) = 0.0975; Cr (WS2) = 0.9025
dist(WS3, Consensus) = 0.8786; Cr (WS3) = 0.1214,

with the credibility estimated using $\lambda = 1$. This estimated credibility can then be used for evidence discounting prior to fusion operations in meaning extraction and inference.

4 MEANING EXTRACTION AND INFERENCE

As mentioned above, DS fusion offers several advantages over other meaning extraction methods, given that it incorporates a more rigorous modeling of uncertainties, and that it allows relevant fusion operations even in the presence of incomplete data. Nevertheless, there are some challenges typically associated with DS-based fusion: (1) limitations when handling information with dissimilar FoDs; (2) sensitivity to contradictory evidence; and (3) computational complexity.

The first of these challenges is particularly magnified when soft information is processed. When dealing with multiple sources of soft data, it is not uncommon to find data generated from non-identical FoDs. For example, the information contained in a public database of vehicles belonging to town residents would have a much larger, but not completely disjoint, scope than the vehicles that had been recorded at a checkpoint. Conventional DS methods are not suitable for such problems. Moreover, conventional DS fusion methods are very sensitive to contradictory evidence, usually rendering counter-intuitive results (which is the second challenge mentioned above). We address challenges (1) and (2) by performing fusion operations based on the CUE (Wickramarathne, et. al., 2010).

Computational complexity (i.e., challenge (3)) of DS methods exponentially increases with increasing cardinality of the FoD. As a result, in many DS-based applications, even the most common and fundamental task of conditioning can quickly become computationally prohibitive, especially in the presence of FoDs with high cardinality. To reduce computational complexity we make use of the Conditional Core Theorem (CCT) (Wickramarathne, Premaratne, and Murthi, 2010). The CCT identifies the propositions that will receive a positive mass after conditioning without any numerical computations. The advantage of such a result is that it is possible to avoid the computation of all the $2^{|\Theta|}$ propositions that otherwise would have to be computed to evaluate the conditional masses. In real application settings, the CCT may yield computational savings of 80% or more.

It is worth noting that CUE-based operations are embedded into the credibility estimation method described in Section 3 above. Then, in our example, meaning extraction could be obtained from analyzing the consensus BoE defined by $m_{consensus}(\ \{white\} \times \{sedan\}\ ) = 1.00$. In this case, the result indicates that there is total certainty that the suspect was driving a white sedan. Inference and meaning extraction in more complex scenarios can be done by following the methodology introduced in (Wickramarathne, et. al., 2011).

5 CONCLUSIONS

In this paper we introduced a general framework that allows automatic and efficient processing of both hard and soft information, leading to enhanced decision-making in multi-source environments. In this framework, soft data is converted into FOL constructs by a NLP module. This module consists of part-of-speech tagging, dependency parsing, coreference resolution, and a conversion to semantics. The logic constructs resulting from this module are quantified and mathematically modeled. In particular, DS theoretic models are generated based on methods arising from the CUE properly tuned for consistency with logic operations. The mathematical models are then used for assessing credibility. The latter is estimated based on the distance to the consensus among information sources. The credibility measure is then used for discounting BoEs in DS-based information fusion. The overall framework could be directly applied for solving problems in varied areas such as neuroscience, business analytics, military intelligence, and sociology, among others.

ACKNOWLEDGMENTS

This work is based on research supported by the US Office of Naval Research (ONR) via grants #N00014-10-1-0140 and #N00014-11-1-0493, and the US National Science Foundation (NSF) via grant #1038257.

REFERENCES

Benavoli, A., L. Chisci, A. Farina, B. Ristic, 2008. Modelling Uncertain Implication Rules in Evidence Theory. 11th International Conference on Information Fusion, pp.1-7, June 30 2008-July 3 2008

Cantrell, R. 2009. Mink: An Incremental Data-Driven Dependency Parser with Integrated Conversion to Semantics. Student Workshop at RANLP, Borovets, Bulgaria.

Genesereth, M. R., N. Nilsson, 1987. Logical Foundations of Artificial Intelligence. Morgan Kaufmann Publishers: San Mateo, CA.

Hewawasam, K.K.R., K. Premaratne, S. Mei-Ling, 2007. Rule Mining and Classification in a Situation Assessment Application: A Belief-Theoretic Approach for Handling Data

Imperfections, IEEE Transactions on Systems, Man, and Cybernetics, Part B: Cybernetics, vol.37, no.6, pp.1446-1459, Dec. 2007

Kübler, S., M. Scheutz, E. Baucom, and R. Israel. 2010. Adding Context Information to Part Of Speech Tagging for Dialogues. Proceedings of the Ninth International Workshop on Treebanks and Linguistic Theories (TLT). Tartu, Estonia.

Kübler, S., E. Baucom. 2011. Fast Domain Adaptation for Part of Speech Tagging for Dialogues. Poceedings of the International Conference on Recent Advances in NLP (RANLP), Hissar, Bulgaria.

Kübler, S., R. Cantrell, M. Scheutz, 2011. Actions Speak Louder than Words: Evaluating Parsers in the Context of Natural Language Understanding Systems for Human-Robot Interaction. Proceedings of the International Conference on Recent Advances in NLP (RANLP), Hissar, Bulgaria.

Marcus, M., B. Santorini, and M. A. Murthi. 1993. Building a Large Annotated Corpus of English: The Penn Treebank. Computational Linguistics 19(2), pp. 313-330.

Nivre, J., J Hall, J. Nilsson, A. Chanev, G. Eryigit, S. Kübler, S. Marinov, and E. Marsi. 2007. MaltParser: A Language-Independent System for Data-Driven Dependency Parsing. Natural Language Engineering 13(2), 95-135.

Pravia, M.A., R. K. Prasanth, P. O. Arambel, C. Sidner, C. Chee-Yee, 2008. Generation of a Fundamental Data Set for Hard/Soft Information Fusion. 11th International Conference on Information Fusion, pp.1-8, June 30 2008-July 3 2008

Premaratne, K., M. N. Murthi, J. Zhang, M. Scheutz, P. H. Bauer, 2009. A Dempster-Shafer theoretic conditional approach to evidence updating for fusion of hard and soft data. Procedings of the International Conference on Information Fusion (ICIF'09), pp. 2122–2129. Seattle, WA 2009

Ristic, B., P. Smets, 2005. Target Identification Using Belief Functions and Implication Rules. IEEE Transactions on Aerospace and Electronic Systems, vol.41, no.3, pp.1097-1103, July 2005

Shafer, G. 1976. A Mathematical Theory of Evidence. Princeton, NJ: Princeton University Press.

Wickramarathne, T. L., K. Premaratne, M. N. Murthi, and M. Scheutz, 2010. A Dempster-Shafer theoretic evidence updating strategy for non-identical frames of discernment, In Proc. Workshop on the Theory of Belief Functions (WTBF), Brest, France, Apr. 2010.

Wickramarathne, T. L., K Premaratne, M. N. Murthi, 2010. Focal elements generated by the Dempster-Shafer theoretic conditionals: a complete characterization. International Conference on Information Fusion (FUSION), Edinburgh, UK, pp. 1-8, July 2010

Wickramarathne, T.L., K. Premaratne, M. N. Murthi, M. Scheutz, S. Kübler, M. Pravia, 2011. Belief Theoretic Methods for Soft and Hard Data Fusion. IEEE International Conference on Acoustics, Speech and Signal Processing (ICASSP), pp.2388-2391, 22-27 May 2011

Wickramarathne, T. L., K. Premaratne, and M. N. Murthi. 2012. Consensus-Based Credibility Estimation of Soft Evidence for Robust Data Fusion. 2nd International Conference on Belief Functions, Compiegne, France, 2012.

Zhekova, D. and S. Kübler. 2010. UBIU: A Language-Independent System for Coreference Resolution. Proceedings of the 5th International Workshop on Semantic Evaluation (SemEval), Uppsala, Sweden.

Zhekova, D. and S. Kübler. 2011. UBIU: A Robust System for Resolving Unrestricted Coreference. Proceedings of the Fifteenth Conference on Computational Natural Language Learning (CoNLL), Shared Task, Portland, OR.

CHAPTER 11

Economic and Civic Engagement: Indicators Derived from Imagery

John M. Irvine, John R. Regan, Janet A. Lepanto
Draper Laboratory
Cambridge, MA, 02139
jirvine@draper.com

ABSTRACT

The application of remote sensing to the social sciences is an emerging research area. Recognizing that people's behavior and values shape, and are shaped by, the environment in which they live, analysis of overhead imagery can characterize geographic factors related to economic status and levels of social connectivity in a region. Observables associated with economic well being include the presence of commercial infrastructure, house size, number and types of livestock, presence of vehicles, and access to transportation. The transportation and communication infrastructure also indicates the expected level of interactions among elements of the society. Other important factors may be inferred from indicators derived from the imagery, such as level and types of agricultural production, population density, access to improved roads, distances to schools and businesses, and attributes of communities. Using imagery data collected over sub-Saharan Africa, we present an initial exploration of the direct and indirect indicators derived from the imagery. We demonstrate a methodology for extracting relevant measures from the imagery, using a combination of human-guided and machine learning methods. Using two regions for comparison, we present an initial image-based characterization of the levels of urbanization. Plans for extending and validating these methods are discussed.

Keywords: remote sensing, imagery, economic indicators, governance

1 INTRODUCTION

The application of remote sensing and geographic information systems (GIS) to the social sciences is an emerging research area (Goodchild, 2000). "Remote sensing can provide measures for a number of dependent variables associated with human activity—particularly regarding the environmental consequences of various social, economic, and demographic processes." (NRC, 1998) Recognizing that people's behavior and values shape, and are shaped by, the environment in which they live, researchers have explored a number of issues, including socio-economic attributes, ethnography, and land use (Jenson and Cowan 1999; Weeks, 2003; Jiang, 2003).

Numerous image processing tools are available for extracting traditional geospatial features (roads, buildings, landcover, water bodies) from overhead imagery (O'Brien and Irvine, 2004). Most of this research focuses on land use, land cover, and similar environmental issues for which remote sensing is ideal. Our research explores the extension of remote sensing to the analysis of economic and civil indicators. Because various conditions (e.g., active insurgency, civil war, challenging terrain, scarce local ground resources) can impede direct observation and measurement of these indicators, remote sensing could provide surrogate measures that are more easily acquired.

2 APPROACH AND METHODOLOGY

Our approach combines political, socio-cultural and economic theory with rich sources of survey data and overhead imagery to analyze the relationships between measurements acquired from direct surveys and features inferred from remotely-sensed data. The goal is to develop models that predict the values of specific indicators using the features extracted from the satellite imagery (figure 1). The approach consists of six steps (figure 1):

1. *Study plan*: Based on relevant social science theory and related literature we identify hypotheses to explore.
2. *Analysis of survey data*: Analysis of the survey questions and responses produces indicators of local attitudes about economic well being and attitudes towards democratic ideals and practice.
3. *Imagery analysis*: Using image processing techniques to extract relevant features from the imagery, we construct measures of the local conditions.
4. *Model development*: Statistical analysis of the relationships between the survey-based indicators and imagery-derived features yields possible models. The overhead imagery covers the same geographic region as the survey data and the geolocation of survey responses is the link between the two data sources. Statistical analysis of the model parameters will provide an empirical test of the hypotheses articulated in the study plan.
5. *Model validation and analysis*: Model validation requires a separate set of survey and imagery data that is held back (i.e., sequestered) during the model development process. Using this new set of data, we will compute survey-based indicators and imagery-derived features. Comparing the observed survey-based indicators to model predictions provides a method for validating the model. Because the research is in early stages, this step has not been performed

6. *Applications of the model*: The result of this process is a model for predicting the survey-based indicators from imagery-derived features, an assessment of the model-based methods, and recommended further research.

This paper explores initial relationships based on available survey data and a limited imagery sample of several regions in sub-Saharan Africa. Since our research is in early stages of development, this paper addresses only steps 1 through 4.

3 ISSUES TO EXPLORE

Social science research suggests several ways in which local attitudes and behaviors will correspond with phenomena that are observable in overhead imagery. Although this set is not exhaustive, the initial research issues we propose to investigate have a foundation in social science research and a logical connection to observable phenomena (Table 1):

Income and Economic Development: Numerous indicators of economic status (housing, vehicles, crop land, livestock, and infrastructure) are evident from overhead imagery and studies have explored the relationship between remote sensing data and the economy (Elvidge, 1997).

Centrality and Decision Authority: Both higher income and equitable distribution of income are associated with good governance. Observables associated with economic well being, including measures of wealth distribution, are also relevant indicators of governance. The transportation and communication infrastructure provide indicators of expected level of the levels of social interactions.

Social Capital: High social capital has been linked to good governance (Bowles, 2002). Observing meaningful measures that correlate with social capital is a challenge. Evidence of economic growth can be associated with higher levels of social capital (Knack, 1997). Although direct measurement of social cohesion and connectedness is not possible, durable institutions (e.g., schools, places of worship) and infrastructure (e.g. roads, cell towers) are indirect indicators of social connectedness.

4 PUBLIC OPINION DATA

This initial study leverages extensive public opinion data that have been collected in sub-Saharan Africa over several years under the Afrobarometer Program. The Afrobarometer is a collaborative enterprise of the Centre for Democratic Development (CDD, Ghana), the Institute for Democracy in South Africa (IDASA), and the Institute for Empirical Research in Political Economy (IREEP) with support from Michigan State University (MSU) and the University of Cape Town, Center of Social Science Research (UCT/CSSR). Each Afrobarometer survey collects data about individual attitudes and behavior, including indicators relevant to developing societies. These issues are summarized in Table 2.

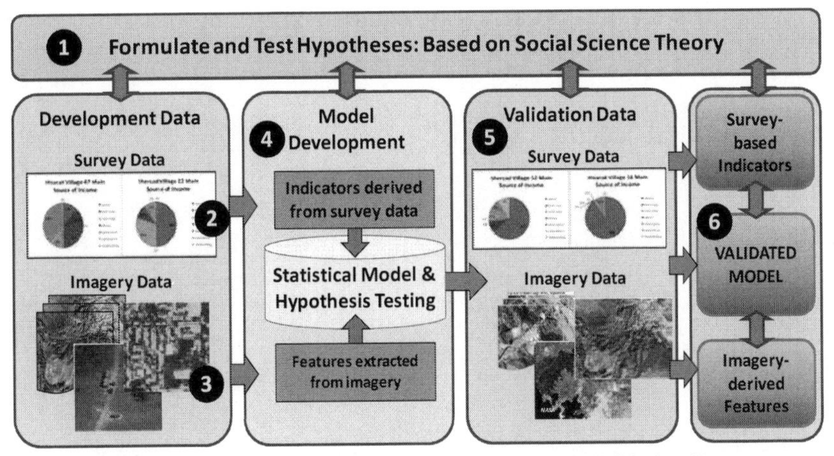

Figure 1 Overview of the Model Development and Validation Process

Table 1. Research Issues and Potential Observables to Explore

Research Issue	Potential Observables
Income and Economic Development:	House sizes: average size, range of sizes, materials Numbers and types of motor vehicles Physical infrastructure (electrical power lines, wells, paved roads, access to major transportation centers, e.g., rail, air) Agriculture: Extent and mix of cultivation, crop health, presence and extent of livestock
Centrality and decision authority	Road network; Lines of communication Physical infrastructure (bridges, paved roads, schools, mosques)
Social Capital	Community infrastructure and prevalence of meeting places and institutions (schools, places of worship). Communications infrastructure (e.g. cell towers, roads)

The survey data provides a rich portrait of societal attitudes across several countries and multiple regions within each country. For some countries, surveys have been repeated over multiple years, giving a temporal characterization of shifting attitudes and opinions. For instance, one survey question explored attitudes concerning the importance of having a democratic society and the freedom to criticize the government. Responses show distinct patterns by country and, to a lesser extent, within country (figure 2). The research challenge is to identify observables in the overhead imagery that are predictive of the patterns of attitudes evident in the survey data.

Table 2. Issues addressed in Afrobarometer Surveys

Democracy	Popular understanding of, support for, and satisfaction with democracy, as well as any desire to return to, or experiment with, authoritarian alternatives.
Governance	The demand for, and satisfaction with, effective, accountable and clean government; judgments of overall governance performance and social service delivery.
Livelihoods	How do families survive? What variety of formal and informal means do they use to gain access to food, shelter, water, health, employment and money?
Macro-economics and Markets	Citizen understandings of market principles and market reforms and their assessments of economic conditions and government performance at economic management.
Social Capital	Whom do people trust? To what extent do they rely on informal networks and associations? What are their evaluations of the trustworthiness of various institutions?
Conflict and Crime	How safe do people feel? What has been their experience with crime and violence?
Participation	The extent to which ordinary people join in development efforts, comply with the laws of the land, vote in elections, contact elected representatives, and engage in protest. The quality of electoral representation.
National Identity	How do people see themselves in relation to ethnic and class identities? Does a shared sense of national identity exist?

5 OVERHEAD IMAGERY

Cultural, social and economic factors that are critical to understanding the societal attitudes are associated with a number of phenomena that are observable from overhead imagery. Distinguishing among industrial, commercial, and residential areas, for example, is a standard use for imagery. Measures of socio-economic status (e.g., lot size, house size, number and types of livestock, presence of vehicles) can also be extracted from high-quality imagery. Other factors may also be inferred from indicators derived from the imagery such as level and types of agricultural production, population density, access to improved roads, distances to schools and businesses, and attributes of communities and domiciles. Patterns which emerge from the correlation of the geospatial analysis with the survey results can suggest phenomena observable in imagery that are reasonable surrogates for the direct measurements of public opinions.

We are exploring a number of image processing techniques to extract and identify specific indicators that can be directly inferred from the imagery (table 3). Certain measures immediately derivable from the imagery indicate the size and spatial distribution of the population, the level of infrastructure development, access to transportation, and other indicators of well-being. Other attributes will require exploration and may only be indirectly inferred from the imagery.

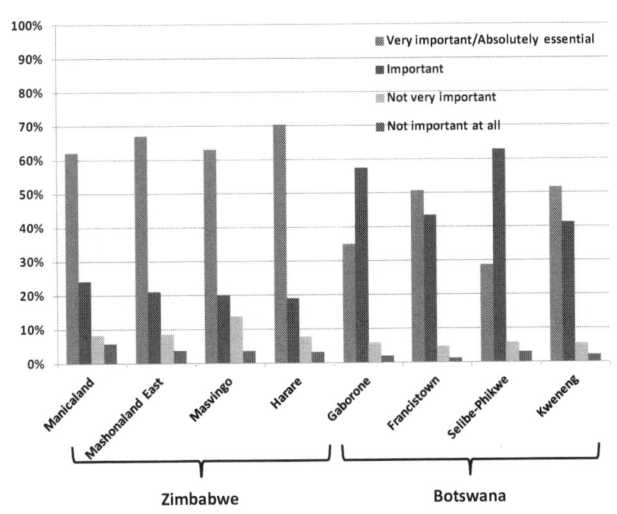

Figure 2. "Democratic society and freedom to criticize the government" -- Responses by Country and Region

A simple example of an imagery-derived feature is the ratio of built up areas to open vegetation (figure 3). In this example road centerlines have been extracted and the image has been classified according to built up areas (residential, commercial, industrial) and open vegetation. We constructed this classification using a machine learning technique based on a genetic algorithm (Harvey, *et al*, 2002). The ratio of the number of pixels classified as a built up area to the number classified as vegetation is one feature that indicates the degree of urbanization. This imagery-derived feature is a candidate explanatory variable for predicting indicators of economic development. Similarly, image features related to communication and transportation infrastructure are potential explanatory variables for predicting indicators of community involvement and connectedness, which relate to social capital.

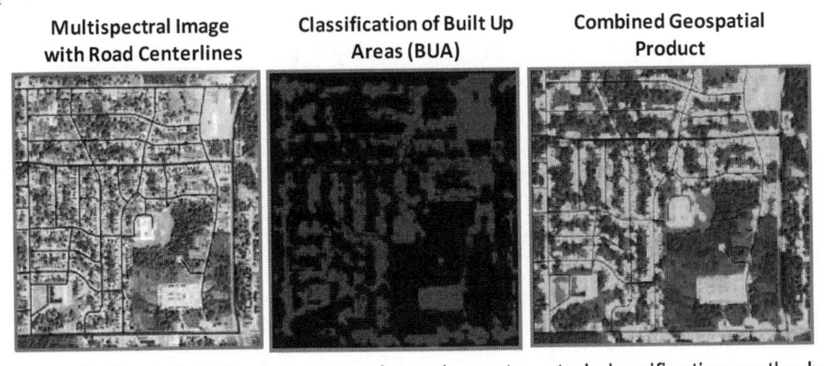

Figure 3. Example of image processing using automated classification methods

Table 3. Illustrative Imagery Observables and Derived Indicators

Feature Class	Observables	Derived Features
Landcover	• Arable land • Area under cultivation • Crop types, quantities, health • Presence /extent of livestock • Deforestation	• Level of commercial agriculture • Opium production • Expected food supply
Buildings	• Types of buildings (e.g., schools, residential, industrial) • Number of types of buildings • Size and character of residential buildings • Damage to buildings	• Degree of urbanization or industrialization • Population density and distribution • Average size of residential buildings
Lines of Communication	• Road network (paved & unimproved) • Rail and air transportation access • Numbers and types of vehicles • Distance from major commercial or industrial areas	• Level of communication • Infrastructure to support local trade and commerce • Access to transportation
Other Infrastructure	• Bridges • Reservoirs and/or wells • Cell towers • Power lines & sub-stations	• Access to potable water, irrigation • Access to communication • Support for appliances, power tools

6. COMPARISON AND PRELIMINARY FINDINGS

A preliminary investigation of available imagery suggests some issues to explore. Visual inspection of the imagery indicated regional differences in the level of urbanization and infrastructure. Consider, for example, Zimbabwe and Botswana (figure 4). We saw in figure 2 a distinct distribution of opinion about democratization for Zimbabwe vs. Botswana. There is also a clear difference in the level of urbanization, as seen in figure 4. The relationship suggests an avenue of further exploration. Survey data for Botswana show a large number of respondents indicating that "Democratic society and freedom to criticize the government" are important, but fewer respondents indicate it is *very important*. Is the degree of urbanization a factor in reducing the desire or tempering expectations? The survey data includes an extensive set of questions that can support deeper analysis.

Using a random sample of imagery transects, we estimated the relative level of urbanization in the imagery. This technique classifies pixels as belonging to one of two classes: built-up or natural. For the initial investigation, areas of well-tended cultivation we considered built up. This simple classification produces a score indicating the estimated percent of development for a region. Comparison of this imagery-derived score to survey responses suggests some specific relationships. For example, consider one question about ties to ones identity group:

> ***Please tell me whether you disagree, neither disagree nor agree, or agree with these statements: You feel much stronger ties to [members of your identity group] than to other [citizens]?***

The responses (strong agree, agree, neutral, disagree, strongly disagree) were coded numerically, such that higher values indicate respondents who feel closer ties to their own identity group. There is a statistical relationship between this score and the percent of development. Although the relationship is weak, higher development is associated with lower identification with one's own identity group within each country (figure 5).

Figure 4. Selected imagery for Botswana and Zimbabwe indicates the levels of urbanization and agriculture in specific study regions

Another interesting indicator from the survey data is derived from respondents' attitudes about job creation. Low values indicate a belief that the government should promote job creation through support for new businesses, whereas high values indicate a belief that the government should create jobs through direct hiring by the government. When we modeled this indicator as a function of the imagery-derived measure of development, an interesting relationship emerged (table 4). Lower levels of development correspond to higher belief that the government should hire people directly and more developed regions exhibit greater support for job creation through support for business. Furthermore, there is a systematic difference between Zimbabwe and Botswana, with greater support for development of new business in Botswana.

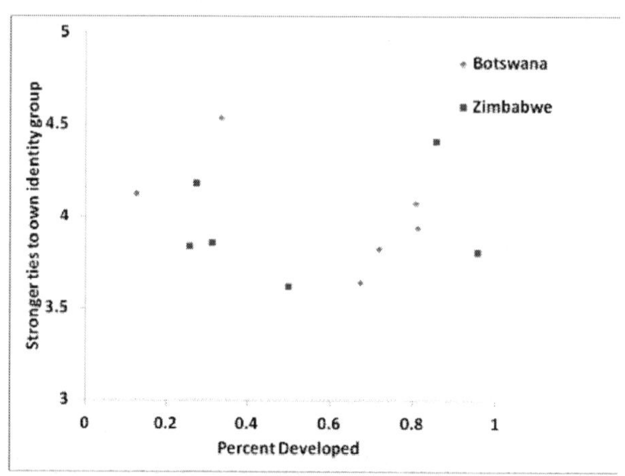

Figure 5. Relationship between estimated percent developed and mean survey response regarding ties to own identity group

Table 4. Linear Model of attitude toward job creation

Parameter	Coefficient	t-statistic	Significance level
Intercept	3.137	26.5	< 0.0005
Country (Botswana)	-0.461	-4.66	0.001
Development	-0.471	-2.59	0.029

7. CONCLUSIONS AND FUTURE INVESTIGATIONS

This paper presents a methodology for inferring indicators of economic well-being and governance through analysis of overhead imagery. We presented a brief discussion of the social science theory that supports this investigation. Our initial investigation is leveraging available opinion survey data for several countries in sub-Saharan Africa. This rich set of data quantifies attitudes in multiple regions which can be associated at a coarse level with imagery-derived information. This initial investigation considered simple measures of development/urbanization and their relationship to attitudes about democratic government, freedom of expression, and group identity. The findings are suggestive at this stage. More extensive and rigorous analysis is needed to assess the validity of indicators that can be derived from imagery.

REFERENCES

Bowles, S. and H. Gintis. 2002. "Social Capital and Community Governance" *Economic Journal*, Vol. 112, pp. 419-436, 2002

Elvidge C.D., Baugh KE, Kihn EA, Kroehl HW, Davis ER, Davis CW. 1997. "Relation between satellite observed visible-near infrared emissions, population, economic activity and electric power consumption." *International Journal of Remote Sensing* 1997; 18: 1373-9.

Goodchild, M.F., L. Anselin, R. P. Appelbaum and B. H. Harthorn. 2000. "Toward Spatially Integrated Social Science," *International Regional Science Review* April 2000 vol. 23 no. 2 139-159

N.R. Harvey, J. Theiler, S.P. Brumby, S. Perkins, J.J. Szymanski, J.J. Block, R.B. Porter, M. Galassi, and A.C. Young, 2002. "Comparison of GENIE and Conventional Supervised Classifiers for Multispectral Image Feature Extraction", *IEEE Transactions in Geoscence. and Remote Sensing*.40, 2002, pp. 393-404

Jenson, J. R. and D. C. Cowen. 1999. Remote Sensing of Urban Infrastructure and Socio-economic attributes, *Photogrammetric Engineering and Remote Sensing*, vol.65, no. 5, pp.611-622, 1999.

Jiang, H. 2003. "Stories Remote Sensing Images Can Tell: Integrating Remote Sensing Analysis with Ethnographic Research in the Study of Cultural Landscapes," *Human Ecology*, Volume 31, Number 2, 215-232, June 2003.

Knack, S. and P. Keefer. 1997. "Does Social Capital Have an Economic Payoff? A Cross-Country Investigation," Quarterly Journal of Economics, 112(4), pp. 1251-1288

NRC. 1998. *People and Pixels: Linking Remote Sensing and Social Science*, Commmittee on the Human Dimensions of Global Change, National Research Council, 1998.

O'Brien, M. A. and J. M. Irvine. 2004. "Information Fusion for Feature Extraction and the Development of Geospatial Information" *International Conference on Information Fusion, Stockholm*, Sweden, 30 June – 2 July 2004

Weeks, J.R. 2003. Using Remote Sensing and Geographic Information Systems to Identify the Underlying Properties of Urban Environments, In Tony Champion and Graeme Hugo, Editors. *New Forms of Urbanization: Beyond the Urban-Rural Dichotomy*. Aldershot, UK: Ashgate Publishing Co., 2003

CHAPTER 12

Soldier Skills to Develop, Enhance, and Support Host-Nation Governance

Jeffrey E. Fite,[1] *Trevor M. Conrad,*[2] *Rebecca Mulvaney,*[2] *Alice Hirzel,*[2] *Jennifer Stern,*[2] *Steven Aude,*[2] *and Lauren Tindall*[2]

[1]U.S. Army Research Institute for the Behavioral and Social Sciences, Fort Hood, TX, USA
jeffrey.e.fite.civ@mail.mil

[2]ICF International
Fairfax, VA 22031 USA

ABSTRACT

During interviews with U.S. Army commanders, researchers from the U.S. Army Research Institute for the Behavioral and Social Sciences (ARI) learned of a need for training that would equip unit-level Soldiers – particularly junior officers and senior noncommissioned officers (NCOs) – with the knowledge, skills, and abilities needed to carry out and support missions to develop, enhance, and support local level host nation governance. At that time, however, the specific governance-related skill sets required of Soldiers were not clear, as it was largely believed that only the most senior officers and NCOs were regularly engaged in governance-related activities. Furthermore, Army doctrine offered only broad guidance on governance-related activities that are critical for fostering good governance – it did not provide detailed guidance on the best practices for unit-level Soldiers carrying out those activities or training them. In this report we describe our initial research efforts to identify those skill sets. Through interviews and literature searches we identified more than 30 governance activities frequently required of Soldiers, and shed light on the different roles of officers, NCOs, and enlisted Soldiers in the conduct of governance missions. The research also revealed aspects of governance operations that Soldiers found to be most challenging, and those that Soldiers found to be most important for mission success. We conclude the report with a description of the design and development of the Host-Nation Operations: Soldier Training on

Governance (HOST-G) Training Support Package (TSP), which contains a variety of training tools and job aids to support Soldiers tasked with local level governance missions. We also describe ongoing research to evaluate and refine the TSP, and to identify the skills required to meet one of the key challenges for Soldiers tasked with governance missions – that is, the challenge of operating in joint, interagency, intergovernmental, and multi-national (JIIM) environments.

Keywords: U.S. Army; Soldiers; Governance; JIIM; Training; HOST-G

1 INTRODUCTION

According to U.S. Army doctrine, a primary objective of counterinsurgency and stability operations is to foster the development of effective governance – the process, systems, institutions, and actors that enable a state to function (Department of the Army, 2008b, pp. 2-11) – by a legitimate government (Department of the Army 2006, pp. 1-21; 2008b, pp. 1-7). Legitimate governments rule with the consent of the people and are therefore better able than "illegitimate" governments – governments that rule primarily through coercion – to carry out key functions (Department of the Army, 2006, pp. 1-21). Legitimate governments can more effectively utilize their authority to regulate social relationships, extract resources, and take actions in the public's name. They are also better able to manage, coordinate, and sustain collective security, as well as economic, political, and social development. For those reasons, legitimate governments are stable governments.

To foster the development of legitimate host-nation governance, the Army makes use of both military and nonmilitary means. Historically, forces that have learned to effectively enhance and develop host-nation governance have engaged in a number of key local-level learning and training tasks that include: developing doctrine and best practices locally; soliciting and evaluating advice from the local people in the conflict zone; learning about the broader world outside of the military and requesting outside assistance in understanding political, cultural, and social situations beyond their experience; and establishing local training centers during operations (Department of the Army, 2006, p. x). These are not easy practices for any organization to establish and, for a military actively engaged in conflict, one would assume even more challenging to accomplish.

In early 2009, during interviews with battalion and brigade commanders, researchers from the U.S. Army Research Institute for the Behavioral and Social Sciences (ARI) learned of a need for training to support missions aimed at developing, enhancing, and supporting legitimate host-nation governance (Fite, Briedert and Shadrick, 2009). Commanders spoke of their desire for training that would equip Soldiers – especially junior officers and senior noncommissioned officers (NCOs) – with the knowledge, skills, and abilities required to carry out and support governance missions at a local level. More specifically, they requested training that would prepare junior officers and senior NCOs to measure and monitor the quality of local governance; gain situational awareness of (personal and professional) local-level politics (Figure 1(A)); interact effectively with local spheres of influence (Figure 1(B)); teach others about government and how to solve problems through government processes (Figure 1(C)); lessen or eliminate the influence of shadow governments; and, enhance partnerships between host nation and joint, interagency, intergovernmental, and multinational (JIIM) agencies and organizations (Figure 1(D)).

At that time, however, the specific governance-related skill sets required of unit-level Soldiers were not clear. Prior to our interviews with Army commanders, it was widely believed that only the most senior officers and NCOs were regularly engaged in governance-related activities. Furthermore, Army doctrine offered only broad guidance on governance-related activities that are critical for fostering good governance – it did not provide detailed guidance on the best practices for unit-level Soldiers responsible for carrying out those activities or training them (Fite, Briedert and Shadrick, 2009). Governance is a complicated concept, and definitions, dimensions, and measures of governance have remained elusive, even to the academicians and economists who have spent decades studying the topic (see reviews in Arndt and Oman, 2006; Besançon, 2003; International Bank for Reconstruction and Development/World Bank, 2006; Landman and Häusermann, 2003; Munck and Verkuilen, 2002). Thus, in order to meet the Army's need for training on these difficult concepts, and to facilitate the development of training plans and strategies, research was required.

In August of 2009, ARI initiated research on the Soldier skills required to successfully conduct governance missions. The specific goals of that research were to identify governance-related activities that are not tied to a particular nation, culture, or area of operations; design and develop prototype training materials that would introduce Soldiers at all levels to the concept and significance of legitimate governments; and, design and develop training tools and job aids that would assist the efforts of junior officers and senior NCOs to develop, support, and enhance local level host-nation governance.

2 METHODS

2.1 Literature Review

Army doctrine was searched for conceptual information regarding governance, and any guidance that Soldiers could use to measure, monitor, and foster legitimate governance locally. Included in those searches were *FM 3-0 Operations*, (Department of the Army, 2008a), *FM 3-07 Stability Operations* (Department of the Army, 2008b), *FM 3-24, Counterinsurgency* (Department of the Army, 2006); and, *FM 3-24.2, Tactics in Counterinsurgency* (Department of the Army, 2009). We also reviewed manuals and publications from military centers (e.g., Center for Army Lessons Learned, Army Peacekeeping and Stability Operations Institute), U.S. government organizations (e.g., U.S. Department of State, U.S. Institute of Peace, U.S. Agency for International Development), and research and policy organizations (e.g., The Brookings Institution, The RAND Corporation). Additionally, academic and professional journals were searched, to include military journals such as *Military Psychology*, *Military Review*, and *Small Wars Journal*.

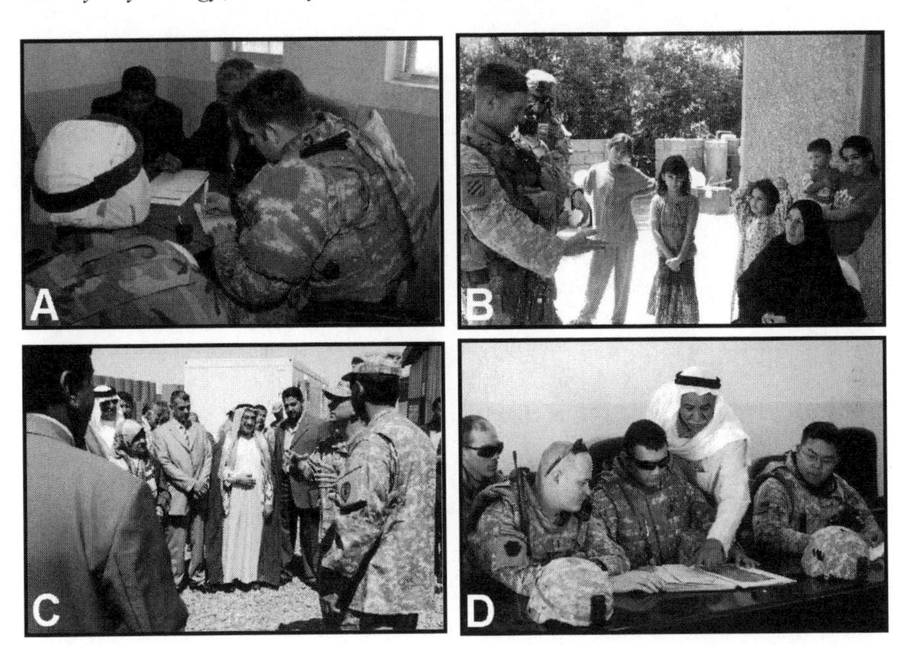

FIGURE 1: A) Image of Soldiers gaining situational awareness of local-level politics. A captain takes notes during a neighborhood advisory council meeting. Photo by Staff Sergeant Bronco Suzuki. B) Image of Soldiers teaching others about government and how to solve problems through government processes. A 1st lieutenant speaks with the representative of a Women's Council outside her home. Photo by Sergeant Kevin Stabinsky. C) Image of Soldiers enhancing partnerships between coalition and host-nation agencies and organizations. A captain explains through an interpreter how local Iraqis are learning job skills and receive paid employment and experience working in carpentry, plumbing, and electrical. Photo by Specialist Jazz Burney. D) Image of Soldiers interacting with local spheres of influence. A 1st lieutenant speaks with a sheik at a key leader meeting. Photo by Mass Communications Specialist 2nd Class Robert Whelan. All images were retrieved from http://www.defenselink.mil/multimedia/multimedia.aspx.

Data from the literature review were used to create preliminary governance-related activity lists for officers, NCOs, and enlisted Soldiers. The literature review also informed the list of individuals we sought out for interviews and our interview protocols. An annotated bibliography of relevant sources that provides the resource, the focus of research, and the resource type (e.g., government document, internet or media, etc.) was also produced (Conrad, et al., 2011a).

2.2 Interviews

Interviews were conducted with subject matter experts (SME) representing a broad range of perspectives. The interviews were intended to solicit current and direct information from individuals who have demonstrated expertise and experience in fostering local level host-nation governance. We also sought out individuals who had witnessed governance efforts within their home countries.

Interviews were conducted with U.S. Army officers (captains, $n = 2$; majors, $n = 3$; lieutenant colonels, $n = 3$; colonels, $n = 2$; brigadier general, $n = 1$) and NCOs (command sergeant majors ($n = 4$)), as well as civilians with recent international development experience ($n = 6$). Interviews were also conducted with civilians representing the host nations of Iraq and Afghanistan (n = 4). Those participants included a former Iraqi mayor, a former Iraqi citizen, and a former aide to President Hamid Karzai. Soldiers were recruited primarily through research team and ARI contacts. Host-nation representatives were recruited with assistance from the University of Nebraska at Omaha's Center for Afghan Studies, National Defense University, and the U.S. Army's Human Terrain System program. Civilians were identified through our literature search and through points of contact at development agencies.

Interviews ranged from one to two hours in length. Most were conducted over the phone, though some were conducted in-person. Participants received a brief "read ahead" by e-mail well in advance of their interviews. The Soldier and civilian versions of the interview protocol (Conrad, et al., 2011a) focused on how the SME defined governance and government legitimacy, typical governance activities, indicators of progress, challenges and lessons learned, and governance training needs and existing resources. The host nation SME protocol (Conrad, et al., 2011a) focused on defining governance and governmental legitimacy, broad activities required to establish local governance and leadership, and perception of governance-related outcomes. In all protocols, we asked for governance-related critical incidents that could potentially be used in later training development.

2.3 Analysis

After the interviews were completed and transcribed, we reviewed the interview content and organized the data according to the four topic areas targeted by all three interview protocols – definitions of legitimate government, key and common governance-related activities of U.S. Army Soldiers, common challenges for Soldiers, and drivers of success (i.e., best practices). We then used a modified grounded theory approach (Strauss, 1987) to analyze the data. First, the three interviewers came to a consensus on major themes that emerged from the interview data for each question or set of questions. Then, interview transcripts were systematically analyzed to determine the frequency of each theme's mention and to categorize quotes and illustrations appropriately for training development purposes.

3 RESULTS

3.1 Governance-Related Activities Commonly Required of Soldiers

Our research revealed more than 30 governance activities frequently required of Soldiers. Seven common activities were related to assessing and understanding the social, political, and economic environments, such as interacting with and engaging the local population in conversation to learn and obtain governance-related information, identifying unmet governance needs (e.g., processes to interface with constituents, basic administrative processes, and mechanisms to select and vet leaders and civil service staff), and observing and recording functional and dysfunctional indicators of local governance. Seven activities were related to monitoring the sentiment of, and building cooperation amongst, the population. Those activities included treating the indigenous populace with respect, interacting with and engaging the local population to communicate governance-related accomplishments and themes, and conducting deliberate outreach activities in conjunction with local officials to strategically communicate governance-related issues and progress. Nine common activities were related to establishing basic services and governance processes, such as identifying and applying knowledge of current indigenous methods and systems of governance in an area of operations, participating in indigenous local meetings and councils, and supporting and providing essential services until local governance systems can be restored. Eleven activities were specifically related to advising and coaching host nation leaders. Those activities included assisting local leadership in the creation of a long-range community development plan, coaching host nation civil servants in facilities/essential services management, and transitioning authority and responsibility for leadership and decision-making back to indigenous leadership.

3.2 Challenging and Important Tasks

Two aspects of local level governance operations were found to be particularly challenging for Soldiers – maintaining continuity of operations between incoming and outgoing units, and learning to identify the most effective local leadership. Yet of all the governance tasks identified, participants considered the monitoring of the social, political, and economic environments – and teaching local host-nation leaders to do the same – to be the most important for mission success. In terms of pre-deployment training for Soldiers, interpersonal and relationship-building skills, as well as a basic understanding of public administration and public services, were thought to be critical.

3.3 Roles of Officers, NCOs, and Enlisted Soldiers

The research also elucidated the different roles of officers, NCOs, and enlisted Soldiers in the conduct of governance missions. Officers were most often responsible for identifying and assessing local leaders, replacing unreliable leadership, providing feedback to the local government on population sentiment, and fostering common thinking among JIIM partners within the area of operations (e.g., civil affairs teams, provincial reconstruction teams, non-governmental organizations, etc.). NCOs tended to have less interaction with local leaders than officers, but played a critical role in governance operations by performing daily, on-the-ground assessments of population sentiment. NCOs, along with junior enlisted Soldiers, played an important role in gaining situational awareness of the area. Everyone – officers, NCOs, and enlisted Soldiers – was responsible for re-establishing security and essential services.

3.4 Design and Development of Prototype Training, Training Tools, and Job Aids

Based on our findings, we designed and developed the *Host-Nation Operations: Soldier Training on Governance* (HOST-G) Training Support Package (TSP) (Figure 2A; Conrad, et al., 2011a), and associated Training Tools and Job Aids (Conrad, et al., 2011b). The TSP includes the following materials:

- **HOST-G Computer-Based Training** (Figure 2B) – Introduces Soldiers at all levels to the concept of governance and the skills necessary for long-term success of host-nation governance. Designed for use at the company and battalion level.
- **Governance Activity List** – Identifies more than 30 key activities for the development and assessment of governance-specific training. Can be used at the company or battalion level.
- **Governance Situational Training Exercise** – Serves as a tool for training platoon-level leaders on individual and collective governance tasks.
- **LEGIT Assessment Tool** – Provides Soldiers at all levels with questions, observations, and tips to guide them through the challenges of collecting information about local governance processes.
- **Governance Annex to Continuity Books** – Provides in-country guidance for documenting governance-specific information that will be helpful to an incoming unit. For use at the company level.
- **Governance Metrics Workbook** – Aids battalion and brigade-level leadership in the measurement and evaluation of progress within governance lines of effort during stability operations.
- **Governance BOLO (Be On the Look Out) Worksheet** – Serves as a tool for gathering information that can be used by higher headquarters to measure and evaluate progress within governance lines of effort during stability operations. Can be used at platoon, company, and battalion levels.

In June and July of 2010, the TSP was reviewed by junior officers (n = 4) and NCOs (n = 10). Those leaders provided favorable feedback regarding the utility and necessity of the training.

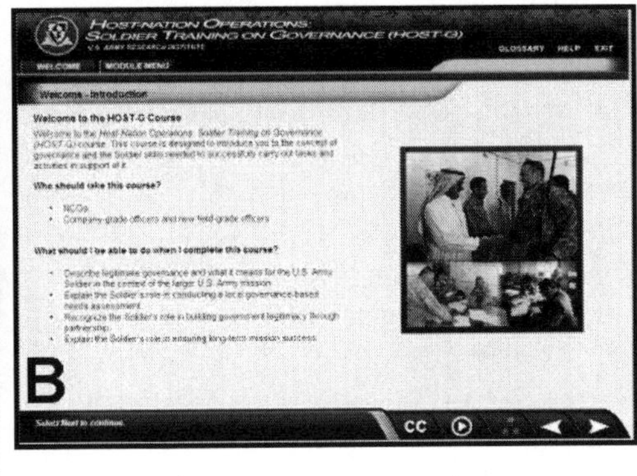

FIGURE 2: A) Cover of the HOST-G Training Support Package, which contains a variety of training tools and job aids for Soldiers at all levels. B) Screen capture from the HOST-G Computer-Based Training, which was designed to introduce Soldiers at all levels to the concept of governance and the Soldier skills necessary for long-term success of local level host-nation governance.

4 DISCUSSION

The research described here revealed a broad array of governance-related activities with which Soldiers are frequently tasked. It also shed light on governance activities that Soldiers find to be particularly challenging, and those that Soldiers have found to be critical for the mission success. Of great interest to us was the finding that officers, NCOs, and enlisted Soldiers all play key roles in the successful development, support, and enhancement of host nation governance. With an improved understanding of what Soldiers do to support governance missions, we were able develop prototype training, training tools, and job aids to prepare Soldiers for future operations, and to assist them once they have deployed. Taken together, the efforts described here provide a foundation upon which we intend to build a research and development program to address in greater depth and detail the Army's need for governance-skills training.

In 2011, ARI began a formative evaluation of the HOST-G TSP with Soldiers, leaders, and trainers to determine the usefulness of the TSP to the operational Army, and to assess the relevance and efficacy of the TSP across different geographical/cultural areas and areas of operation. At the same time, ARI initiated research on a closely related topic that operational units view as a complicating factor in the conduct of governance missions – that is, the challenge of operating in JIIM environments.

Operations in JIIM environments are difficult affairs, representing a nexus of sociocultural, political, and organizational challenges in already demanding environments. During interviews for the research detailed in this report, Army leaders frequently reported poor unity of effort with and among partner agencies, governments, and nations, and difficulty garnering their support and cooperation. They reported difficulty in understanding the organizational cultures of Army partners, what they do, and how they do it. They also reported difficulty in communicating Army doctrine and military operations. Research now underway seeks to identify the best practices for overcoming those obstacles, and the specific JIIM-related skill sets required of junior officers and NCOs.

5 CONCLUSIONS

Governance is a complicated concept. For the U.S. Army, the already complex issue of developing legitimate host-nation governance has been made even more difficult by cultural and language barriers, and being actively engaged in conflict. Moreover, as the research presented here indicates, governance missions often entail the efforts of Soldiers at all levels – officers, NCOs, and enlisted Soldiers alike. Nonetheless, the Army appears to have done a remarkably good job of navigating the complexities of governance in very challenging physical and socio-cultural environments – even with limited formalized training in place to support those efforts at the unit level.

Because governance is such a complicated concept, the development of governance training for the Army must be realistic and focused. It will not be possible to resolve the long-standing difficulties of defining and measuring governance, or solve all of the challenges related to working with a multitude of JIIM partners. Instead, we feel that the primary goal of such work should be to develop training for Soldiers at all levels that is relevant to the missions that they carry out and support – as we have shown here, governance missions are not the sole responsibility of senior officers and NCOs. We also recommend that future research and training development in this area not be limited to specific nations, cultures, or area of operations. The U.S. military has been, and is expected to be, the primary U.S. instrument of nation- and state-building (Chiarelli, 2007). Therefore, we feel that future research and development efforts should provide tools that can be used by, and/or modified for, Soldiers faced with governance-related missions anywhere in the world.

6 ACKNOWLEDGEMENTS

The authors wish to acknowledge the U.S. Army leaders and Soldiers who took time out of their busy schedules to share their needs and experiences with us. We also wish to acknowledge the civilians who graciously shared their international development and host-nation expertise. We thank Dr. Brian Crabb, SMA (R) Bill Gates, Dr. Scott Shadrick, and Cindy Underwood for their continued support of, and assistance with, this research.

7 REFERENCES

Arndt, C. and Oman, C., 2006. *Uses and abuses of governance indicators*. Paris: Development Centre of the Organisation for Economic Co-Operation and Development.

Besançon, M., 2003. *Good governance rankings: The art of measurement*. Cambridge, MA: World Peace Foundation.

Chiarelli, P.W., 2007. Learning from our modern wars: The imperatives of preparing for a dangerous future. *Military Review*, September-October, pp. 2-15.

Conrad, T.M., Mulvaney, R., Hirzel, A., Stern, J., Aude, S. and Tindall, L., 2011a. *Host-Nation Operations: Soldier Training on Governance (HOST-G) Training Support Package (ARI Research Product 2011-05)*. Arlington, VA: U.S. Army Research Institute for the Behavioral and Social Sciences.

Conrad, T.M., Mulvaney, R., Hirzel, A., Stern, J., Aude, S. and Tindall, L., 2011b. *Host-Nation Operations: Soldier Training on Governance (HOST-G) Training Tools and Job Aids (ARI Research Product 2011-06)*. Arlington, VA: U.S. Army Research Institute for the Behavioral and Social Sciences.

Department of the Army, 2006. *Counterinsurgency (Field Manual 3-24)*. Washington, DC: Author.

Department of the Army, 2008a. *Operations (Field Manual 3-0)*. Washington, DC: Author.

Department of the Army, 2008b. *Stability Operations (Field Manual 3-07)*. Washington, DC: Author.

Department of the Army, 2009. *Tactics in Counterinsurgency (Field Manual 3-24.2)*. Washington, DC: Author.

Fite, J.E., Breidert, J.T. and Shadrick, S.B., 2009. *Initial evaluation of a U.S. Army training need: Soldier skills to develop, enhance, and support local level host-nation governance (ARI Research Report 1912)*. Arlington, VA: U.S. Army Research Institute for the Behavioral and Social Sciences.

International Bank for Reconstruction and Development/World Bank, 2006. *A decade of measuring the quality of governance*. Washington, DC: Author.

Landman, T. and Häusermann, J., 2003. *Map-making and analysis of the main international initiatives on developing indicators on democracy and good governance*. Cambridge, MA: World Peace Foundation.

Munck, G.L. and Verkuilen, J., 2002. Conceptualizing and measuring democracy: Evaluating indices. *Comparative Political Studies*, 35(1), pp. 5-34.

CHAPTER 13

Towards a Taxonomy of Socio-Cultural Factors That Influence Decision Making

Charneta Samms, Asisat Animashaun, Shanell Henry, Susan Hill, Debra Patton and Diane Ungvarsky

U.S. Army Research Laboratory
Aberdeen Proving Ground, MD

ABSTRACT

The Army Research Laboratory has launched a research program, Relevant Information for Social Cultural Depiction (RISC-D) to understand and model the socio-cultural factors that affect Soldier/Commander decision making. Through synthetic analysis of the literature, we have developed a draft taxonomy of socio-cultural factors that are potentially influential to decision making. This paper discusses our taxonomy development process, introduces the taxonomy framework and lays out our next steps for this line of research.

Keywords: decision-making, socio-cultural, taxonomy, Soldier

1 INTRODUCTION

As the missions of U.S. Soldiers expand and the variety of cultural environments in which they work increase, it is essential that scientific research and technology development result in enhanced understanding of various social and cultural differences Soldiers will face. To support the Soldier in mission accomplishment, we need to translate this socio-cultural layer into information. There are many research programs focused on understanding and modeling our adversaries and host populations of the countries we occupy. The Minerva Initiative (http://minerva.dtic.mil/) and the Human Social, Culture and Behavior (HSCB) Modeling Program (http://www.onr.navy.mil/Science-Technology/Departments/Code-30/All-Programs/Human-Behavioral-Sciences.aspx) are examples of programs whose focus is on external socio-cultural phenomena. Increasing emphasis on operations such as stability and the defense support of civil authorities requires Soldiers to make more complex decisions when dealing in environments where non-combatants reside and combatants may hide. Unfortunately, the research conducted to understand decision making has not focused on the inclusion of culture (Guss, 2004) and more specifically on decision making in a military environment including the effects of culture. Even less of this research is focused on the Soldier's own socio-cultural attributes and how they affect his or her decision making. In this vein, the Army Research Laboratory has launched a research program, Relevant Information for Social Cultural Depiction (RISC-D). RISC-D focuses on understanding and modeling the socio-cultural factors that affect Soldier/Commander decision-making. This research seeks to address how a Soldier's own cultural background influences his or her decision-making.

To begin this line of research, a synthetic analysis of the current literature was conducted to develop a taxonomy of socio-cultural factors believed to influence how people make decisions. Synthetic analysis is the process of conducting an extensive literature review to gain a synoptic view of a complex problem as a whole and detect links, commonalities and gaps in a given research area. A taxonomy provides a classification of key characteristics that focus on the general principles that describe a particular phenomena (Scherperell, 2005). Our taxonomy will serve as the base of our research by functioning as the foundation of a framework to develop a socio-culturally influenced model of decision making. It will also assist as a way to categorize the research in an organized fashion. This paper discusses our taxonomy development process, introduces the taxonomy framework and lays out our next steps for this line of research.

2 DEVELOPING THE TAXONOMY [1]

This research effort began with a synthetic analysis of research literature dealing with the broad categories of decision making, culture and military operations. While the area of decision making has been extensively examined, much of the research is focused on decision making in static environments and in the analysis of decision making in role playing game-type environments. These task environments are not always generalizable to a military context but do point to areas where research could be tailored to examine more dynamic and military-like environments. To support identification of initial factors, we examined research across a wide variety of domains including business, management, economics and healthcare. Because decision making is so common to so many parts of life, we knew that while research in these areas may not be directly applicable, it would

---

[1] The concepts used for our taxonomy development process were derived from Whittaker & Breininger (2008) and DeRue & Morgeson (2005).

provide a good start at identifying socio-cultural factors of interest to military decision making.

Next, we convened a brainstorming session where all the researchers used sticky notes to document every socio-cultural factor or idea that they believed could affect decision making. These ideas were based on expertise developed from the synthetic analysis and other research they had identified as useful to understanding the problem. Once everyone had exhausted their lists, the sticky notes were placed on the wall and reviewed by the group. Duplicate concepts were combined. As a group, we discussed the meaning of each concept and agreed as to whether or not we believed it to be a potentially influential construct of decision making. As the discussion continued, similar items were grouped together and items that did not seem to fit a group were put aside for further discussion. Out of this process, five major areas were identified: demographics, personality, experience, values, and context. These five categories grouped approximately fifty socio-cultural factors. A sixth category emerged from the items that were set aside and we labeled them external factors. The groupings serve as a basis of the taxonomy and for continued research into how these factors may affect a person's decision-making process.

As next steps, individual researchers were assigned a section of the taxonomy to conduct continued examination of existing literature. Their job was to identify if there was any specific current research that supports the notion of that socio-cultural factor being linked to decision making. They then needed to develop a definition of the concept for inclusion in the draft taxonomy paper. The draft taxonomy paper will be used in the next phase of taxonomy development.

2.1 Purpose

We identified several uses for this type of taxonomy. Our main purpose is to focus the research into areas that will allow us to develop a qualitative and informed cognitive model of decision-making. The taxonomy also supports the binning of research into categories that allow for the identification of research gaps and areas to focus new research. We also think that the taxonomy can indicate skills that can be developed by the training community and serve as a basis of what is important when modeling host populations.

This taxonomy is focused on the socio-cultural influences that effect the decision making of the individual Soldier, small unit leader and Commander during their operation in a given environment. While our focus is on socio-culturally diverse environments, the taxonomy should support the understanding of military decision making in any environment since we are focused on the factors of the Soldier. We predict this taxonomy will be useful for force designers, military researchers, analysts and trainers across the services and academia.

2.2 Assumptions

Several assumptions were made to frame the taxonomy development. In the ever-changing military environment, it is very certain that the context in which decisions need to be made will change and as such a discussion of context and its elements are included with the taxonomy. Also, we want to ensure that we focus on decision making in an operational context, so our focus is on those decisions that need to be made real time in the operation of missions and not necessarily in the initial or strategic planning phases of operation. We also wanted to define what we meant by factors. Factors, for our purposes are socio-cultural constructs that can be measured or identified. As the taxonomy has evolved, we have found that some items could clearly be labeled as a factor by our current definition. We have also found other items may be constructs that are influential but not necessarily measureable.

3 THE TAXONOMY

The taxonomy is currently divided into five major areas: demographics, personality, experience, values and context. Approximately fifty initial socio-cultural factors are grouped across these five categories. A sixth category of external factors contains items that are influential and are still under consideration for inclusion into one of the five main categories. The following section discusses each category and highlights some of the identified factors.

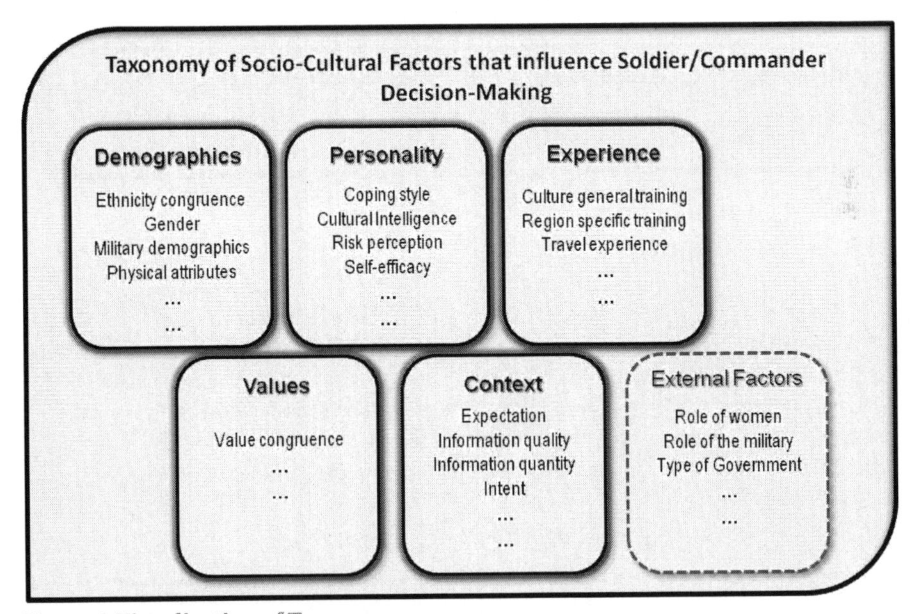

Figure 1 Visualization of Taxonomy

3.1 Demographics

Demographics represent characteristics of the Soldier that potentially influence his or her decision making. Socioeconomic status as an example, depends on a combination of variables, including occupation, education, income, wealth, and place of residence and is often used by sociologists as a means of predicting behavior. Table 1 identifies a few other demographic factors included in the draft taxonomy.

Table 1 Examples of factors from demographics category

Factor	Definition
Ethnicity congruence	quality of agreeing with, relating to, or conforming to values and behaviors for a particular large group of people classed according to common racial, national, tribal, religious, linguistic, or cultural origin or background
Family structure/ upbringing	the composition and membership of the family and the organization and patterning of relationships among individual family; the way in which a family is organized according to roles, rules, power, and hierarchies
Gender	the behavioral, cultural, or psychological traits typically associated with one sex (i.e., male or female)
Military demographics	characteristics unique to a military environment such as rank, position, role, time in service, etc.
Physical attributes	appearance; individual physical characteristics of a person (height, weight, hair color, skin color, stamina, strength)
Political affiliation	being formally or informally connected, joined, associated with, allied with a particular political party or message

3.2 Personality

Personality represents the totality of an individual's behavioral and emotional characteristics. Research demonstrates that the "Big Five" personality traits (openness, conscientiousness, extraversion, agreeableness, and neuroticism) strongly predict work behavior across time, contexts, and cultures—in domestic settings and in overseas assignments (Ang, 2006; Costa & McCrae, 1992). Table 2 identifies a few of the personality factors included in the draft taxonomy.

Table 2 Examples of factors from personality category

Element	Definition
Coping Style	a person's characteristic strategies used in response to life problems or traumas. Coping styles include problem-focused thoughts or behaviors, seeking social support, wishful thinking, blaming self, and avoidance (Vitaliano, 1987)

Element	Definition
Cultural Intelligence (CQ)	a state-like individual difference that describes an individual's malleable capability to deal effectively with people from other cultures. (Earley & Ang, 2003)
Risk Perception	A subjective judgment that people make about the characteristics and severity of a risk driven by the need for arousal that can be provided by varied, novel, complex and intense experiences (Lauriola & Levin, 2001)
Self-efficacy	the belief that people have in their ability to exercise control over events that affect their lives (Bandura, 1977)
Sensation seeking	an individual's desire for varied, complex, novel and intense stimulation (Zuckerman, 1971)
Uncertainty	A lack of certainty. (Greco, 2003). A state of having limited knowledge where it is impossible to exactly describe existing state or future outcome, more than one possible outcome

3.3 Experience

Experience is the understanding of an object, thought, or emotion through the senses or mind, either through direct observation of or participation in events. Experience coupled with relevance informs the basis of knowledge. Hsee and Weber (1999) found that the inherent influence of culture affects basic judgment, decision making, and risk preference. Table 3 provides a few examples of experience factors.

Table 3 Examples of factors from experience category

Element	Definition
Culture General Training	teaches about a culture in general by identifying the points on which cultures vary, providing a framework to consider cultural similarities and differences
Region Specific Training	provides descriptive facts and figures about a locale
Travel Experience	exposure to a variety of different cultural environments

3.4 Values

The common dictionary definition of a value is "something (as a principle or quality) intrinsically valuable or desirable" (http://www.merriam-webster.com/dictionary/value) while the Dictionary of Human Geography defines values as "the principles or standards informing individual or group ideas and beliefs" (Values, 2009). "Cultural differences are reflected in values, which in turn affect behavior, including the way in which decisions are made." (Albaum et al., 2010, p.142). One factor in the value category is value congruence. Value congruence is the extent to which an individual holds values or behaves in a manner consistent with their own self-image such as their self-identification with an ethnicity or group (Simmons, 2010).

3.5 Context

The word context has many meanings, but in reference to this taxonomy, we will define context as "the set of circumstances or facts that surround a particular event, situation, etc." (http://www.merriam-webster.com/dictionary/context). Context provides the background that will influence Soldier/Commander decision-making. One factor included in the context category is the physical environment. The physical environment is identified as a critical issue because climate and terrain can affect the success of the mission (Pirnie & Simons, 1996). Table 4 provides some more example factors for the context category.

Table 4 Examples of factors from context category

Element	Definition
Expectation	the act or state of anticipation of the results of a particular action
Information Quality	goodness of information
Information Quantity	amount of information available to a decision maker
Intent	the purpose motivating a given action
Physical Environment	the external surroundings and conditions in which something exists
Social Power	society's perception of an individual or ones perception of their own influence
Social Pressure	a force that a person feels to behave in a certain way based on society and other people's expectations

3.6 External Factors

During the taxonomy development process, we identified a group of factors that were determined to be relevant to decision making, however, their inclusion in the taxonomy was still uncertain at that point. The factors are the role of women, role of the military, type of government, globalism/isolationism, role of the media, locality, and customs. For the purpose of this paper, these items have been labeled as external factors. The external factors that we believe will ultimately be included in the taxonomy are the role of the military and the role of the media.

The literature discusses the importance of the military as a social institution and how it affects people's lives both directly and indirectly as well as how it effects other institutions (Segal, 1999). The central role of the military is to defend its country by combating threats through the use of lethal force when necessary. Additional roles may include advancing a political agenda, supporting or promoting economic expansion, or offering support externally to respond to non-military crises.

The ultimate role of the media should be to develop a range of diverse mediums and voices that are credible, and to create and strengthen a sector that promotes such outlets. Credible outlets enable citizens to have access to information that they need to make informed decisions and to participate in society. A media sector supportive of democracy would be one that has a degree of editorial independence, is financially viable, has diverse and plural voices, and serves the public interest (Center for Democracy and Governance, 1999). When discussing the role of the media, the term "CNN Effect" is often identified as a relevant phenomenon that can drive foreign policy and decision-making. "CNN effect" is described as the effect that real-time news media has on politics and government during political conflict and natural disasters (Livingston, 1997). The media's role in influencing national and international public opinion has grown immensely over the past decade with the evolution of computers and the creation of the internet. The media is an immeasurable tool in national and military decision making (Marye, 2004).

4 VALIDATION PLANS

This next step of this research effort is to gain insight into the validity of the draft taxonomy. Our first step is to introduce the draft taxonomy to researchers in the area of decision making, more specifically military decision making. These will include social scientists, psychologists, and other experts in the field of decision making who can provide some face validity to the structure and items of our taxonomy. After incorporating suggestions from these experts, we will take the taxonomy to Soldiers and commanders via an online survey. Because we are trying to determine the influence of socio-cultural factors on decision making that may be obscure to the actual decision maker, we plan to design a study that incorporates understanding the levels of these socio-cultural factors for each decision maker. We will then put them into experimental decision-making environments to examine if decision-making performance does in fact vary across these factors. By utilizing this approach, we can gain face and construct validity for the taxonomy before we begin to use it to build our cognitive framework.

5 FUTURE TAXONOMY USAGE

Utilizing the final validated taxonomy, we will build a cognitive framework of how these socio-cultural factors influence decision making. The socio-culturally influenced decision-making model will be used to develop guiding principles for the presentation of relevant socio-cultural information in a format most effective to the Soldiers and commanders using the information and the various environments and missions where they will work. Additionally, these guidelines will be used in the design of future decision support tools for Soldiers and commanders.

6 REFERENCES

Albaum, G., Yu, J., Wiese, N., Herche, J., Evangelista, F., & Murphy, B. 2010. Culture-Based Values and Management Style of Marketing Decision Makers in Six Western Pacific Rim Countries. *Journal of Global Marketing*. 23, 139–151.

Ang, S., Van Dyne, L., & Koh, C. 2006. Personality Correlates of the Four-Factor Model of Cultural Intelligence. *Group & Organization Management*. *31*(1), 100-123.

Bandura, A. (1977). Self-efficacy: Toward a Unifying Theory of Behavioral Change. *Psychological Review*. 84(2), 191-215.

Center for Democracy and Governance. 1999. The Role of Media in Democracy: A Strategic Approach. Technical Publication Series. Washington, D.C.

Costa, P., & McCrae, R. 1992. Revised NEO Personality Inventory (NEO-PI-R) and NEO Five-Factor Inventory (NEO-FFI) Professional Manual. Odessa, FL: Psychological Assessment Resources.

DeRue, S. & Morgeson, F. 2005. Developing a taxonomy of team leadership behavior in self-managing teams. *20th Annual Conference of the Society for Industrial and Organizational Psychology*, April. Los Angeles, CA.

Earley, P., & Ang, S. 2003. *Cultural intelligence: Individual interactions across cultures*. Stanford, CA: Stanford University Press.

Greco, V., & Roger, D. 2003. Uncertainty, stress, and health. *Personality and Individual Differences*. 34, 1057-1068.

Guss, C. 2004. Decision Making in Individualistic and Collectivistic Cultures. *Online Readings in Psychology and Culture*. 4(3). Downloaded 1/3/2012.

Hsee, C. and Weber, E. 1999. Cross National Difference in Risk Preference and Lay Predictions. *Journal of Behavioral Decision Making*. 12 (1999):165-179. Print.

Lauriola, M., & Levin, I. 2001. Personality traits and risky decision-making in a controlled experimental task: an exploratory study. *Personality and Individual Differences.* 31(2), 215-226.

Livingston, Steven. 1997. Clarifying the CNN Effect: An Examination of the Media Effects According to Type of Military Intervention. Research Paper R-18. The Joan Shorenstein Center for Public Policy, Harvard University.

Marye, James M. 2004. The Media and National Security Decision-Making. USAWC Strategy Research Project; U.S. Army War College, Carlisle Barracks, Pennsylvania.

Pirnie, B. & Simons, W. 1996. *Soldiers for Peace: Critical Operational Issues.* National Defense Research Institute, RAND. Santa Monica, CA.

Scherpereel, C. 2006. Decision orders: a decision taxonomy. *Management Decision.* 44(1), 123-126.

Segal, Mary W. 1999. Gender and the Military. *Handbook of the Sociology of Gender.* New York.

Simmons, B. 2012. http://www.bretlsimmons.com/2010-07/leadership-integrity-value-congruence-and-employee-engagement/#ixzz1iazYmq4r.

Values. (2009). In D. Gregory, R. Johnston, G. Pratt, M.J. Watts, and S. Whatmore (Eds.) Dictionary of Human Geography (5th Edition). Hoboken, NJ: Wiley-Blackwell.

Vitaliano, P.P., Maiuro, R.D., Russo, J. and Becker, J. 1987. Raw Versus Relative Scores in the Assessment of Coping Strategies. *Journal of Behavioral Medicine.* 10, 1–18.

Whittaker, M. & Breininger, K. 2008. Taxonomy Development for Knowledge Management. *World Library and Information Congress: 74th International Federation of Library Associations and Institutions (IFLA) General Conference and Council*, 10-14 August 2008. Quebec, Canada.

Zuckerman, M. 1971. Sensation Seeking Scale IV. *Journal of Consulting and Clinical Psychology.* 36(2), 45-52.

Section III

Application of Human, Social, Culture Behavioral Modeling Technology

CHAPTER 14

Enhancing Cognitive Models with Affective State for Improved Understanding and Prediction of Human Socio-Cultural Behavior

Priya Ganapathy[a], *Tejaswi Tamminedi*[a], *Ion Juvina*[b], *Jacob Yadegar*[a]
a: UtopiaCompression Corporation, Los Angeles, CA
b: Carnegie Mellon University, Pittsburgh, PA

ABSTRACT

With an increasing shift towards asymmetric warfare, soldiers must be trained to enhance their social and behavior skills. Models/agents based on Adaptive Control of Thought-Rational (ACT-R) cognitive architecture have been widely used in military simulation platforms to educate and track trainee performance in various tactical domains. However, these models are inept in emulating the full range of human behavior in socio-cognitive tasks. Through this project, we have been successful in building an enhanced ACT-R (E+) model by incorporating an emotion (E) mechanism that produces better performance mapping compared to conventional ACT-R (E-) model for simple socio-cognitive games of Prisoner's dilemma (PD) and Intergroup Prisoner's Dilemma with Intragroup Power Dynamics – (IPD^2). By extending this fundamental work to incorporate various facets that affect decision-making in a social environment, the E+ model/agents could be used to understand and forecast the outcomes of real-world socio-cultural interactions. This study entails developing high-fidelity cognitive models/agents that can be employed for training of soldiers to improve their decision-making skills in scenarios that involve socio-cultural interactions with civilians or allied forces.

Keywords: Adaptive Control of Thought-Rational (ACT-R), cognitive architecture, decision-making, socio-cultural behavior, socio-cognitive games

1 BACKGROUND

1.1. Decision-making influenced by socio-cultural behavior

Today's combat scenario is multi-faceted. A wrong decision at the level of a single individual or team of personnel can cause serious retaliation from the local civilian population, the targeted hostiles and even allies. In accordance there has been a shift in interest within the department of defense (DoD) from conventional to irregular warfare (termed coined by CDR Dylan Schmorrow from OSD) (Schmorrow, 2010). To facilitate this shift, there is a growing demand for a broader and in-depth understanding of socio-cultural dynamics by developing human social behavior tools. This would enable non-traditional warfare planning and decision-making starting from the soldier level all the way to higher ranks of leaders, military analysts and planners. Decision-making tasks significantly depend on the individual's social, cultural, economic underpinnings as well as their personal experiences. As a result, systems that can measure decision-making factors which are directly affected by the social-cultural-behavior will be beneficial to predict not only the outcome of any decision (reaction of an infiltrator) but also the response of a trained soldier in responding without any loss of civilian lives and suffering injury (Farry, Pfautz and Carlson, 2010, Galiardi, 1995, Lee, 2009)

1.2 Meta-cognition models (E+ models) based on ACT-R architecture for training

Of the cognitive architectures that have been developed, three of the oldest and the most widely used of them are Soar, ACT-R and EPIC. In this study, we are interested in building cognitive models based on ACT-R (Adaptive Control of Thought-Rational) which is an architecture which takes a lot of inspiration from neuropsychological ideas, such as neural networks to do adaptation and learning. The 'Theory of mind' studies have demonstrated that diverse HSCB factors such as 'impulsiveness/response to inhibition', 'interest', 'pleasure', 'rationality', etc. influence pure cognition processes (memory, attention) and thereby, affect decision-making (Kohlberg, 1973, Leslie, 1991, Permack and Woodruff, 1978). The stimuli (or tasks) which elicit either of these factors prior to decision-making are coined as meta-cognition and the corresponding system or model generated to predict outcome for these tasks is called a meta-cognition model. Conversely, models which study the effect of stimuli that elicit only pure cognition processes such as memory, attention, and perception prior to any decision-making are called cognition models. To develop better and complete intelligent systems for training, it is necessary to build robust predictive models that capture both cognition and meta-cognition aspects of decision-making.

1.3. Neuroimaging and ACT-R modeling

The pioneer of ACT-R, Dr. Anderson wanted to explore the underlying neural underpinnings of his model and therefore, initiated comparative studies using brain imaging of subjects performing a given task and refined the corresponding ACT-R model to perform the same task with better matching result to that of the subject (Anderson, Bothell, Byrne, Douglass and Lebiere, 2004). These studies of brain localization created many revisions to the ACT-R theory and introduced several concepts of declarative, procedural representations, interactions between the two representations by introducing buffers to hold the current active information, etc. and thereby better simulated the actual cortical activations that can be observed during a functional MRI (fMRI) analysis (Anderson, Bothell, Byrne, Douglass and Lebiere, 2004). The ACT-R architecture can similarly be modified from E- (read E minus, lacks emotions) to E+ (read E plus) by introducing 'affect' parameters (such as 'trust', 'displeasure', etc.) and refined to match the fMRI traces of subjects across different brain regions associated with emotional reactivity. This would provide us with models that match better with subject data both internally and output behavior.

1.4. E+ models for individualized training

The advantage of these models is that once developed they do not require any external inputs such as providing neural data, etc. for classification/prediction. The parameters of the ACT-R model can be varied to simulate different cognitive conditions and thereby, allow the model to perform as humans (errors in making decisions, learning, etc.). By collecting extensive fMRI data, it is possible to accurately personalize the E+ model to match an individual subject's performance which in turn relies on his/her domain-knowledge, domain-independent skill set and predisposition to learning. The objective evaluation retrieved based on predictive performance of E+ model on alternate scenarios will aid experts in providing focused training to subjects. Consequently, we expect such training will result in reduction of training duration for subjects to become experts in a given domain.

To this end, as a feasibility study, we have developed E+ models for a simple socio-economic game of Prisoner's dilemma (PD) and Intergroup Prisoner's Dilemma with intragroup power dynamics (IPD^2) based on behavior and fMRI data collected by Rilling et al (Rilling, Goldsmith and Glenn, 2008). The output of the enhanced model playing the PD/IPD^2 games are compared with an average human player in terms of model accuracy and completeness. We then discuss the results of our study and comment on future direction.

2. METHOD

2.1. ACT-R ARCHITECTURE

ACT-R is a modular cognitive architecture composed of various interacting modules, mainly perceptual systems (vision and audio modules), motor systems (manual and speech modules), and memory coordinated through a production system. This modular view of cognition is in line with the recent advances in neuroscience concerning the localization or mapping of brain functions. The advantage of a symbolic system like ACT-R's production system is that, unlike connectionist systems, it can readily represent and apply symbolic knowledge of the type specified by military doctrine. The fundamental advantage of an integrated architecture like ACT-R is that it provides a framework for modeling basic human cognition and integrating it with specific domain knowledge. In ACT-R, performances defined by the parameters at the sub-symbolic level determine the availability and applicability of symbolic knowledge which makes ACT-R behave more human like (not deterministic). Those parameters underlie ACT-R's theory of memory, providing effects such as decay, priming and strengthening. But they also play a broader role of being cognitively adaptive, stochastic in nature, capable of handling alternate scenarios and being robust in uncertain situations. These qualities provide ACT-R models the capability to reason, learn, plan and make decisions. Finally, because of the continuous nature of the sub-symbolic layer, architecture parameters (example: W-working memory parameter) can be varied to incorporate the gradual degradation of behavior to emulate stress and fatigue (Best, Lebiere and Scarpinatto, 2002) (Figure 1).

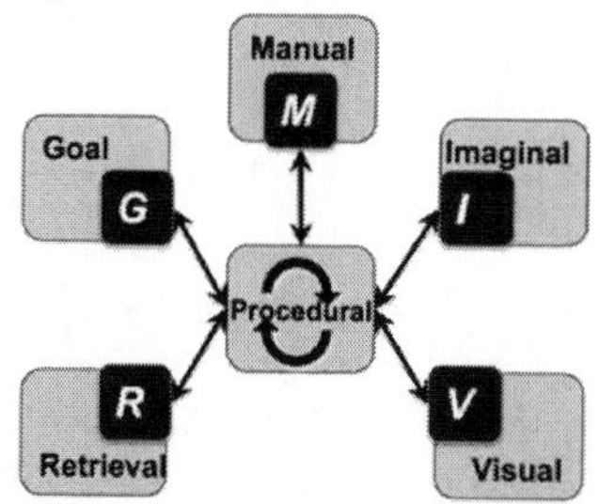

Figure 1: An overview of ACT-R architecture

2.2. IPD^2 game

We present a justification for including emotional biasing in ACT-R models by discussing our work on a social-cognitive task of Intergroup Prisoner's Dilemma with Intragroup Power Dynamics – (IPD^2). The IPD^2 game paradigm is useful for studying human behavior in conflict situations. IPD^2 game offers a richer opportunity to model real-world situations in which intra- and inter-group interactions are always entangled. The power dimension facilitates greater mutual co-operation (trust) while keeping unilateral and mutual defection at low. More information regarding the rationale behind the game and details on power and cumulative payoff calculation is provided in Juvina et al. (Juvina, Lebiere, Martin and Gonzalez, 2011).

2.3. Learning in ACT-R E minus model

First, we briefly introduce a computational model of strategic interaction in IPD^2. We label this model as E- (read E minus) because it is the baseline model that we use later on to argue for the necessity of including emotional biasing. Thus, in short, E- is a "purely rational" model that does not include emotional biasing. E- is a combination between an instance-based learning model (IBL) and a reinforcement-learning model (RL) of IPD^2. These models have been described in detail in Juvina et al. (Juvina, Lebiere, Martin and Gonzalez, 2011) and are only briefly introduced here. We combined IBL and RL in order to take advantage of their strengths and created E-. The E- model starts as an IBL model and creates instances of the game as the game unfolds. Gradually, the model develops context specific rules and relies less and less on retrieval of instances. Thus, as a result of this rule generation, E- becomes a complex RL model with a large number of rules instead of only two rules (corresponding to the two moves cooperate and defect) as in the simple RL model. The IBL model exhibits too much cooperation and the RL model too much defection as compared to the human players. E- also fails to match the human data (Juvina, Lebiere, Martin and Gonzalez, 2011). This failure is essentially related to E-'s inability to maintain mutual cooperation and avoid the risk of unreciprocated cooperation.

2.4. E plus model via neuroimaging studies

In searching for a solution to these limitations of the RL and IBL models, we found behavioral and neuroimaging evidence showing that humans do not always behave in accord to the principles of reinforcement learning. Humans in general have certain propensities to react to previous outcomes and these might interfere with their assessment of the current situation. From a brain imaging perspective, Rilling et al., demonstrated that humans react emotionally to certain outcomes and these emotional reactions bias their subsequent selection of moves (Rilling, Goldsmith and Glen, 2008). These results were the basis of our attempts to develop a model that includes emotional biases in decision-making. E+ (read E plus) is identical to E- and in addition it includes an affective mechanism that interacts with the decision to select a move in a particular context. The main idea in E+ is that each decision is biased by emotional reaction to previous outcomes. There are two types of emotional reactions: appetitive and aversive. The appetitive reactions make it more likely that the previous choice will be repeated. The aversive reactions make it more likely that the model will switch to the alternative choice. The model manifests appetitive emotional reactions in response to the following situations: reciprocated cooperation, unreciprocated defection, and reciprocated defection. The model manifests aversive emotional reactions in response to unreciprocated cooperation and powerlessness. As shown in Figure 2, E+ constructs an "attractive move" based on its own previous move, the outcome of that move, and its emotional reaction to that outcome. The attractive move might be different from the "reasoned" move that was constructed by the IBL and RL mechanisms. In this case they both act as sources of activation that spreads towards the memory representations of the two available moves (cooperate and defect). A retrieval process that is stochastic in nature determines the move that is eventually executed. There is a body of evidence from the brain imaging literature supporting such dissociation between reinforcement-guided choice and emotional biasing in strategic decision-making (e.g., Rushworth, 2008)

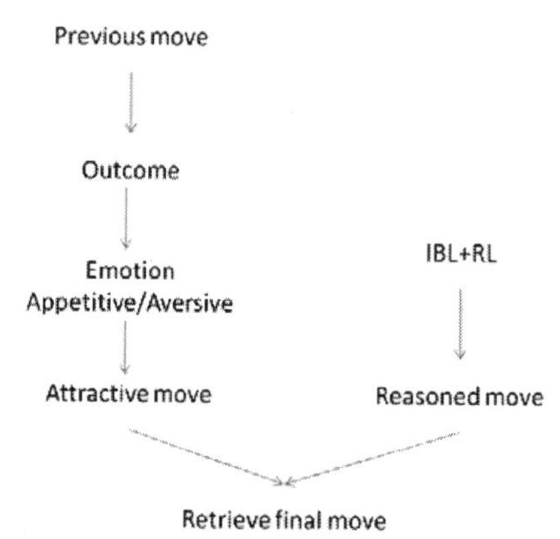

Figure 2: A basic structure of the E+ model. All other buffer modules remain the same including the internal parameter values

2.5. Extraction of activation trace from regions of interest

Using Rilling et al., fMRI data collected on a typical PD task (n = 10 subjects, 5 females, 5 males) and conventional General linear modeling (GLM) analysis, we identified brain regions that are significantly active during various phases of the game (Rilling, Goldsmith and Glen, 2008). A trial stimulus for the PD task is shown in Figure 3.

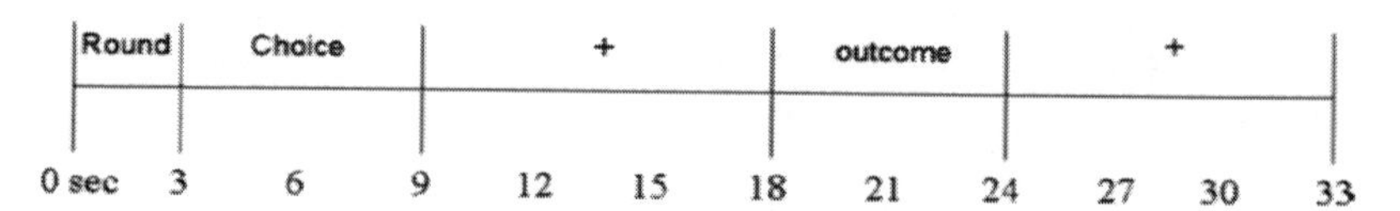

Figure 3: Timeline of a given trial. The two fixation periods are denoted as Fixation1 and Fixation2

We studied the brain regions (activity and duration) associated with vision, motor and retrieval modules that are known to be active at different time points during the trial stimuli. Standard fMRI processing steps including GLM analysis using SPM software were applied on fMRI images. We then applied de-trending and deconvolution steps to extract activity signal from different regions of interest in the brain. Signals from 100 voxels around each module (e.g., lateral inferior prefrontal cortex for retrieval module) in a rectangular area were averaged to get the time series for that module. Following this step, we then compare the activity across defined brain regions with the corresponding modules of the ACT-R model playing the IPD^2 or PD game.

3. RESULTS

We present the results of E- and E+ models playing the IPD^2 game and compare their performance with a small subset of behavior data. We created a graphical interface on MATLAB that communicates with the models developed on LISP. The interface allows us to test multiple computer strategies for a stronger proof of model behavior validation. The level of reciprocity, time course of outcomes and cumulative payoff clearly demonstrate the convergence of E+ model behavior with subject data (Figure 4, 5 and 6).

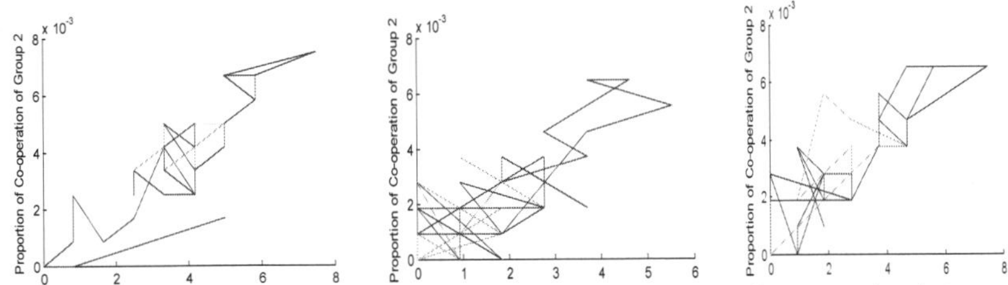

Figure 4: Left: Human player, Center: E-, Right: E+; Reciprocity of co-operation between groups with one group having an ACT-R model (E +, E-) or a human subject. Each IPD^2 game has several trials (here, n = 50). Red: 1 to 15 trials, Green: 16 to 30 trials, Blue: 30 to 50 trials. Note: Reciprocity is computed for cases when the subject (human or models were in majority for group 1 (x-axis)). Increase in mutual co-operation over time in games with human or E+ player. The trend is reversed for E- model.

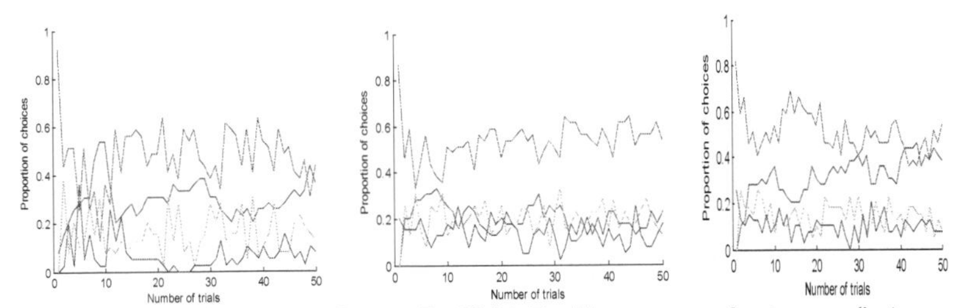

Figure 5: Left: Human player, Center: E-, Right: E+; Time course of outcomes (between groups) - Red: CC, Green: CD, Blue: DC, Magenta: DD. There is an increase in mutual co-operation (red line) between groups with increase in trials for a game with human as player or E+ as player.

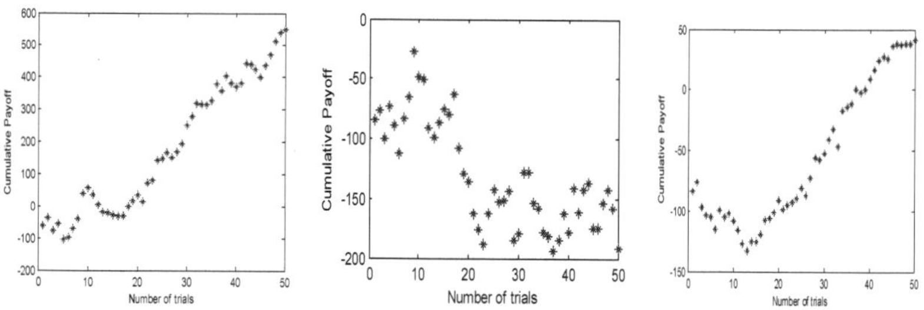

Figure 6: Left: Human player, Center: E-, Right: E+; Cumulative payoff trend. Both E+ and human player make positive payoff over time. The trends are better correlated than E- payoff trend.

Note: The models (E- and E+) play the same game as that of the human subjects

(existing data) in terms of number of sessions, strategies within each session. Each session consists of 50 rounds/trials. Even with a small sample set of subject data (n = 13 subjects) - the overall slope in terms of proportion of mutual co-operation with increase in trials and the cumulative payoff trend remain the same for E+ model and human players. Since we used only a small set of data, the subjects may not be a representative sample of average human behavior; we therefore see a difference in magnitude in the subject and model payoff trend across trials. The conventional E- model and enhanced E+ model provide an average correlation of 0.67 and 0.85 respectively with human subjects in terms of behavior data. In terms of internal matching with neural data, the vision and motor modules of the E+ model provided correlation of 0.91 and 0.85, respectively (Figure 7).

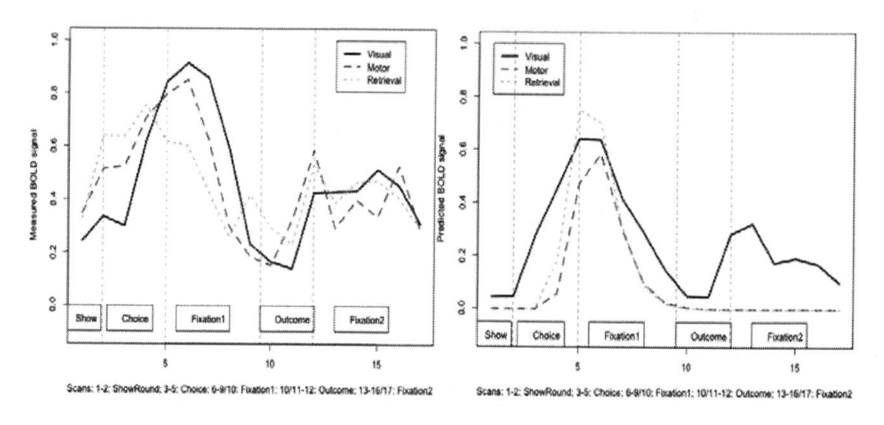

Figure 7: Left: Measured activity of subject 10. Right: Fitted activity of E+ model

We therefore, demonstrate that E+ model is a more complete cognitive model compared to E- model and that the match is achieved not only with subject behavior data but also internal mapping in terms of visual, motor and retrieval modules involved in the IPD^2/PD game.

4. DISCUSSION

The results clearly demonstrate the validity of refining cognitive models based on neural data. The enhanced ACT-R models (E+) match humans in terms of task breakdown involved in social cognition and also the overall response. Although, we demonstrated the feasibility of building better models using fMRI data for a simple socio-cognition task, it is possible to extend this work and develop normative or personalized models for more complex HSCB applications. The advantage of using ACT-R architecture is that once fine tuned using neural data, the resultant models/agents can be used to study outcomes in alternate scenarios and interact with one another without the need to further capture and feed real-time neural signals to the models. As has been established through literature (Bechara, 2004, Naqvi, Shiv and Bechara, 2006, Benedetto, Jumaran, Seymour and Dolan, 2006) and this work, emotion plays a critical role in the decision making process. Our experiments with IPD^2 games established that choice of action is strongly dependent of the perceived trustworthiness of the other player to the extent that individuals will punish the other's uncooperative behavior at an expense to themselves. Individual biases that are socio-cultural found to affect soldier behavior in the field can also be mapped to the emotional spectrum. These factors encourage the development of affective personality traits or emotional dispositions by producing or rewarding specific appraisal biases (Kluas, Scherer and Brosch, 2009). The emotions felt in a particular situation will be recorded in the emotional memory and can be activated when the person faces a similar situation or has to make a difficult decision in a short period of time. Therefore, it is crucial that we identify what emotions will play a dominant role in the training platforms of choice. The simplicity of delineating emotion into positive and negative is attractive but may not suffice as demonstrated by Raghunathan and Pham, 1999 (difference was observed in behavior of anxious and sad subjects, both negative emotions). We will study resolution and choice of emotion to ascertain a good fit with the platforms and their intended training goals. The extent to which ACT-R (E+) can emulate human behavior in both non-social and social contexts is dependent on the resolution of fMRI data collected and the strength of the sample set available to sufficiently model the breakdown of emotion-driven cognitive tasks. Our focus will be to develop modules within ACT-R (E+) which simulate the functions of amygdala, hippocampus, bilateral anterior insula and lateral orbital frontal cortex. Once we are able to understand and incorporate the causal modeling of these brain regions, we will then build ACT-R module(s) and buffer(s) to mirror this behavior. It is our hypothesis that the ACT-R(E+) predicted BOLD trace across mentioned regions will closely emulate BOLD response from fMRI studies.

5. CONCLUSIONS

As part of UC's future research programs, the E+ model will be further extended to represent a cognitive agent consisting of the primary mental faculties of *rational*, *emotional* and *volition/ethics* including their constituent components. Through a connectionist paradigm to be developed, these faculties will excite or inhibit each other depending on the environment the agent is immersed in. The hope is that such study will set a principled basis for a theory of mind with diverse applications including the current pursuits in the HSCB community, psychology, sociology, culture, history, etc. As an ensuing research program, once the theory and the practice of the aforementioned agents are in place, UC will investigate the sociability of the agents and their aggregated emerging behavior in swarms. A temporal dimension to such swarms will investigate the evolution of such societies and their ensuing consequences. Such an approach may provide a consistent method for understanding cultures.

6. ACKNOWLEDGMENTS

The authors would like to acknowledge our program manager CDR Joseph Victor Cohn, Ph.D and Office of Naval Research for funding this Phase I SBIR work (N00014-11-M-0138).

REFERENCES

Schmorrow D. 2010. Human social culture behavior modeling program, Office Of Secretariat Defense.

Stocco A., and D. Fum. 2008. Implicit emotional biases in decision making: The case of the Gambling Task. *Brain & Cognition*, 66(3), 253-259.

Mysiak J., C. Giupponi, and P. Rosato. 2005. Towards the development of a decision support system for water resource management. *Environmental Modelling and Software* 20: 203–214.

Farry M., J. Pfautz, and E. Carlson. 2010. Trust and Reliance in Human Socio-Cultural Behavior Models, In Proceedings of 3rd International Conference on Applied Human Factors and Ergonomics, Miami, FL.

Gagliardi J.L. 1995. Seeing and knowing: Knowledge attribution versus stimulus control in adult humans (Homo sapiens). *Journal of Comparative Psychology,* 109, 107-114.

Lee. 2009. Game theory and neural basis of social decision making. *Social Decision Making Social Dilemmas, Social Values, and Ethical Judgments*, 200-210.

Kohlberg L. 1973. The Claim to Moral Adequacy of a Highest Stage of Moral Judgment. *The Journal of Philosophy*, 70 (18): 630–646.

Leslie, A. M. 1991. Theory of mind impairment in autism. In A. Whiten (Ed.), *Natural theories of mind: Evolution, development and simulation of everyday mindreading*, pp. 63-77. Oxford: Basil Blackwell.

Premack D. G. and Woodruff G. 1978. Does the chimpanzee have a theory of mind? *Behavioral and Brain Sciences*, 1, 515-526.

Anderson J., D. Bothell, D. Byrne, S. Douglass, C. Lebiere and Y. Qin. 2004. An Integrated theory of the mind. *Psychological Review*.

Best B., C. Lebiere, and C. Scarpinatto. 2002. A model of synthetic opponents in MOUT training simulations using the ACT-R cognitive architecture. In Proceedings of the Eleventh Conference on Computer Generated Forces and Behavior Representation. Orlando, FL.

Juvina I, Lebiere C, Martin J. and Gonzalez C. 2011. Intergroup Prisoner's Dilemma with Intragroup Power Dynamics, *Games*, 2, 21-51.

Rilling J, D. Goldsmith and R. Glenn. 2008. The neural correlates of the affective response to unreciprocated cooperation, *Neuropsychologia*,

Rushworth M. 2008. Intention, Choice, and the Medial Frontal Cortex, *Annals of the New York Academy of Sciences*, 1124, 181–207.

Bechara A. 2004. The role of emotion in decision-making: Evidence from neurological patients with orbitofrontal damage, *Brain and Cognition,* 55, 30–40.

Naqvi N., B. Shiv and A. Bechara. 2006. The role of emotion in decision making: a cognitive neuroscience perspective. *Curr. Dir. Psychol.* Sci., 15, 260–264.

De Martino B., D. Kumaran, B. Seymour and J. Raymond. 2006. Dolan, Frames, Biases, and Rational Decision-Making in the Human Brain, *Science,* 313 (5787), 684-687.

Kluas R., T. Scherer. T. Brosch. 2009. Culture-Specific Appraisal Biases Contribute to Emotion Dispositions, *European Journal of Personality*, 23: 265–288.

Raghunathan, R. and M. Tuan Pham. 1999. All negative moods are not equal: Motivational influences of anxiety and sadness on decision making. *Organizational Behavior and Human Decision Processes*, 79 (1), pp. 56–77.

CHAPTER 15

Simulating Civilization Change with Nexus Cognitive Agents

Deborah Duong, Jerry Pearman

Agent Based Learning Systems
Birmingham, Alabama, USA
dduong@agentBasedLearningSystems.com

Augustine Consulting Inc.
Monterey, California, USA
jerry@aciedge.com

ABSTRACT

Civilization change includes not only the changes of material life but also changes in the social structures that go along with material change. Changes in the economic conditions of a country, such as an increase in the amount of corruption, are accompanied by changes in social relations. For example, corruption in Africa, an economic condition, is said to be the result of globalization, caused by the conflict between the roles of modern bureaucracies and traditional patron-client roles. The Nexus cognitive agent based simulation models the changing nature of role behaviors, and how these social norms co-evolve with economic and materialistic changes, based on cultural goals. This presentation is about the application of Nexus to the civilization change of corruption in Africa.

The Nexus cognitive agent model has been used in three major analysis community studies of Irregular Warfare (IW) at the Department of Defense. It was used to simulate corruption for the first IW analytic baseline, the Africa Study, at the Office of the Secretary of Defense (OSD). It was used to simulate Key Leader networks and Key Leader Engagements for the 2010 Tactical War Game, and additionally HVI networks for the 2011 Tactical War Game at the US Army Training and Doctrine Command Analysis Center. Nexus is unique in that it can simulate both short term micro-level role behaviors, as one might find between individuals in Key Leader, HVI, and corruption networks, and it can also be run out for macro-level analysis at the level of civilization change over the long run. This is an important quality for the study of Irregular Warfare because we would like to anticipate the long term effects of international interventions on state stability, and especially in terms of whether we would have to return to a country in the near future.

Nexus works at the level of individual motivation so that role based behaviors such as bribing and stealing, and role based choices such as nepotism are learned by individuals in an attempt to optimize their cultural goals. For example, in the Africa study, an agent from a Matrilocal tribe may have the goal to maximize the transactions in which its maternal relatives obtain goods and services. However, whether an agent's individually learned choices of bribing, stealing, and nepotism increase or decrease the number of times a maternal grandmother buys goods at the market is a social matter. It is social because in order for a plan to bribe to work, it must be likely that a person you offer a bribe to will accept it. In order for a plan to reject employers who steal from paychecks to work, there have to be some employers who don't steal from paychecks. Whether these behaviors result in advantages ultimately relies on greater social trends, also called institutions. Nexus computes from first principles, introducing the co-evolution between individually motivated behaviors and social institutions to a realistic social scenario. Every agent has its own evolutionary algorithm to judge the effectiveness of its behavior plans for achieving its goals in an environment where other agents are also learning to adjust their behaviors. Agents not only react to international interventions, but to other agent's reactions to those interventions, and so on. Nexus has a way to introduce realism in that the distribution of agent's behaviors is controlled so that it stays within the bounds expected by its class (based on ethnicity, tribe, or whatever else is important to an agent's behavior). The coevolution is a type of reinforcement learning that can start with what presently exists, due to special properties of the evolutionary algorithm, the Bayesian Optimization Algorithm. These properties, under co-evolution, allow Nexus agents to absorb and mimic real world scenarios. Results for the Africa corruption scenario are presented.

Keywords: agent-based simulation, Bayesian networks, Bayesian optimization algorithm, cognitive agents, corruption, terrorist networks, data adaptation

1 A SYMBOLIC INTERACTIONIST SIMULATION

Nexus Network Learner is part of the author's research program on Symbolic Interactionist Simulation, that started with the first Cognitive Agent Social Simulation, the Sociological Dynamical System Simulation (SDSS) (Duong 1991, Duong and Reilly 1995). Symbolic Interactionist Simulation addresses the fundamental social aggregation challenge with coevolving autonomous agents that maximize the utility of their actions based on their individual perceptions, acting solely in their own self-interest. From these individual actions emerge institutions, or expected behaviors that are more than the sum of their parts. Institutions in Symbolic Interactionist Simulations are dissipative (dynamic) structures: vicious or virtuous cycles of corresponding behaviors that symbolic interactionist agents develop as they adjust to each other. SDSS concentrates on vicious cycles such as racism, social class, and status symbols. The Symbolic Interactionist Simulation of Trade and Emergent Roles (SISTER) demonstrates Adam Smith's invisible hand in virtuous cycles such as price, a standard of trade (money), and a role-based division of labor (Smith 1994, Duong 1996, Duong and Grefenstette 2005). Symbolic Interactionist Simulations, as adaptive systems, model not only the emergence of social institutions from individual motivation, but the effect of institutions on individual motivation, a process known as immergence (Andrighetto et al. 2007) (See Figure 1).

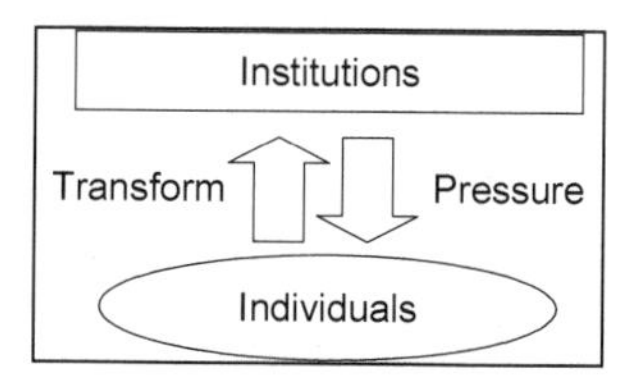

Figure 1. Emergence with Cross-scale Dynamics in Symbolic Interactionist Simulation.

However, both of these Symbolic Interactionist Simulations are more theory based than simulations of real world scenarios. SDSS, SISTER, and most agent based simulation are general representations of processes, that don't exist in the same way in the real world. For example, SISTER agents can form a grid of stores that have some things in common with the real world, such as more expensive local convenience stores vice less expensive centrally located markets (Duong 2010). However, they are not the markets that actually exist on any street. Nexus was made to bridge the gap between theory based simulation and simulations that can anticipate real world outcomes, for the purpose of Irregular Warfare analysis at the Department of Defense.

Getting phenomena to emerge accurately from a simulation that is initialized to real world data is a difficult task. The social structures that form from individual agent motives are dissipative (dynamic) structures, and to anticipate what will happen to them with any accuracy, we need to recreate the cycles that derived the data in the first place. We need to reverse engineer the data by taking correlations in the data and explaining them with agent motivation causes. This is what Nexus was designed to do: every agent in Nexus has its own Bayesian Optimization Algorithm (BOA), an estimated distribution algorithm (EDA) that is very similar to a Genetic Algorithm, except that it can be initialized from an existing scenario and change from that initialization point. Nexus agents use the BOA to learn strategies to behave within roles and to choose role partners in a way that benefits their cultural goals. Through coevolution, agents initialized to behave in the expected distribution of behaviors in accord with their attributes and role come to behave in that distribution through motivated strategies, performing the desired "reverse engineering" from statistical correlation to cause. This capability is the most important contribution of Nexus Network Learner.

2 NEXUS NETWORK LEARNER CORRUPTION SCENARIO

The Nexus Network Learner corruption scenario possesses a theoretical basis in interpretive social science, which is expressed in economics in the "New Institutional Economics" (NIE) theory (North 2005). NIE is the theory that institutions, which are social and legal norms and rules, underlie economic activity and constitute economic incentive structures. These institutions come from the efforts of individuals to understand their environment, so as to reduce their uncertainty, given their limited perception. However, when some uncertainties are reduced, others arise, causing economic change. To find the levers of social change, NIE would look at the actor's definition of their environment, and how this changes incentives and thus institutions. Interpretive social science is expressed in sociology in Symbolic Interactionism (SI), in which roles and role relations are learned, and created through the display and interpretation of signs. In Nexus Network Learner examples of roles and role relations are "Consumer" and "Vendor," and examples of signs are social markers such as gender and ethnicity. Symbolic Interaction provides the basis for individual interaction in the model.

Nexus Network Learner is a model of corruption based on the theory that corruption is a result of globalization. Many social scientists assert that corruption is the result of conflict between the roles and role relations of the kin network and the bureaucratic network, two separate social structures with their own institutions forced together because of globalization (Smith 2007). Nexus Network Learner models the kin network and the bureaucratic network (in the context of the trade network or network of economic activity), as well as the role behaviors which result in corruption, and the capacity of individuals to learn new behaviors resulting in new institutions, based on their cultural motivations (Duong et al. 2010) (See Figure 2).

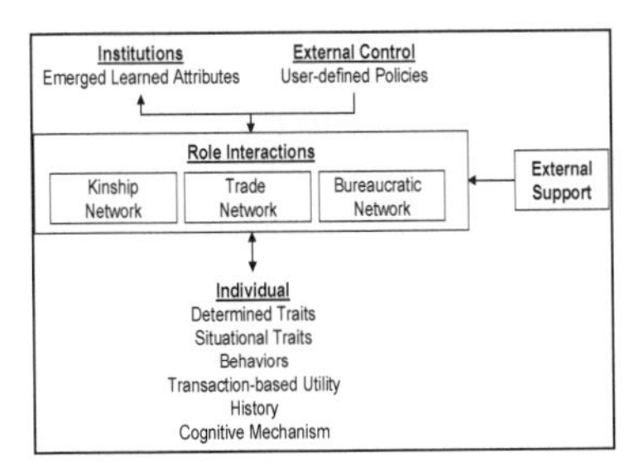

Figure 2. Nexus Network Learner Conceptual Model.

In order to model institutions from first principles, a model of 'cognition' and 'agency' is needed, including 'cognitive agents' (or individuals) that perceive their environment and act based on their motivations. Institutions 'stabilize' when the individuals converge upon an agreement on how to act with each other, at a point when no agent can improve its situation by changing its behavior (Nash Equillibrium). Changes to institutions, such as the reform of corruption, may be studied by applying policy changes which change the motivations of the actors. New institutions form, perhaps repairing the problem, as individuals react not only to the policy changes but to each other as well.

Nexus Network Learner individuals learn to navigate their environment according to their individual incentives. As they learn they affect each other's incentives. As they change each other's incentives, the choices they make become new social structures (institutions). The choices include both "actions" and "perceptions". For example, individuals learn the 'type of persons' to include in their social network, including their kinship, ethnicity, and bribing behavior. Individuals learn whether to divert funds across networks through bribing and stealing. With incentives modeled, the effects of different government actions such as increased penalties for behaviors, foreign aid, or actions which affect the price of natural resources can be objectively evaluated. Changes in the habits of individuals change the prevalence and types of corruption as they evolve. These changes are driven by evolving incentives from government actions, and individuals' reactions to those actions and to each other.

3 COMPUTATIONAL MODEL

Nexus Network Learner was created with the REPAST Simphony agent based simulation (North and Macal 2007). It was used in the first large study of Irregular Warfare at the Office of the Secretary of Defense (OSD), the Africa Study. In this study, the effects of international interventions on corruption were examined.

The Nexus Network Learner uses a Bayesian Network to characterize the demographic data of a country, which generates the initial agents of the scenario. The Bayesian Network describes characteristics that agents cannot change, for example, social markers such as ethnicity or gender. It also describes other characteristics that agents can change on an individual basis during the simulation, for example, behavioral characteristics, such as bribing or stealing, or preferences for choices of others in social networks (based on social markers or behavioral characteristics). Finally, the Bayesian Network describes demographic characteristics which individual agents do not learn, but are rather the output of the computations made during the simulation, such as unemployment statistics. (See Figure 3).

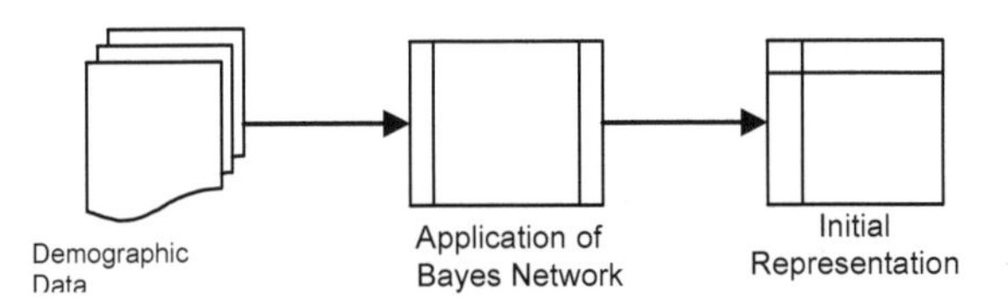

Figure 3. Model Initialization Process.

In Nexus Network Learner, agents are adaptive, that is, they are able to learn and adapt to new role behaviors through the use of evolutionary computation techniques of artificial intelligence, also known as genetic algorithms. Agents start out with different propensities to behave that are generated by a Bayesian Network, but they learn other behaviors based on utility, that is, what they care about. Utility is in the trade interactions (transaction-based utility) of the kin that an agent cares about. As the simulation runs, agents learn how to navigate their environment according to their individual traits and experience through the BOA. These choices include both actions and perceptions. Agents learn the type of persons to include in their social network, including kinship, ethnicity, and bribing behavior. They also learn whether to divert funds across networks through bribing and stealing. With incentives modeled, the effects of different whole of government actions such as increased penalties for behaviors, foreign aid, or actions which affect the price of natural resources may be studied. Corruption behavior changes through synchronous alterations in the habits of individuals, driven by new incentive structures that come both from whole of government actions, in agent's reactions to those actions, and to each other. (See Figure 4).

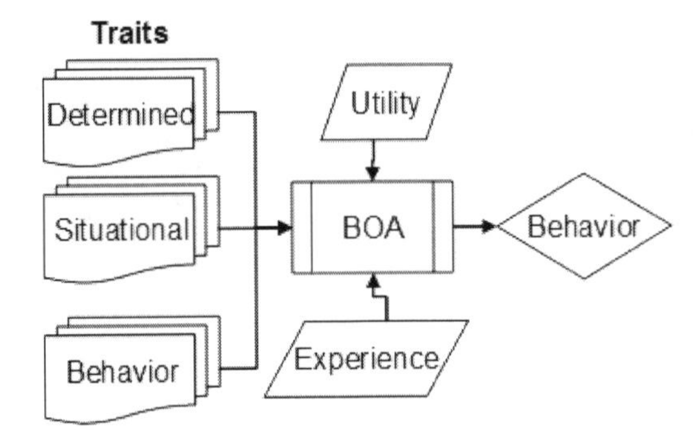

Figure 4. Agent Components.

There are eighty-two different roles in the three networks. Typical Kin roles include Father, paternal grandmother, maternal cousin. Typical Trade roles are retailer and vendor. Typical Bureaucratic roles include government employee and government employer. There are eight types of corruption relations possible:

1. Stealing/Trade Network (Scam)
2. Bribing/ Trade Network (Gratuity)
3. Hiring Kin/ Trade Network (Nepotism)
4. Bribing to be hired/ Trade Network (Misappropriation)
5. Stealing/Government Network (Levy, Toll, Sidelining)
6. Bribing/Government Network (Unwarrented Payment)
7. Hiring Kin/Government Network (Nepotism)
8. Bribing to be hired/Government Network (Misappropriation)

In a Nexus Network Learner run, agents periodically choose their role partners with traits according to what they learned to be their optimized goals, accept offers for partnership according to their goals, and periodically distribute funds using corruption behaviors that have served their goals in the past. The distribution of agent behaviors as well as their intended behaviors may be outputted. (See Figure 5).

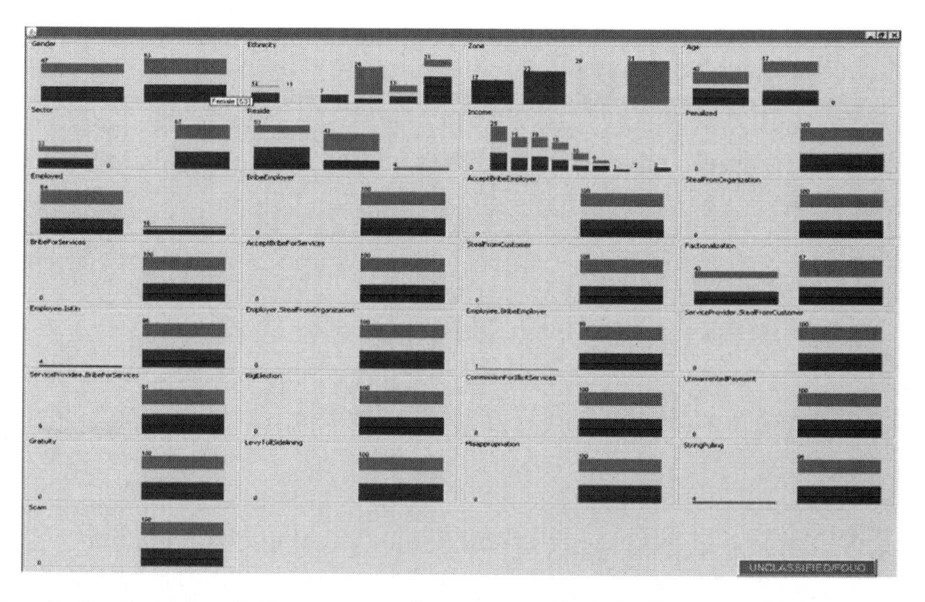

Figure 5. Sample output "arff" file in Weka, showing percentage of persons of each zone having corrupt and other attributes.

Each agent implements the BOA to assess results of 20 rated strategies after 400 days of simulated time. Specifically, the BOA captures the 10 best strategies for future use and discards the 10 worst strategies. The number of strategies for future use and the number of strategies discarded are simulation parameters set by the user. The BOA utilizes the 10 best strategies to reformulate 10 additional strategies for further testing by the agent. The output in Figure 6 shows and excerpt of person (agent) #11's scored strategies after approximately two years of simulated time. The output shows 20 scored strategies ranked worst (-30.75) to best (49.75).

Person 11: Strategy Score	Employee.IsKin	Employer.StealFromOrganization	Employee.BribeEmployer
-30.75	IsKin	StealFromOrganization	BribeEmployerNot
-9.5	IsKin	StealFromOrganizationNot	BribeEmployerNot
0	IsKin	StealFromOrganizationNot	BribeEmployerNot
0	IsKin	StealFromOrganization	BribeEmployerNot
0	IsKin	StealFromOrganizationNot	BribeEmployerNot
0	IsKin	StealFromOrganizationNot	BribeEmployerNot
0	IsKin	StealFromOrganization	BribeEmployerNot
0	IsKin	StealFromOrganizationNot	BribeEmployerNot
0	IsKin	StealFromOrganization	BribeEmployer
0	IsKin	StealFromOrganizationNot	BribeEmployerNot
7.5	IsKin	StealFromOrganizationNot	BribeEmployerNot
8.25	IsKin	StealFromOrganization	BribeEmployerNot
9.75	IsKin	StealFromOrganizationNot	BribeEmployerNot
12	IsKin	StealFromOrganization	BribeEmployerNot
12	IsKin	StealFromOrganizationNot	BribeEmployerNot
18	IsKin	StealFromOrganization	BribeEmployerNot
21.75	IsKin	StealFromOrganization	BribeEmployerNot
31.75	IsKin	StealFromOrganizationNot	BribeEmployerNot
39.75	IsKin	StealFromOrganization	BribeEmployerNot
49.75	IsKin	StealFromOrganizationNot	BribeEmployerNot

Figure 6. Excerpt of Scored Strategy for Agent #11 After Two Years Simulated Time

4 NEXUS NETWORK LEARNER TERRORIST NETWORK SCENARIO

Learning occurs in Nexus slowly, but because agents are initialized to a distribution of behaviors expected by their demographic class, Nexus can give reasonable outputs upon initialization. Nexus' networks are dynamic, changing at a reasonable statistical rate even before learning and change take hold. The realistic rate of attrition of relationships and representation of roles make for networks with a high enough fidelity that they were used to test the skill of intelligence analysts in the US Army Irregular Warfare Tactical Wargame 2011.

In this wargame of terrorist tracking in Afghanistan, a base case of companies without a dedicated Company Intelligence Support Team (COIST) was tested against an enabled case of companies that had COIST intelligence officers. Their job was to study rumors from the population to infer roles, identities and locations in terrorist networks. Knowledge of behaviors and rumors travel through Nexus' social networks in what one would expect based on the role of the person. These rumors could be about everyday role behaviors such as having tea together, or more telling clues such as the sale of weapons, as well as being distorted with frequent lying. The scenario is realistic in that it represents all the role relations that are important in the culture as well as their typical role behaviors, and they attrit at the frequency expected the real world. Because the dynamism is realistic, the COIST encountered a similarly hard, ambiguous problem in discerning the meaning of rumors for roles in a terrorist network. The fidelity was sufficient enough to separate the COIST equipped units from units that did not have COIST assets.

SUMMARY

Nexus Network Learner is a Symbolic Interactionist Program designed to anticipate real world outcomes in social simulations. As a cognitive model, it simulates basic principles of interpretive social science: in sociology, roles and role relationships that develop in the process of symbolic interaction, and in economics, individual self interested transactions that form corresponding trade plans. Institutions emerge and at the same time affect the motivations of individual agents. Coevolution fuels this engine of society, as well as enabling Nexus Network Learner to adapt to and explain data, "reverse engineering" the correlative data relations into the causes and motivations that underlie it.

REFERENCES

Andrighetto, G., Campennì, M., Conte, R., Paolucci, M. 2007. On the Immergence of Norms: a Normative Agent Architecture. In Proceedings of AAAI Symposium, Social and Organizational Aspects of Intelligence, Washington DC.

Duong D. and Pearman, G. 2012. Nexus Scenario Development Guide, US Army Training and Doctrine Analysis Center. http://www.scs.gmu.edu/~dduong/NexusTechnicalReport.pdf.

Duong, D.V, Turner R. and Selke K. 2010. "Crime and Corruption," ESTIMATING IMPACT: A Handbook of Computational Methods and Models for Anticipating Economic, Social, Political and Security Effects in International Intervention, Kott and Citrenbaum, eds. New York: Springer Verlag. http://www.scs.gmu.edu/~dduong/Estimating_Impact_Corruption.pdf.

Duong, D. V. 2010. "Autonomous Tags: Language as Generative of Culture" Second World Conference on Social Simulation, Takadama, Keiki; Revilla, Claudio Cioffi; Deffuant, Guillaume, eds. New York: Springer Verlag. http://www.scs.gmu.edu/~dduong/WCSS08.pdf.

Duong, D.V. and Grefenstette, J. 2005. "SISTER: A Symbolic Interactionist Simulation of Trade and Emergent Roles". Journal of Artificial Societies and Social Simulation, January. http://jasss.soc.surrey.ac.uk/8/1/1.html.

Duong, D. V. 1996."Symbolic Interactionist Modeling: The Coevolution of Symbols and Institutions". Reprinted in Gilbert, ed. Computational Social Science, London: Sage Publications, 2010. http://www.scs.gmu.edu/~dduong/symbolicInteractionistModeling.pdf.

Duong, D.V. and Kevin D. Reilly. 1995. "A System of IAC Neural Networks as the Basis for Self-Organization in a Sociological Dynamical System Simulation". Behavioral Science,40,4,275-303. http://www.scs.gmu.edu/~dduong/behavioralScience.pdf.

Duong, D. V. 1991. A System of IAC Neural Networks as the Basis for Self-Organization in a Sociological Dynamical System Simulation. Master's Thesis, The University of Alabama at Birmingham. http://www.scs.gmu.edu/~dduong/behavioralScience.pdf.

Messer, K. 2009. The Africa Study. HSCB Focus 2010 Conference.

North, D. 2005. Understanding the Process of Economic Change. Princeton: Princeton University Press.

North, M. and Macal, C. 2007. Managing Business Complexity: Discovering Strategic Solutions with Agent Based Modeling and Simulation. New York: Oxford University Press.

Smith, A. 1994. The Wealth of Nations. New York: The Modern Library.

Smith, D. 2007. A Culture of Corruption. Princeton: Princeton University Press.

Witten, I., Frank, E. 2000. Data Mining: Practical Machine Learning Tools and Techniques with Java Implementations. NewYork: Morgan Kaufman.

CHAPTER 16

Cosmopolis: A Massively Multiplayer Online Game for Social and Behavioral Research

Marc Spraragen[1], Peter Landwehr[2], Balakrishnan Ranganathan[1], Michael Zyda[1], Kathleen Carley[2], Yu-Han Chang[3], and Rajiv Maheswaran[3]
[1]University of Southern California, GamePipe Laboratory
[2]Carnegie Mellon University, CASOS Center
[3]USC Information Sciences Institute
{sprarage,brangana,zyda}@usc.edu,{plandweh,Kathleen.carley}@cs.cmu.edu

ABSTRACT

The community aspects of Massively Multiplayer Online Games (MMOGs) are an exciting opportunity for studying social and behavioral models. For that purpose we have developed *Cosmopolis,* an MMOG designed to appeal to a wide variety of player types, and that contains several key research-oriented features. The course of development has revealed several challenges in integrating behavioral models with an MMOG test bed. However, the Human Social, Cultural, and Behavioral (HSCB) research value of Cosmopolis has been demonstrated with a number of prototype studies, and based on these studies and challenges we propose an ongoing experimental plan largely driven by collaboration with HSCB researchers.

Keywords: social, behavioral, modeling, massively, multiplayer, online, game

1 INTRODUCTION AND MOTIVATION

A 2008 study by the National Research Council entitled "Behavioral Modeling and Simulation – from Individuals to Societies" (NRC, 2008) discusses how we need to expand research in modeling and simulation to include models of individual and societal behaviors. In the study, it is pointed out that a technological infrastructure needs to be developed for behavioral modeling such that we can properly develop, test and then deploy such models. The study, in fact, suggests the development of a massively multiplayer online game (MMOG) for that infrastructure. Such an MMOG can be utilized as a test bed for models of individual and group phenomena.

Cosmopolis is an MMOG we have developed for this purpose. In designing the game, we have been motivated by the need to balance the diverse interests of players and researchers: players want an engaging game experience, while Human Social, Cultural, and Behavioral (HSCB) researchers need the flexibility to perform various experiments of their own design. To accomplish these goals, we've designed the game to incorporate specific features to work towards these ends.

For players, *Cosmopolis* is an MMOG built around an outer game world and a collection of subgames; these subgames may be of any genre (action, puzzles, sports, etc.). For researchers, *Cosmopolis* is a unique test bed and data source for studying social and behavioral models, particularly via custom-designed experimental subgames. The models can be of individual players or multiple players over time, as well as of non-player (AI/Artificial Intelligence) characters (NPCs), or combinations of the above. *Cosmopolis* provides various and flexible methods to facilitate these needs. *Cosmopolis* also has a novel approach to information channels, with multiple real-world and game world sources being combined to create effects on the game AI, and customized output for players.

Section 2 of this paper will discuss the theoretical framework of Cosmopolis from game design and social/behavioral modeling perspectives. Section 3 will cover the approach to development of the game engine, in terms of both design and engineering. In Section 4, the first set of experiments in Cosmopolis will be described and their results analyzed. Section 5 will summarize and draw conclusions from our findings during the Cosmopolis project.

2 THEORETICAL FRAMEWORK

Research into video games and their scientific uses has currently taken several paths. Work that has taken a Human-Computer Interaction (HCI) perspective has looked at issues of real world reaction to virtual appearance, (Yee, 2007). Other researchers have begun investigating the broader issue of how real world social phenomena translate into virtual spaces. Castronova specifically proposed that virtual worlds might serve as ideal platforms for experimenting with a wide variety of individual, organizational, and societal (IOS) models (Castronova, 2006). To demonstrate this point, he carried out a small scale experiment demonstrating that the real world concepts of supply and demand mapped reasonably to a virtual space (Castronova, 2008).

A variety of researchers have also looked at the social structures that form in games and their relative strengths and weaknesses. Such analyses have been derived from qualitative and ethnographic observation of player interactions, from surveys of player opinions, and from social network analyses of the strength of ties between different players. One notable ethnography-based analysis is Pearce's long term study of "Uruvian expatriates" (players of *Myst: Uru Online* who migrated to *There.com* after the first game's abrupt closure), and the roles that emerged among them. (Pearce2009) Social network analyses have been specifically conducted using both *Everquest* and *World of Warcraft*, (Ducheneaut, 2006, Huang, 2009, Huffaker, 2009, Williams et al., 2006) demonstrating the relatively small levels of interaction among players within the same guild structures, while Johnson et al. have developed a model of guild formation patterns that also helps to explain the formation patterns of offline gangs. (Johnson, 2009)

All of these efforts fall into the general category of *mapping*, as described by Williams: researchers want to know how virtual actions and representations serve as analogs of real world actions and representations. (Williams, 2009) By implementing a new MMOG, as opposed to relying on working in the diversity of extant MMOGs and virtual spaces, we can establish a unified mapping environment in which a variety of different phenomena can be explored, linked across a single space. Then, as our understanding of mapping principles develops, we will be able to implement and test different mappings with autonomy unavailable to a corporation beholden to a much more fixed game structure.

Mapping is a difficult phenomenon to deal with in MMOG development because it is difficult to predict exactly how different experiences will map for different individuals. That said, game designers have done considerable work to try and understand the differing natures of play styles practiced by different individuals in online environments. Designers have long been aware of the emergent values and behaviors of different MMOG communities and attempted to foster a broader awareness of this fact in the community at large. Raph Koster, one of the designers of *Ultima Online* and lead designer of *Star Wars Galaxies*, famously formulated that "(An MMOG is) a community. Not a game. Anyone who says, 'it's just a game' is missing the point." (Koster, 2009) Morningstar and Farmer, developers of LucasArts's social game *Habitat,* encountered the same phenomenon and noted that "a cyberspace is defined more by the interactions among the actors within it than by the technology with which it is implemented" and that from a design standpoint "detailed central planning is impossible." (Farmer, 1990) While the rules and incentive structures for certain behaviors can be incorporated into MMOGs, players will be driven by their own motivations as well. Instead of looking at player growth as a process opposed to these rule structures, however, community development should be considered in tandem to them. Players' reactions to the IOS models as implemented in the game environments will help to evolve our understanding of these models themselves.

That said, to apply these ideas to our specific development of an MMOG for the study of individual, organizational, and societal (IOS) models, it is still necessary to develop both an understanding of the community that will play the game and a method for allowing investigators to translate the salient features of IOS models into game dynamics. While measuring players' reactions to model implementations is essential, it is impossible to engage in accurate study without any theory of the base population. Bartle notably broke down players into four types based on discussion within a game's forum about what people want out of a multi-user dungeon or MUD (a precursor of modern MMOGs): Achievers,

Explorers, Socializers, and Killers. Bartle posited that these player groups can exist in various stable states of flux, determined by the type of MUD that had been created (Bartle, 1996). Yee later followed up this work with an attempt at a multi-factor analysis of the survey results from players of different MMOGs, identifying three salient factors in players' motivations for play: Achievement, Socialization, and Immersion (into a virtual environment). Yee also noted that these factors did not suppress each other, but might actuallycoexist within an individual at equal intensities, bolstering each other. (Yee, 2006) As noted earlier, significant research has already been done to determine the demographics of several conventional MMOGs, though Aschbacher's report on *Whyville* demonstrates the possibility for considerable demographic variability based on design, a phenomenon also at the heart of Pearce's study. (Aschbacher, 2004, Pearce, 2009) Game designers can appeal to all or some subset of these perspectives via design choices, and in creating *Cosmopolis* we have sought to provide a framework that would support multiple combinations of desires. Additionally, given our expected ability to segment players based on their play habits and associations, we can hope to provide a more refined breakdown of play habits than has been previously found.

Another important aspect of creating and instrumenting a game, particularly an MMOG, as a tool for social research is that of using competition among players as a means to reduce search complexity over a large-scale problem space. Our "human heuristic hypothesis" is the assumption that competing human players in increasing numbers and at growing levels of expertise will be able to find better solutions in large-scale social scenarios than would brute-force AI methods assigned the same problem. This is particularly true if the conditions discussed above hold in order to keep the players engaged and on task.

3 APPROACH

As a research test bed, *Cosmopolis* offers a critical degree of experimental flexibility beyond the data-logging capability of the standard MMOG. Our overall design comprises a federated model architecture: each subgame is a potential lab for a different social and behavioral model, maintaining interoperability with the outer game world model. Subgames may be added, and gameplay of the outer world can be tweaked, all to meet the needs of researchers who use our game to validate or collect data for their models. While all in-game events will be logged, we will be specifically providing appropriate data export, reporting, and visualization capabilities so that researchers can easily analyze the experiments that they design and conduct in the game environment. Exported data would include player characteristics and activities, relational information such as who played with whom, performance outcomes, geo-temporal activity sequences and so on.

From an engineering standpoint, the subgames' content and logic is completely isolated from the outer game except for a controlled data access pipeline. This enables administrators and researchers to determine exactly what parts of the outer world (or other subgames) that a subgame can modify, to prevent any unexpected behavior. This also means that in the event of bugs or design inconsistencies in a subgame, it can safely be taken down without affecting the rest of the game.

The event-based networking model enables efficient logging management, which is vital for researchers using *Cosmopolis* as a data source. The logging parameters can vary from player to player, subgame to subgame, based on the needs of various researchers. The current networking model has a separate gameplay and analysis server. The analysis server can be tasked to perform near real-time processing in addition to logging the data.

To separate the account management from gameplay logic, *Cosmopolis* uses web services to perform authentication and initiate game connection. This also paves the way for enabling researchers to control different parameters of the game from the browser—anything from adding characters to changing the weather.

4 *TRIBES* EXPERIMENT SYNOPSIS

One of the key ideas in the development of Cosmopolis was to have an AI system underlying the world and providing feedback for player actions. Such an AI would open the world for possible large-scale experiments involving the game's entire population. As a first step towards this, we began work on a basic experiment using player interaction with a computational model to investigate the possibility of crowdsourcing policy development.

The research question underlying *Tribes* was whether individual players, when presented with a computational model with too many configurations to be completely solved by a computer, could find optimal configurations through reasoning. In the same way that researchers have built custom scenarios in *SimCity*, accepted the model's assumptions, and analyzed the results (Peschon, 1996) we sought to build a model, have individuals play it, and would then analyze their results.

In the case of this game, we developed a model of inter-tribal relationships in Sudan. Our model was coded using Construct (Schreiber, 2004), a simulation engine that has been used for a variety of different social simulations. Construct posits that agents interact and exchange facts; these facts have positive or negative associations with particular beliefs. Correspondingly, agents choose to interact with each other based in part on defined affinities and in part based on overall similarity of belief.

Figure 1: Player comparing tribal beliefs in the Sudan Tribes subgame

We modeled Sudan as a set of 14 prominent tribes, each with a different degree of affinity to the others, and each tribe with a leader. The tribal prominence and affinity data is based on a semi-automated analysis of the corpus of the Sudan Tribune newspaper over the period of 2003-2010. The newspaper transcripts are reduced to a pre-defined list of relevant entities that are then linked together based on their relative proximity. Our data thus assumes closer relationships between tribes discussed in the same context and more distant relationships between those that are not mentioned in the same context. We correspondingly assume that the most important Sudanese tribes are those mentioned most often in the paper. The process of vetting and analyzing this data was not part of our own research and is more thoroughly covered in (Van Holt, 2011). Tables describing the key components of the model are available on request to authors.

The Sudanese are not only defined in terms of tribe membership but also as possessing a set of eight beliefs, each consisting of ten facts – five positive and five negative. The set of beliefs and their distribution across the different tribes were derived from selected readings on Sudan and consultation with a Subject Matter Expert (SME) on Sudan (Carley2010,ICS2007,Johnson2006). We posit that a central cause of tension in Sudan to be that different tribes possess markedly different beliefs, and that to rectify this it is important to bring the tribes' beliefs into alignment. We measure hostility by considering the net difference in average beliefs between each tribe pair. If at least four of the beliefs of the tribe in the pair differ more than a particular threshold, we consider the tribes to be hostile. All of these hostilities are then normalized across the entire set of possible pairs, providing an overall hostility indicator.

To support the propagation of beliefs and the corresponding reduction of hostility, our model contains two different interventions: one where tribal or national leaders give speeches to their particular regions and one where leaders meet in conference to increase their own knowledge (approximating the *Tamazuj* forums which occurred in Sudan before the referendum (ICS, 2010)). To "win" the model, a sequence of interventions needs to be chosen that will successfully reduce hostility below a particular threshold. (Our theoretical minimum is 0.2, but this is not based on a fully play-tested implementation of the game.) While it may seem that by incorporating only two forms of intervention our model keeps the complexity limited, this is not the case. Given that our model incorporates 14 tribal leaders and six national leaders, eight beliefs that can be construed either positively or negatively, and the ability of leaders to meet in groups of two, three, or four, there are:

$$(6+14)\cdot(8\cdot 2)+\left[\binom{6+14}{2}+\binom{6+14}{3}+\binom{6+14}{4}\right]\cdot 8$$
$$=20\cdot 16+23180\cdot 8=185760$$

possible interventions at any time period. This number grows exponentially over time, making complete computation impossible.

Successful solutions to the Sudan game can be considered high-level policy solutions for Sudan - an ordered mixture of conferences and interventions with specific tribes. While one player's successful game should not be a sole determiner of policy, a winning combination of choices that is validated with multiple simulation runs outside of the game could be used to make a recommendation about an appropriate course of action. Players thus serve as individual policy analysts, their opinions and choices informed by the information that is included within the scope of the simulation and the game that is wrapped around it.

The *Tribes* game is a graphic interface to this model. A player takes on a role as a member of the UN, talking to the different tribal leaders mentioned above, and choosing interventions based on their understanding of the current situation in Sudan. This understanding is cultivated in several ways. First, all leaders provide a modicum of contextual text explaining their relationship to their tribe or the nation as a whole. The player can also ask the leaders about the precise effects of their interventions on Sudan. Secondly, and more importantly, the player can access visual information about the beliefs of the different Sudanese tribes, the beliefs about which a leader is informed (that is, about which they could give an effective speech), and the current hostility in the country.

Beyond tracking players' solutions and their relative effectiveness, we plan to look at players' dialogue in *Cosmopolis* relating to *Tribes*, as well as their interactions in the section of *Cosmopolis*'s online forum dedicated to *Tribes*. By doing so, we hope to better understand players' motivations for why they make the choices that they do when playing. Players will be developing their solutions based on their understanding of Sudan as presented in the game. It is not essential that they develop a deeper understanding of the policies of the country -the game is based on the assumption that such an understanding *should not* be required of the players- but to not have any insight into their process is anathema to common sense. As such, we will log and analyze this conversation in order to better understand not simply players' decisions but also why they make them.

Because *Tribes* has not been launched, no formal results currently exist for it. However, we have tested out the model in simulation and as such have some understanding of the game's difficulty. To prepare, we have carried out two alternate "greedy" test simulations of the game. By "greedy", we mean that in analyzing the simulation we commit to the best choice for a particular time period and then determine the best choice for the next time period, thus building a complete set of best choices. We do not, however, limit our determination of the best possible option by only looking at the current time period. Rather, we look at the long-term consequences of an intervention (as if no future interventions occurred) and thus choose interventions based on net performance over time. If no intervention can be considered the best, we choose randomly between them.

A downside of the greedy approach is that it makes it impossible to study the impact of conferences; a conference between leaders will have no immediate positive impact on the beliefs of the Sudanese. To compensate for this, we have run two alternate sets of simulations, one in which all of the leaders have their natural knowledge, and one in which all leaders have perfect knowledge – the maximum possible outcome from a set of conferences. These two set of simulations thus provide bounds for the problem space – the maximum and minimum impact that can be had by the best intervention at a particular time period.

Our results from these tests are shown in Table 1 below. They suggest several key elements that will need to be taken into account when actually releasing the game. The first of these is that the difficulty level of the game needs to be significantly refined. Our initial plan was to ask the player to reduce hostility to a level of 0.2; this was not possible in even our optimal case, where the lowest hostility value seen was 0.46153885 and occurred because of the first intervention. That is, we never see a hostility value lower than this initial case, suggesting that it may be impossible to dramatically alter the hostility levels in the country. Any model that is this difficult to manipulate needs to be reworked to be made amenable to gameplay. The player needs to think that they are making headway on a problem, not running into a brick wall.

We also need to develop other methods of probing the model. This initial map provides us with some boundaries for the space, but a better map could be made by using AI agents to make greedy decisions on the fly as opposed to by locking in a particular solution at each time period. Our current method works best to find a single, optimal solution for each time period. Given that in both the imperfect and perfect knowledge cases we ran into a situation where more than 200 possible choices performed equally well at one time period, randomly choosing a particular path will not necessarily guarantee any long term measure of success. Looking at the output of a host of AI agents will certainly not yield a single best path, but will definitely provide better bounds on the space.

Table 1 Impact of interventions in Tribes game

Inter-vention	Imperfect Knowledge			Perfect Knowledge		
	Min. Hostility Seen	# of Minima	Max Hostility Seen	Min. Hostility Seen	# of Minima	Max Hostility Seen
1	0.4725275	1	0.5274725	0.4615385	2	0.5274725
2	0.4615385	1	0.5274725	0.4725275	6	0.5274725
3	0.4725275	2	0.5274725	0.4725275	1	0.5274725
4	0.4725275	1	0.5384615	0.4725275	2	0.5164835
5	0.4615385	1	0.5274725	0.4835165	290	0.5274725
6	0.4725275	4	0.5384615			
7	0.4725275	1	0.5274725			
8	0.4725275	3	0.5274725			
9	0.4835165	283	0.5274725			
10	0.4725275	1	0.5274725			

5 CONCLUSION

MMOGs are widespread and popular online communities; *World of Warcraft* alone boasts millions of subscribed players. The significance of *Cosmopolis* is its uniqueness as an MMOG designed specifically as a research testbed for social and behavioral models, with a correspondingly high degree of researcher control over experiments performed in and data gleaned from the game world. A few of the key features that *Cosmopolis* incorporates are a multi-genre system of subgames, a dynamically modifiable outer world, and a channel-based information system featuring real-world feeds and game-world effects. While these features help to make the game novel and engaging, they also have specific applications for scientists opting to use *Cosmopolis* as a research platform: subgames are a way for researchers to conduct isolated experiments; the modifiable nature of the game world allows for events to occur that may dramatically alter the main game environment, providing fodder for scientists interested in the evolution of online communities; and the information broadcasting systems will allow different messages to be broadcast to different portions of the community to help manage experiments conducted on the entire player community. Also, any and all *Cosmopolis* actions and interactions (including the internal processes of AI-based non-player characters) may be logged into our databases, and may be used to explore the mappings between game world and real world societies. Ready access to a high-fidelity data set means that researchers will have an easier time determining the impacts of different experiments on the community in *Cosmopolis* than do researchers of more closed gaming environments. It is impossible for one MMOG to be considered the definitive online game, and *Cosmopolis* is not intended to be that. But it is an important step in opening up game environments for use by researchers, and one that can help support the work of scientists interested in studying game environments and how different social and behavioral phenomena manifest within them. We hope that the public presence of *Cosmopolis* will encourage other researchers to look to our game environment as an avenue for research into human behavior.

REFERENCES

Aschbacher, P. Gender Differences in the Perceptions and Use of an Informal Science Learning Web Site. *National Science Foundation, Arlington, VA, 2003.*

Bartle, R.A. Players Who Suit MUDs, 1996. http://www.mud.co.uk/richard/hcds.htm.

Bavelas, A. Communication Patterns in Task-Oriented Groups, *The Journal of the Acoustical Society of America*, vol. 22, Nov. 1950, pp. 725-730.

Carley, K.C. Answers on Sudan, 26-Jul-2010.

Carley, K., Moon, I., Schneider, M., and Shigiltchoff, O. Detailed Analysis of Factors Affecting Team Success and Failure in the America's Army Game, *Pittsburgh, PA: Carnegie Mellon University,* 2005.

Castronova, E. Synthetic Worlds. *University of Chicago.*

Castronova, E. A Test of the Law of Demand in a Virtual World: Exploring the Petri Dish Approach to Social Science. *Indiana University, 2008.*

Christ, R.E., and Evans, K.L. Radio Communications and Situation Awareness of Infantry Squads During Urban Operations, *United States Army Research Institute, 2002.*

Claypool, M., Claypool, K., and Damaa, F. The effect of frame rate and resolution on users playing First Person Shooter Games, *Proceedings of SPIE 2006*, 2006.

Ducheneaut, N., Yee, N., Nickell, E., and Moore, R.J. "Alone together?": exploring the social dynamics of massively multiplayer online games. *CHI '06: Proceedings of the SIGCHI conference on Human Factors in computing systems*, ACM (2006), 407–416.

Farmer, F.R. and Morningstar, C. The Lessons of Lucasfilm's Habitat. In M. Benedikt, ed., *Cyberspace: First Steps.* MIT Press, Cambridge, MA, 1990.

Gibbs, M., Wadley, G., and Benda, P. Proximity-based chat in a first person shooter: using a novel voice communication system for online play. *Proceedings of the 3rd Australasian conference on Interactive entertainment,* Perth, Australia: Murdoch University, 2006, pp. 96-102.

Huang, Y., Shen, C., and Contractor, N. Virtually There: Exploring Proximity and Homophily in a Virtual World. *Proceedings of IEEE SocialCom-09*, IEEE (2009).

Huffaker, D., Wang, J., Treem, J., et al. The Social Behaviors of Experts in Massive Multiplayer Online Role-playing Games. *Proceedings of IEEE SocialCom-09*, IEEE (2009).

International Crisis Group, Sudan: Defining the North-South Border. Juba / Khartoum / Nairobi / Brussels. International Crisis Group, 2010.

International Crisis Group. Sudan: Breaking The Abyei Deadlock. Juba / Khartoum / Nairobi / Brussels. International Crisis Group, 2007.

Johnson, D.H. The Root Causes of Sudan's Civil Wars, 3rd ed. Bloomington, Indiana: Indiana University Press, 2006.

Johnson, N.F., Xu, C., Zhenyuan, Z., et al. Human group formation in online guilds and offline gangs driven by a common team dynamic. *Physical Review E 79*, 6 (2009), 1-11.

Koster, R. The Laws of Online World Design. *Raph Koster's Home Page.* http://www.raphkoster.com/gaming/laws.shtml.

Moon, I., Carley, K., Schneider, M., and Shigiltchoff, O. Detailed Analysis of Team Movement and Communication in the Americas Army Game, *Pittsburgh, PA: Carnegie Mellon University, 2005.*

Landwehr, P., Karley, C., Spraragen, M., Ranganathan, B., and Zyda, M. Planning a Cosmopolis: Key Features of an MMOG for Social Science Research. *In Proceedings of CHI-10, Atlanta, GA.*

National Research Council. Behavioral Modeling and Simulation: from Individuals to Societies, *Committee on Human Factors, Division of Behavioral and Social Sciences and Education, National Research Council, National Academies Press, Washington, DC, 2008, ISBN 0-309-11862-X.*

Pearce, C. and Artemesia. Communities of Play: Emergent Cultures in Multiplayer Games and Virtual Worlds. *MIT Press, 2009.*

Peschon, J., Isaksen, L., and Tyler, B. The growth, accretion, and decay of cities. *Proceedings of 1996 International Symposium on Technology and Society Technical Expertise and Public Decisions,* 1996, pp. 301-310.

Redden, E.S., and Blackwell, C.L. Situational Awareness and Communication Experiment for Military Operations in Urban Terrain: Experiment 1, *Aberdeen Proving Ground, MD: Army Research Laboratory, 2001.*

Schneider, M., Carley, K., and Moon, I. Detailed Comparison of America's Army Game and Unit of Action Experiments, Pittsburgh, PA: Carnegie Mellon University, 2005.

Schreiber, C., Singh, S., and Carley, K.M. Construct - A Multi-agent network model for the co-evolution of agents and socio-cultural environments. Pittsburgh, PA: CASOS, Carnegie Mellon University, 2004.

Slater, M., and Sadagic, A. Small-Group Behavior in a Virtual and Real Environment: A Comparative Study, *Presence,* vol. 9, Feb. 2000, pp. 38-51.

Van Holt, T. and Johnson, J.C. A Text and Network Analysis of Natural Resource Conflict in Sudan. In *Proceedings Sunbelt XXXI*, St. Petersburg Beach, Florida, USA, 2011.

Williams, D. The mapping principle, and a research framework for virtual worlds. *University of Southern California, Los Angeles, California, 2009.*

Williams, D., Ducheneaut, N., Xiong, L., Zhang, Y., Yee, N., and Nickell, E. From Tree House to Barracks: The Social Life of Guilds in World of Warcraft. *Games and Culture 1*, 4 (2006), 338-361.

Wright, T., Boria, E., Breidenbach, P. Creative Player Actions in FPS Online Video Games, *Game Studies*, vol. 2, Dec. 2002.

Yee, N. Motivations for Play in Online Games. *CyberPsychology & Behavior 9*, 6 (2006), 772-775.

Yee, N. and Bailenson, J. The Proteus Effect: Implications of transformed digital self-representation on online and offline behavior. *Human Communication Research 33*, 3 (2007), 271-290.

Zyda, M., Spraragen, M., Ranganathan, B., Arnason, B., and Liu, H. Information Channels in MMOGs: Implementation and Effects. *In proceedings of AHFE 2010, Miami, FL.*

Zyda, M., Spraragen, M., Ranganathan, B., Arnason, B, and Landwehr, P. Designing a Massively Multiplayer Online Game / Research Testbed Featuring AI-Driven NPC Communities. *In Proceedings of AIIDE-10, Palo Alto, CA.*

CHAPTER 17

A Decision Support Capability for the National Operational Environment Model

John Salerno[1], *Jason Smith*[2], *Adam Kwiat*[3]

[1]AFRL, Rome Research Site
Rome, NY 13442
john.salerno@rl.af.mil

[2]ITT Corporation, Advanced Engineering & Sciences
474 Phoenix Drive
Rome, NY 13441
jason.E.smith.ctr@rl.af.mil

[3]CUBRC
725 Daedalian Drive
Rome NY 13441
adam.kwiat.ctr@rl.af.mil

ABSTRACT

In this paper, we describe an integrated set of capabilities currently in development that provides Decision Support functionality to the National Operational Environment Model (NOEM) program. Decision Support functionality allows a NOEM user to define an objective, investigate/compare various policy set options, obtain the "best" set of policies based on the defined objective(s), and provide an English-like explanation of the results. A summary of the NOEM and its components is provided along with a description of the decision process envisioned by the NOEM team, exploring how the algorithms presented are combined to provide analysts and policy makers with success/failure and consequences of possible policies for the region/nation of interest. We conclude the paper using an example to demonstrate this added capability.

Keywords: Decision Support System, sensitivity analysis, decision trees, explanation systems, automated rule generation

1. INTRODUCTION

The NOEM is a large-scale stochastic simulator of models representing the environments of nation-states or regions along with a set of capabilities which allow one to exercise/explore the models. The NOEM supports two mission flows: (1) a prospective look at the future and investigation into reducing potential country/region instabilities and (2) ramifications of actions or events. The NOEM enables the user to identify potential problem regions within the environment, test a wide variety of policy options on a national or regional basis, determine suitable courses of action(s) and to investigate resource allocation levels that will best improve overall country or regional stability. The different policy options or actions are simulated, revealing potential unforeseen effects and general trends. The NOEM also allows users to identify points of impact and rings of degradation to examine the ramifications or impacts that have been or could be created by actions or events.

The objective of this paper is to demonstrate how the NOEM and a set of decision analytics can be used to investigate potential unstable national or regional conditions and potential policies that can be implemented to resolve the problems. We begin our discussion with an overview of the NOEM components. We then discuss the overall decision process and introduce an example as a use case. We conclude the paper by demonstrating how the tools discussed are used by providing results for our use case.

1.1. Architecture Overview

The NOEM infrastructure consists of three main components the: (1) Model Development Environment (MDE), (2) Baseline Forecaster (BF), and (3) the Experiment Manager (EM). The MDE facilitates the creation and management of the model and their components. The MDE's Model Builder is used to create a new nation-state model or edit an existing one. Users may create new regions within a nation using the Model Builder and populate those regions with various, existing system modules. The BF continually and automatically provides an updated set of projected futures for any nation-state model. A Baseline Forecast tells us where a nation of interest is headed over time if the nation remains on its current course. The EM enables intelligence analysts and decision makers to perform "What If" analyses in order to compare policy sets and their effect on a nation or region's stability over time, as well as reveal any potential short- to long-term ramifications of pursuing any one policy set. The EM allows one to examine/assess a multitude of policy options while the BF supports the assessment of whether the actions taken are moving the situation in the desired direction.

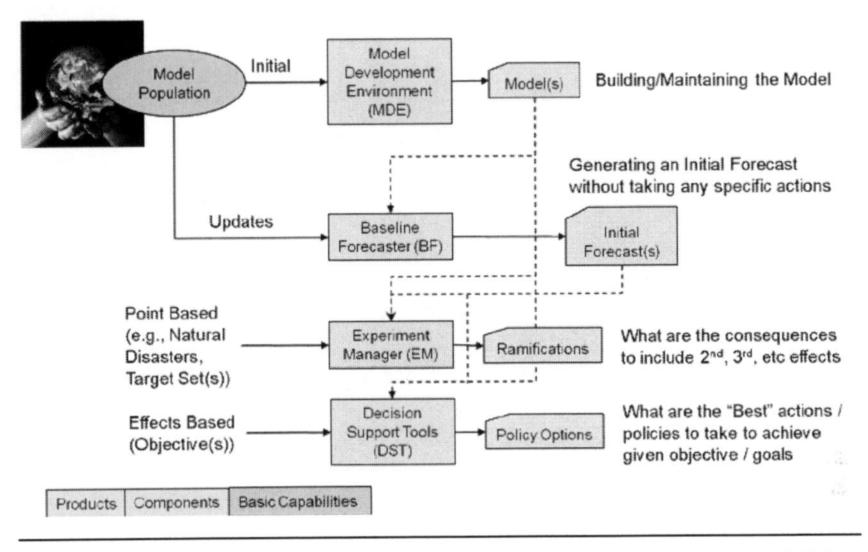

Figure 1 provides an overview of the major components that form the NOEM

The Decision Support Tools allows one to define an objective, investigate policy set alternatives and obtain English-like explanations of the resulting policy sets. It sits alongside the NOEM components, operates concurrently and interacts with them during the course of operations. We will be discussing this capability in much more detail in the following sections; however before we do, let us provide an overview of the NOEM model.

1.2. The Model

The NOEM model supports the simulation and the analysis of a nation-state's operational environment. Within the MDE, the nation models can be configured for one or more regions, where each region is composed of a group of highly interconnected modules which simulate subsystems such as a region's demographics, economy, or critical infrastructure. These modules, in essence, relate to the major pillars of a nation-state based on stability operations theory (Governance, Security/Rule of Law, Economy and Social Well-Being).

Security/Rule of Law is defined as protecting the lives of the populace from immediate and large-scale violence, and ensuring the state's ability to maintain territorial integrity. The Security/Rule of Law pillar is currently comprised of three modules: Indigenous Security Institutions (ISI), Crime, and Police. The ISI module is divided into: Border Patrol, Civil Defense, Facility Protection Services, and the Indigenous Military. Economy is defined as a system made up of various economic policies, macroeconomics fundamentals, free market, and international trade that exchanges wealth, goods, and resources mostly free of economic criminal activity. The Economy pillar is composed of two modules: Economics and Finance & Debt. Governance is defined as a public management process that involves a constituting process, governmental capabilities, and participation of citizens. Social Well-Being is defined as sustenance of life and relieving of suffering by way of humanitarian aid, best practices, human rights, essential services, and emergency response systems. The Social Well-Being pillar is composed of the majority of the modules and includes: Demographics, Health, Migration, Food, and fundamental Utilities (Electric Power, Telecommunications, Natural Gas, Oil, Transportation, and Water & Sanitation).

The agent-based Populace Behavior Module forms the heart of the NOEM model in the sense that all other modules (resources and security) are in place to support the populace. If the populace is not happy or satisfied to a certain degree, they could become activists and rebel against the host-nation government. Whether or not segments of the populace become activists depends on many factors, including their perceived hardship, the legitimacy of (or belief in) their government, their level of risk aversion, and the amount/visibility of security forces. Insurgents, Coalition forces, NGOs, and Host Nation Governments within the NOEM are not modeled as agents, but are characterized by the policies or strategies that they implement. Policies implemented by such groups will affect either the overall security within the environment or the services/resources provided to the people. As such, the NOEM team is researching the use of various Gaming Engine Techniques/Technologies that will allow one to play off these various strategies.

2. USING THE NOEM TOOLS

Our concept begins with a problem at hand. The problem can be identified in a number of ways. The BF can identify it as either an existing or potential problem (based on a set of metrics and their bounds for normalcy for that region/nation) or one can be identified ad hoc. Once the problem is understood, the next step is to define success; the metrics or outputs that will allow us to determine whether a given policy successfully moves the region/nation in the desired direction. If the problem was identified by the BF, then the objective becomes those metrics that are deemed out of normalcy. The goal in this case would be to restore the metrics to their values within the acceptable region. Given the outputs, we can then determine which actions or inputs are necessary to perturb in order to achieve the desired objective. These inputs therefore become the initial policy sets and can be run against the NOEM model to see how close they come to meeting that objective. If any policy set successfully achieve the desired results, we are done. Otherwise, we can revise the input values and develop additional/revised policy sets. This process can be iterated to find the "best" collection of inputs (i.e., policy set(s)) that produce results that are closest to the desired objective. We are also currently developing a new set of tools under the term 'Policy Set Optimization' to automate the generation of a set of policies, with each member of the set having the characteristic that no other policy exists that can outperform it on all of its objectives. This is called the Pareto or non-dominated set of policies.

Before we continue describing the tools, let us introduce a simple example. Our example will be of a fictitious country. We call it ImagiNation. A baseline forecast was run based on the current situation. The baseline forecast indicated that

ImagiNation will be undergoing financial instability in the near future if it does not quickly change its current fiscal policies. The country's borrowing ability will soon be denied since their maximum allowable debt (based loosely on the IMF's policies) will exceed 150% of their GDP. Based on the forecast this will occur around day 28 after our simulation starts. The country is going broke. Therefore, our objective will be to minimize the amount of borrowing (i.e., the debt). One way to accomplish this would be to stop spending or increase revenues (i.e., taxes). However, we know that there are also many downfalls if we attempt to do this. So, while our primary objective is to minimize the debt, we add a secondary objective: minimize the number of people becoming upset (or in our case, the number of people willing to rise up).

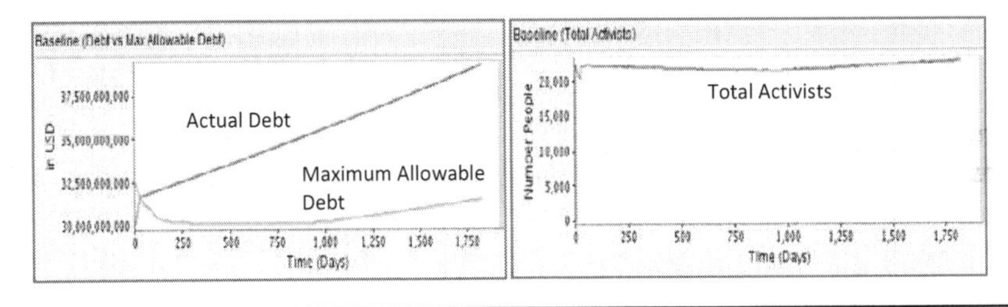

Figure 2 provides the results of a baseline run, indicating that the projected forecast of debt will soon exceed the maximum allowable debt. When this happens the IMF will deny any further borrowing.

Now that we have defined our problem and have identified our goal or measure of effectiveness, we next generate an initial set of policies that we believe would likely provide us the "best" solution. Typically, these potential solutions or policy sets are debated within a group, revised, and then an agreed upon way forward is decided. However, how do we know whether any given policy set chosen was the best one (having taken the right actions), the most robust, or the most resilient? How does one know it is the lowest cost, how long until one starts seeing appreciable results and how long until the full effects can be seen? How does one even know what actions one should take to achieve the most effective solution? These are the types of questions that our policy makers attempt to answer every day. These are the types of questions that the NOEM and its suite of Decision Support Tools attempt to support.

In our example, our primary objective is to minimize the amount of long term debt accrued by our country, Imagination, has. A concern, though, is that any actions we take do not incite riot or a revolt, i.e., any actions we take should not increase (and hopefully decrease) the discontent of the populace (measured in terms of the number of populace willing to become activists). Thus, our initial goal was to minimize debt AND minimize the number of activists, but we quickly found out that we could not simply minimize the debt alone. In fact, what we needed to do was to maximize the difference between the maximum allowable debt and the actual debt. This is due to the fact that the maximum allowable debt is a function of the GDP and only provides an upper ceiling as to the amount a country can borrow.

Driving solely the debt lower does not provide us an insight into whether we are below the maximum allowable debt since the GDP could be going down and the maximum allowable debt will, also. We weight them both equally. The next question is to determine to what extent we can or want to achieve our objective. Of course, we can set the goal for both the debt and the number of activists to be zero, but this is unrealistic. So what are realistic values? One way is to simply state a number for each based on expert opinion, a second option might be to derive them based on a set of runs to determine what is achievable. We opted to use the second approach here and do not specify the actual values for our objective function. In summary, our objective function can be written as:

GOAL = MAXIMIZE(MAXIMUM ALLOWABLE DEBT – ACTUAL DEBT) AND MINIMIZE(TOTAL ACTIVISTS)

After identifying the outputs of interest we next need to define what inputs or actions that we believe are necessary to achieve our desired objective. To identify the actions we have developed a query tool. Bringing up the tool allows the user to select a given output. Once the output is selected, the output the tool will provide a list of inputs that are believed to most directly affect it. The number of inputs displayed will be based on the degree of significance that the input contributes to the selected output (user selectable). Highlighting the inputs will automatically select them for the next step in the process – creating an initial set of policies. For our example, fifteen inputs have been identified. Table 1 provides this list along with their default values and ranges (to be used in the next step).

This next step is to create an initial set of policies. This is accomplished by using the inputs/outputs defined in the previous steps, their specified ranges and space filling techniques such as Latin Hypercubes Sampling (LHS) or Leap Halton. The goal is to create an initial set of policies that uniformly cover the design space. This design space is based on the number of inputs, their acceptable ranges, and the total number of allowable runs (user selectable). Each simulation run consists of one input or policy set along with the list of desired outputs. These are sent to the NOEM, which executes the given simulation and returns the results. The results of each run can now be compared to the objective function to examine how close to "optimal" the given policy set has come. If one or more of the results or policy sets are within a given tolerance, we are done. Otherwise, we can use the closest policy set so far and evolve or mutate it to generate a series of new policies that can be run and evaluated. This is an iterative process and can be continued until the desired results are found or after a specified maximum number of iterations.

Table 1 - Input Variables

Inputs	Default Value	Range
Adjudication Rate	0.7	0.1 – 0.9
Government Corruption Theft Percentage	0.1	0.05 – 0.15
Government Infrastructure Spending Percent	0	0 – 1.0
Government Services Spending Percent	1.0	0 – 1.0

Government Stimulus Spending Percent	0	0 – 1.0
Government Wages	2,835.84	1,400 – 4,200
Initial Police Forces	22,000	112,500 – 137,500
Interest Rate	0.045	0.0225 – 0.0675
Jail Term	30	3 – 100
Long Term Government Share Of Employed	0.001	0.001 – 0.05
Mean Adjudication Processing Time	0.5	0.1 – 0.9
Military Acting As Police Percent	0.1	0.0 – 0.2
Police Forces Goal	22,000	112,500 – 187,500
Stimulus	10,000,000	0 – 6,000,000
Tax_Rates	0.0176	0.03 – 0.50

For our example, a total of 1,000 runs were conducted. In this case, since we are minimizing only two values, we can simply plot them as a two dimensional scatter plot - debt versus total activists. Figure 3 provides this plot for a one year and five year projection. We can see that, as time progresses, the dispersion of the results increases.

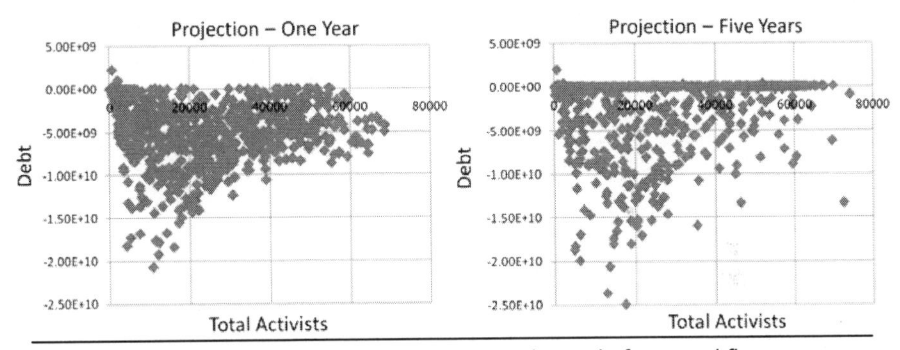

Figure 3 provides the results of 1000 runs at the end of one and five years

Based on these runs we can now find our goal values (the best that can be done independent) for both the delta between the maximum allowable debt and the actual debt, and for total activists. This is done by going through each run for a given day and finding the minimum (or maximum) value for that output. In our example we have:

Table 2 - Goal Values

Output	One Year	Five Years
Maximum Allowable Debt – Actual Debt	2.3E+09	2.00E+09
Minimum Number of Total Activist	622	716

In a two variable world, we can simply find the policy set that is closet to our objective. In a multi-dimensional world it is not so simple. So the question becomes "How do we determine what constituents a "good" policy?" To determine this we can still use the concept of closeness, but generalized for an n-dimensional space. We use the goal as the target point and compute a distance between it and each policy set result. Table 3 provides a listing of the closest ten policy sets after one- and five-year projections.

Table 3 – "Best" Policies

Rank	1 Year		5 Year	
	Run No.	Distance	Run No.	Distance
0	160	0.006	160	0.007
1	984	0.063	984	0.062
2	142	0.087	142	0.064
3	145	0.095	49	0.067
4	411	0.096	630	0.070
5	214	0.096	867	0.070
6	810	0.096	500	0.070
7	581	0.096	389	0.070
8	419	0.096	766	0.070
9	515	0.096	681	0.070

Comparing the two projections we can see that the top three runs are common, Run 160, 984 and 142. We next examine these top policies for each of our fifteen variables. Table 4 provides these values:

Table 4 – "Best" Policies (Input Values)

Inputs		Run 160	Run 984	Run 142
Adjudication Rate [0.7]		0.03	0.08	0.07
Government Corruption Theft Percentage [0.1]		0.06	0.08	0.07
Government Infrastructure Spending Percent [0.0]		0.24	0.75	0.96
Government Services Spending Percent [1.0]		0.007	0.009	0.033
Government Stimulus Spending Percent [0.0]		0.807	0.985	0.503
Government Wages [2,836]		1,479	4,062	1,417
Initial Police Forces [22,000]		124,897	136,228	107,183
Interest Rate [0.045]		0.034	0.058	0.0398
Jail Term [30]		46	96	74
Long Term Government Share Of Employed [0.001]		0.0457	0.0372	0.0079
Mean Adjudication Processing Time [0.5]		0.23	0.48	0.57
Military Acting As Police Percent [0.1]		0.17	0.03	0.095
Police Forces Goal [22,000]		113,976	168,146	142,586
Stimulus [10,000,000]		471,071	42,325	565,747
Tax_Rates [0.0176]		0.4549	0.1699	0.42498
Total Activist	**[End of Year 1]**	**622**	**2057**	**872**
	[End of Year 5]	**716**	**2299**	**985**
Debt	**[End of Year 1]**	**3.47E+10**	**3.15E+10**	**3.37E+10**
	[End of Year 5]	**5.21E+10**	**4.18E+10**	**5.05E+10**
Maximum Allowable Debt	**[End of Year 1]**	**3.70E+10**	**3.25E+10**	**3.40E+10**
	[End of Year 5]	**5.41E+10**	**4.36E+10**	**5.08E+10**

We can now display the results of our policy sets through the NOEM and see how well we did. Figure 4 provides the results for runs 160, 984 and 142.

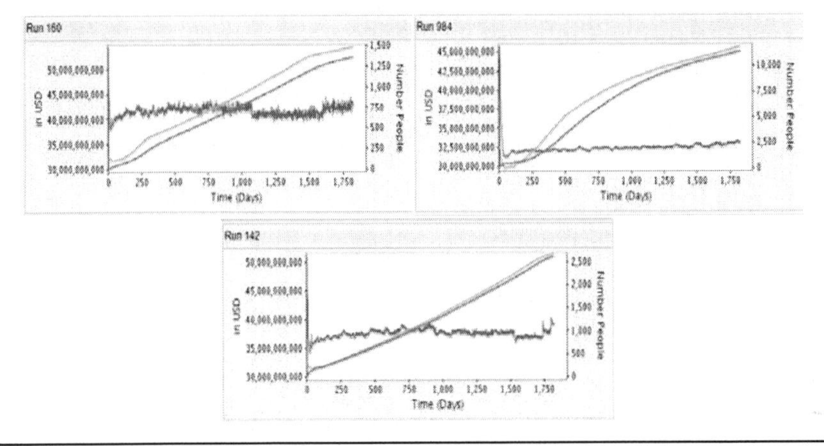

Figure 4 provides the results for runs 160, 984 and 142. Green line indicates the maximum allowable debt, the blue line is the actual debt, while the purple line displays the number of activists.

A second way to view the results is through the use of a Decision Tree. To use a Decision Tree algorithm we must first label which policies are good vs bad. Based on this decision tree, rules can be then be derived and English-like explanations can be generated, describing what a good policy for the given scenario is composed of. Figure 5 provides a sample tree for our one year projection while Figure 6 is a text view of the same graph. The goal would be to form the text view and to generate a set of rules or English-like explanation based on the derived rules.

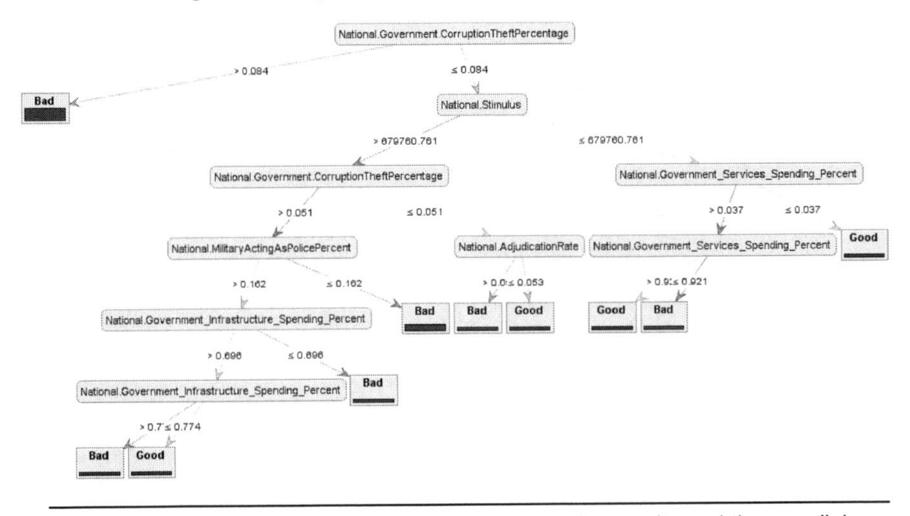

Figure 5 is a Decision Tree generated based on the results and those policies (top 10) identified as “Good” for the one year projection

```
National.Government.CorruptionTheftPercentage > 0.084: Bad {Good=0, Bad=658}
National.Government.CorruptionTheftPercentage ≤ 0.084
| National.Stimulus > 679760.761
| | National.Government.CorruptionTheftPercentage > 0.051
| | | National.MilitaryActingAsPolicePercent > 0.162
| | | | National.Government_Infrastructure_Spending_Percent > 0.696
| | | | | National.Government_Infrastructure_Spending_Percent > 0.774: Bad {Good=0, Bad=12}
| | | | | National.Government_Infrastructure_Spending_Percent ≤ 0.774: Good {Good=2, Bad=1}
| | | | National.Government_Infrastructure_Spending_Percent ≤ 0.696: Bad {Good=0, Bad=42}
| | | National.MilitaryActingAsPolicePercent ≤ 0.162: Bad {Good=0, Bad=243}
| | National.Government.CorruptionTheftPercentage ≤ 0.051
| | | National.AdjudicationRate > 0.053: Bad {Good=0, Bad=4}
| | | National.AdjudicationRate ≤ 0.053: Good {Good=2, Bad=0}
| National.Stimulus ≤ 679760.761
| | National.Government_Services_Spending_Percent > 0.037
| | | National.Government_Services_Spending_Percent > 0.921: Good {Good=2, Bad=0}
| | | National.Government_Services_Spending_Percent ≤ 0.921: Bad {Good=1, Bad=30}
| | National.Government_Services_Spending_Percent ≤ 0.037: Good {Good=3, Bad=0}
```

Figure 6 is a text view of the Decision Tree in Figure 5

3. CONCLUSION

In this paper, we provided an overview of the NOEM and its model. We then described a decision support tool/process that works in conjunction with the NOEM and which aids analysts and policy makers in the investigation of potential policy sets that address a given problem/situation. In combination with the NOEM components, the MDE, BF and EM, these tools provide a unique and unprecedented capability.

ACKNOWLEDGEMENTS

The authors would like to thank a number of additional members of the NOEM Team for their comments and continued support - from Air Force Research Laboratory, Aleksey Panasyuk, Warren Geiler, and Pat McCabe; and Dr Moises Sudit from CUBRC.

REFERENCES

Salerno, John J.; Romano, Brian; Geiler, Warren; The National Operational Environment Model (NOEM), Proceedings of Information Systems and Networks: Processing, Fusion, and Knowledge Generation, Defense & Security, Orlando, FL, 25-29 April 2011.

Salerno, John J.; Romano, Brian; Geiler, Warren; Smith, Jason; Hudson, Brian; Thron, Christopher; The National Operational Environment Model (NOEM) – A Focus on Understanding the Populace, Proceedings of the Modeling & Simulation Conference, MODSIM World 2011 Conference and Expo, Virginia Beach, VA, 11- 14 October 2011

CHAPTER 18

Supporting Situation Understanding (Past, Present, and Implications on the Future) in the USSTRATCOM ISPAN Program of Record

James Starz, Mark Hoffman, Joe Roberts, Jason Losco, Kevin Spivey, Jennifer Lautenschlager, Rachel Hingst

Lockheed Martin Advanced Technology Laboratories
{jstarz, mhoffma, jroberts, jlosco, kspivey, jlautens, rhingst}@atl.lmco.com

ABSTRACT

The DARPA Integrated Crisis Early Warning System (ICEWS) has been a four year effort to develop a comprehensive, integrated, automated, generalizable, and validated system to monitor, assess, and forecast national, sub-national, and international crises in a way that supports decisions on how to allocate resources to mitigate them. Lockheed Martin Advanced Technology Laboratories has been leading a multidisciplinary team to address this objective. This paper chronicles the DARPA ICEWS effort, specifically the transition of research components to operational users and describes the process followed for testing and evaluation with both USPACOM and USSOUTHCOM and the acceptance and transition to the Integrated Strategic Planning and Analysis Network (ISPAN) program of record at USSTRATCOM in spring of 2012. The paper also gives suggestions and lessons learned for transitioning future human, social, behavioral, and cultural systems.

Keywords: transition, computational social science, lessons learned

1 INTRODUCTION TO THE INTEGRATED CRISIS EARLY WARNING SYSTEM

The DARPA Integrated Crisis Early Warning System (ICEWS) has been an effort to develop a comprehensive, integrated, automated, generalizable, and validated system to monitor, assess, and forecast national, sub-national, and international crises in a way that supports decisions on how to allocate resources to mitigate them (Kettler and Hoffman, 2012). Lockheed Martin Advanced Technology Laboratories has been leading a multidisciplinary team of computer and social scientists, including several universities and small businesses to address this objective. Major ICEWS products include a mixed-methods suite of statistical and agent-based models for over 50 countries that produces highly accurate forecasts; an extensible model integration framework that provisions, executes, and manages models; innovative forecast aggregation methods; and tools for the ingest and processing of dynamic data feeds such as over 16 million English and foreign language news stories.

The system is comprised of three major functional components. The first is ICEWS Trending and Recognition and Assessment of Current Events (iTRACE) that provides analytical visualizations for the vast amount of data collected for the ICEWS program. The second is the forecasting capability, iCAST, supporting what-if analysis of six-month forecasts across five events of interest. The third component is iSENT, providing automated sentiment analysis of social media data such as blogs.

One of the key thrusts of the ICEWS program was transitioning the successful research into an existing program that would allow ICEWS capabilities to be widely available to the operational community. Part of this effort has been the deployment of an experimental version of ICEWS at USPACOM and USSOUTHCOM. Our lessons learned and experiences working with operational users significantly shaped the software components and enabled further adoption of the individual ICEWS components. ICEWS is a better system and provides better capability to operational users because of these lessons. USPACOM has demonstrated that an analyst using ICEWS can perform an assessment of real world events in a fraction of the time and with similar results in terms of Measure of Effectiveness (MOEs) and Strategic Effects (SE) as five analysts working without ICEWS. At USSOUTHCOM, ICEWS products are being incorporated into the commander's daily and weekly status briefs. The interest and support from these Combatant Commands helped facilitate the transition of technologies from a research system to a program of record. Significant effort went into aligning the research system with the needs of operators.

Under the DARPA ICEWS and the Office of Naval Research's "Worldwide ICEWS" (W-ICEWS) efforts, we are expanding all ICEWS capabilities to cover the entire world, enabling the transition of ICEWS components to the Integrated Strategic Planning and Analysis Network (ISPAN) program at USSTRATCOM. In early 2012, the iTRACE component was transitioned to ISPAN with the other two components to be transitioned soon thereafter.

1.1 ICEWS Trending, Recognition, and Assessment of Current Events (iTRACE)

The primary goal of the iTRACE capabilities in ICEWS is to provide a fully automated capability to monitor political activity around the globe. This is accomplished by automatically converting over 16 million news reports from over 200 news sources for the past 11 years into a structured form that reflect the character and intensity of interactions between key leaders, organizations, and countries. iTRACE taps into this wealth of data to provide a 360 degree view of who is doing what to whom, when, where, and how in areas of responsibility of interest, based on open-source reporting from major and regional news services. The intuitive layering of charts, graphs and storylines provides sufficient information to quickly focus the analyst on those reports most relevant to an issue, region, or time period of interest.

iTRACE presents the coded event data to users through the ICEWS portal via an extensible suite of analytics to aggregate event data in various ways, while enabling drilldown to the original news stories. The ICEWS framework is modeled on iGoogle™ and allows users to fully customize the organization of charts, graphs, and alerts. The system is also designed to be fully transparent and traceable: From any graph or chart, the user is never more than three mouse clicks away from the actual time-stamped news story used to generate a particular data point or set of points. The ICEWS Portal, shown in Figure 1, is the gateway into the various components of the system, including iTRACE. Via the portal, users can customize the iTRACE interface with specific portlets, small views in a web page, that help them in their mission.

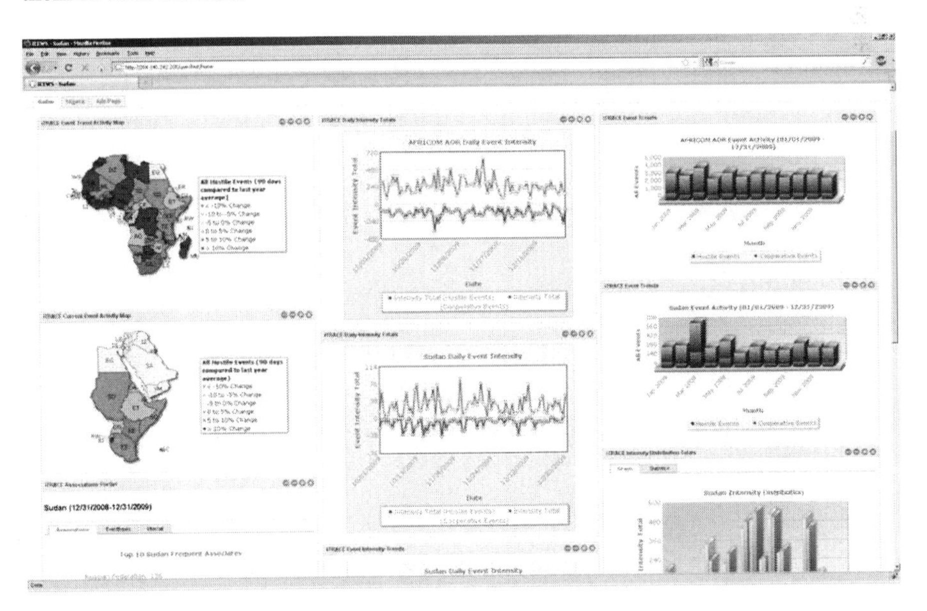

Figure 1 iTRACE portal page for configured for analysis of Sudan and neighboring countries

The iTRACE visualizations primarily show aggregations of events for one or more actors, or class of actors, over a historical timeline. By associating events on a scale (Goldstein, 1992), a user can quickly look at the visualization and see a positive or negative trend in events, if such a trend exists. This analysis suits recognition and assessment of both short-term spikes in activity and long-term changes in behavior. The system can also be useful comparing current trends, with past activities that may be similar. There are a diverse range of visualizations including graphs, tables, timelines, and maps that users can customize for their needs. There is also an advanced querying interface, shown in Figure 2. For all of the displays, the users are also able to download the raw information and do their own analysis in other tools. It is up to the analyst to interpret the information, such as to determine why a specific trend is seen and the implications (good or bad) of those trends.

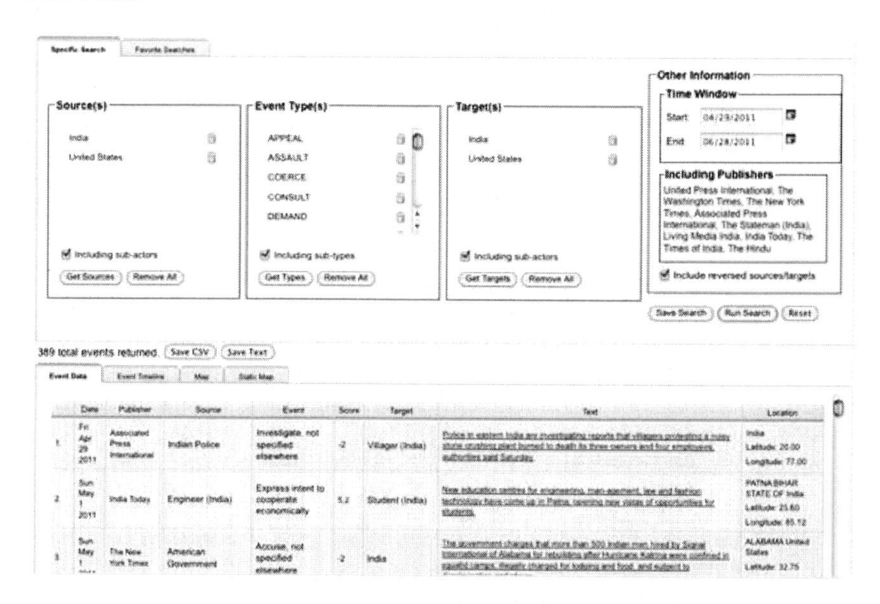

Figure 2 iTRACE Event Browser used to search interactions between two countries

2 USER FEEDBACK AND ADOPTION

ICEWS was very aggressive engaging potential users. This helped shape the design of ICEWS, specifically the transparency of the forecasting models.

2.1 De-mystifying Social Science for Operators

One of the foremost challenges in human, social, cultural, behavioral systems is providing users with the confidence that they both understand how the system works and trust the results of the system. This is especially true of a system based on the "softer sciences." End users may disagree with elements of information presented in ICEWS, such as individual forecasts. With this in mind, it was critically important to allow users to see the formal, scientific, repeatable way the information was derived.

The iTRACE system has numerous visualizations to show historical trends in activity. A key system design goal was that all source information should be no more than three clicks away from any aggregation of the original data. For instance a user may see a graphical display and question how that information was derived. iTRACE provided the ability to drill into visualizations to see partial data aggregations and drill down to the raw data, textual stories and the corresponding coded events.

Both iTRACE and iCAST rely heavily on automated event coding from open sources news media. This requires natural language processing technologies to determine the source, type, and target of events as well as determining the timeframe and location [Van Brackle, 2012]. Natural language processing technologies are frequently imprecise and can lead to confusion by users. Event coding by humans is orders of magnitude slower, highly subjective, and suffers from challenges such as fatigue and inter-coder consistency. Initial versions of the ICEWS coding system, sufficient for forecasting input, contained some egregious errors that caused users to question the results of the system. Three steps were taken to remedy the problem. First, we refocused significant effort to improve the reliability of the coder [Shilliday and Lautenschlager, 2012]. One particular tradeoff was that users preferred increased precision at the sacrifice of recall. It was critically important that users did not see many situations where they would lose faith in the system. This conservative approach was also used by web search engines when they initially became prevalent. To foster trust, we educated users about what the system is doing "under the hood" and some of the limitations and tradeoffs made for our particular implementation. Finally, sessions were held where a small number of candidate users coded items by hand. Our users saw the value in the automated event coding process, and when they actually coded data by hand they were sympathetic to the challenges faced and became advocates of the system.

2.2 Operator Engagement

One of the keys to the success of ICEWS was engaging with operators early so that they could help shape the system. Quickly responding to their feedback was challenging, but it showed this information was valued by the team. With the help of subject matter experts, we used a number of strategies to engage users including demonstrations and supporting a military exercise. Each operator engagement provided feedback that was tracked and reviewed by the team. Though we were unable to act upon every suggestion, this tracking allowed us to prioritize changes and gave end users the confidence that their suggestions were shaping and improving the system.

Typical user engagement began with a system briefing and demonstration. It was very important for us to frame the system as being a work in progress and openly describe its limitations. After doing a number of these sessions, it became obvious what questions and concerns would be raised. We would modify our briefs accordingly and attempt to address the issues in the software. In our first series of engagements with users it was not unusual to receive dozens of feature suggestions. In one session, we received over 100 requests.

Once users were aware of the system and expressed interest in learning more, we would perform training. The user interaction with the system does not require significant training, but the social science underpinnings are important to understand the behavior of the system. After a rudimentary introduction to the system, the training sessions would often focus on how the system was generating calculations and visualizations as well as describing how the system would address specific problems the individual operator faced. When the training was done, it was our intention that each user would have iTRACE portal pages configured for their specific job function and challenges.

A technique that was introduced to the ICEWS team during the transition process was the Joint Application Design/Rapid Application Development (JADRAD). These interactions occurred with experts, managers and subject matter experts, from the ISPAN system. These users had limited expertise in ICEWS, but were experts in the current ISPAN system. A number of these JADRADs were performed allowing the final transitioned version of iTRACE to align better with the existing ISPAN system operations and look and feel.

3 TRANSITION CONCEPT OF OPERATIONS

3.1 Military Exercises

Participation in a military exercise at USPACOM brought iTRACE's concept operations issues to the forefront. To that point, we had a small number of infrequent uses the system, and the exercise was an opportunity to demonstrate operational relevance. The operators worked with us to tune existing visualizations and make entirely new visualizations that could be used to assess ongoing activities during the exercise. While there was significant value from the participation, it was also apparent that there was inconsistent messaging with ICEWS to the operational leadership. Vocal critics did not understand or accept the various limitations with the existing system. It became apparent that the system needed to be more trustworthy and understandable before we could achieve widespread adoption.

A year later, the ICEWS team participated in a planning exercise with USSTRATCOM. Unlike the previous exercise, this focused more on describing the value of ICEWS with respect to the planning process. The goal was to develop a storyboard described how all of ICEWS would fit into the planning process and ultimately into ISPAN. This was a particularly challenging task that required both subject matter experts and ICEWS developers. The resulting storyboard would ultimately shape the transition and concept of operations for the program.

3.2 Operational Relevance

One factor that allowed us to be successful was the limited amount of data, particularly in the cultural realm, which operators can access. Though many operators are faced with information overload, our experience showed that users were willing to investigate new data sources to determine their value. In ICEWS, the open source news and social media data gave us inroads to provide more visionary forecasting functionality in iCAST. This was not the initial intention, but it became obvious that the more timely and widespread information our system provided, the more likely operators were to have interest in it.

The original goal for ICEWS was to support personnel responsible for monitoring and coordinating activities in countries. As the individual components of ICEWS were developed, it became evident that this view could be narrow-minded. The inference was that ICEWS could be impactful for a large number of functions that are performed by operational communities. For instance, ICEWS could easily provide intelligence or assessment information. The system needed to get exposure to operators to help shape the development of the system. To do this, a member of the ICEWS development team along with multiple subject matter experts worked on site at USPACOM for 18 months, demonstrating and training potential users of the system. Working with real challenges of the community allowed us to adapt the system to real-world needs and focus our development efforts.

For ISPAN transition, it was initially unclear how ICEWS would fit into the deliberate and contingency planning process. The ICEWS components did not align perfectly with the formal military doctrine supporting planning. In working with operators and the ISPAN team, it was determined that the best fit for the ICEWS components what to support situational awareness. iTRACE would be used to help determine unusual patterns of events compared to historical patterns. iCAST would be used to for considering courses of action by leveraging the "what-if" capability. iSENT could be used to understand popular sentiment towards particular figures and actions as part of the assessment process. These three components provide a socio-cultural situation awareness that is typically beyond the scope of current systems.

Figure 3 shows the various steps in the planning process. The circled area shows the functions where iTRACE capabilities provide utility. There are steps in the

Figure 3 iTRACE supported portion (circled) of Joint Planning Process Doctrine

process, such as plan initiation and mission analysis, where iTRACE provides value but would need to be supplemented by other functions. This level of planning support provided real user benefit allowing us to pursue the transition into ISPAN.

4 TECHNICAL TRANSITION CHALLENGES

4.1 Architecture Considerations

In ICEWS, we were very fortunate that our system had a fairly compatible architecture to our target transition system. Because of the similar architectures, it was deemed appropriate to closely tie into the existing system leveraging their legacy components. Switching among similar functioning products from different vendors was very time consuming. If you use common practices and widely used vendors in your research system, it is likely to streamline the transition effort. We found there was significant value in doing experimental threads to gauge the possibility of integration. A crash development "loose integration" example between iTRACE and ISPAN became the reference for the actual transition. With this we were better able to determine the level of effort for the transition. It also provided a larger audience to shape the direction of the system going forward from an operational perspective.

For ICEWS, there were a number of radically different paradigms getting introduced into ISPAN. While transitioning iTRACE presented some data challenges, the more daunting hurdle was the forthcoming large data requirements associated with transitioning iSENT. iSENT is dependent on large collections of social media. The notion of collecting and storing large collections of data was completely outside the capabilities of the ISPAN system. Supporting this particular need requires a significant lead-time to work within the development cycles of the existing system.

4.2 Information Assurance and License Considerations

As with many research systems, ICEWS relied heavily on free and open source software (FOSS) to provide much of the infrastructure of the system. While this is a great way to jumpstart a research and development effort, it does come with drawbacks. In ICEWS, we ended up backing out a lot of FOSS and commercial products that were viewed as potential information assurance (IA) concerns. While the issues would not be consistent across operational programs, it is extremely important to be cognizant when dealing with third-party tools concerning information assurance and licensing issues.

The success of ICEWS caused us to face IA issues that were not originally part of our focus. Some awareness of the IA problem up front would have gone a long way into building a safer, secure, and more easily transitioned solution. Ultimately, there was significant effort to retrofit the system to meet the minimum IA requirements. One of the primary requirements was to validate all information going into the system. This information could be entered by users or through trusted data sources. The Open Web Application Security Project has an invaluable guide of potential security vulnerabilities and remedies. Another IA requirement we faced was to ensure that our software passed static analysis tests. We have incorporated this into our regular process, as it is often easier to fix potential issues closer to the time, and nearer the developers, that it is written. For nearly all research and development projects, it really does not make sense to build a completely secure solution. It does make sense for architects and developers to be familiar with these paradigms and design solutions that will not require significant modification later. IA requirements will vary from application to application and it is of the utmost importance to understand the requirements when the transition starts, and even earlier to avoid any potential major misalignments when transition "might" be an ultimate objective.

As with the IA issues, software licensing issues did not become a challenge until transition was imminent. In ICEWS, we started the project with aggressive use of open source software using various licensing terms. As we started down the transition path, many of the particular licenses were heavily securitized. A large software program of record obviously does not want to use the wrong license that would cause it to share the entire production system with the world. In the end, we replaced a number of libraries due to license concerns. There is no one size fits all rule for using particular licenses, but we have found it useful to stick with FOSS that use common licenses.

4.3 System Maintenance Considerations

As is the nature of research and development systems, there is a point where the research team disappears and the operational support team takes over management of the system. For ICEWS, there are two particular challenges beyond a typical software system: maintaining the event coding and the social science models.

For event coding, there are two major concerns about system maintenance. The first is connectivity to outside data sources that provide open source data. These are bound to change over time and any open system would likely face this same issue. Of particular interest for ICEWS continually updating dictionaries of actors and events that feed the event coding process. ICEWS can use linguistic clues to determine the roles of individuals in a document, but it is often beneficial for these to be explicitly stated in a dictionary of actors. Even items that appear relatively static like dictionaries of verbs face challenges when new paradigms appear such as an actor "tweeting" their messages on the Internet. In our system, we believed that only administrators should be able to modify dictionaries. We did provide tools to edit these dictionaries and a mechanism for end users to browse and suggest actors.

The challenges are exacerbated when updating social science models is involved. These systems are quite complicated and they take numerous outside data sources as input. Both the model and the format are bound to change as time progresses. We are working towards minimizing operations and maintenance costs for the upcoming transition of the iCAST component.

5 CONCLUSIONS

The path to transitioning research and development is always challenging. The topic of human, social, cultural, and behavior only adds to the complexity of such efforts. The utility and design of ICEWS were critical enablers to the transition process, but there are a number of items that any group could do trying to facilitate transitioning of HSCB technologies to operational systems. From our experience, it was imperative to work early with operators and be receptive to their feedback. Some of the more interesting research results may not always transition to the concept of operations for the users. Additionally, build systems early on with an eye toward ultimate transition. In ICEWS the transition path was paved with bumps, but we were fortunate that we were able to work through the issues and get the software in the hands of operational users. This paper is meant to provide a roadmap to help smooth that path for other future transition activities.

ACKNOWLEDGEMENTS

The authors would like to acknowledge our subject matter experts who had a significant role in making transition possible: Mr. Ed Smith, Dr. Evelyn Dahm, Mr. Steve Sladky, and Mr. Pat Bindl. We especially acknowledge Dr. Sean O'Brien and Dr. Philippe Loustaunau for their vision in creating and guiding the ICEWS Program. We also acknowledge the ICEWS champions at USSOUTHCOM, USPACOM, and USSTRATCOM, as well as the dedication of the ISPAN team.

REFERENCES

Goldstein, J. "A Conflict-Cooperation Scale for WEIS International Events Data," Journal of Conflict Resolution, 36 (2), June 1992: 369-85.

Kettler, B. and Hoffman, M. 2012. Lessons Learned in Instability Modeling, Forecasting, and Mitigation from the DARPA Integrated Crisis Early Warning System (ICEWS) Program. 2nd International Conference on Cross-Cultural Decision Making: Focus 2012.

Shilliday, A., and Lautenschlager, J. 2012. Data for a Global ICEWS and Ongoing Research. *2nd International Conference on Cross-Cultural Decision Making: Focus 2012.*

Van Brackle, D. 2012. Improvements in the Jabari Event Coder. 2nd International Conference on Cross-Cultural Decision Making: Focus 2012.

CHAPTER 19

Civil Society Initiatives in Post-Revolution Tunisia

Jeffry R. Halverson, PhD

Arizona State University
Tempe, AZ, USA
jeffry.halverson@asu.edu

ABSTRACT

Cross-cultural decision-making requires an understanding of the sociopolitical and historical context in which policies are enacted. This analysis evaluates methods designed to support civil society organizations in Tunisia after the revolution, using the Tunisian Scouts as a case study. In particular, it discusses the challenges at play and the necessary features of an effective communication strategy. Fostering support networks for the growth and prosperity of NGOs, such as the Tunisian Scouts, will play an instrumental role in the future sociopolitical direction of the country. The benefits and risks of US government involvement are thus weighed. In September 2011, the author traveled throughout Tunisia assessing conditions and many of his observations are related in this paper.

Keywords: Tunisia, civil society, communication strategy, narrative

1 TUNISIAN CIVIL SOCIETY AND THE SCOUTS

Recognizing that a vibrant civil society is critical to the success of liberalism in a budding democracy, especially when located in a key geopolitical region, the Public Affairs Office of the U.S. Embassy in Tunis has reached out to NGOs in post-revolution Tunisia, all of them lacking in resources and experience. Typically, these efforts follow a bureaucratic management model that tries to direct certain funds through red tape, but seldom involves itself in the delicate nuances that cross-cultural work requires. This observation brings us into the domain of strategic communication and cross-cultural decision making. American support of Tunisian civil society organizations is desperately needed and critical to U.S. interests in the region; however, simply by supporting them, the U.S. government may actually undermine and discredit the efforts and legitimacy of these organizations due to the current sociopolitical climate. As such, one might ask: Is it better for the U.S. government to do nothing at all? To explore this dilemma further, this paper will look at the case of the Tunisian Scouts and Guides.

In September of 2011, I traveled throughout Tunisia as part of the U.S. Embassy's efforts to support civil society organizations in the wake of the revolution that launched the historic Arab Spring. Under the authoritarian regime of Zine al-Abidine Ben Ali, which lasted from 1987 until January 14, 2011, civil society in Tunisia was systematically repressed by the state. As Anna Würth and Claudia Englemann observed in 2010: "In Tunisia, civil society can barely breathe, and activists are regularly slandered, physically assaulted, and under threat of spurious charges." (255) Since the revolution, however, there is new hope and possibilities for the future, even if the status of civil society in Tunisia at the present time remains uncertain. Prior to the revolution, civil society organizations or NGOs were required to receive government approval, which was almost never given, except for a very small handful of exceptions. Today, no approval is necessary. The only requirement is a notification that a group has been established. As such, we can expect a wave of new organizations in the coming months and years

While in Tunisia, I spent the greatest amount of time with the Tunisian Scouts, one of the few NGOs to survive the Ben Ali regime. There are approximately 40,000 official members of the Tunisian Scouts at present. Anecdotal claims suggest that the numbers are growing rapidly and the organization is ill-equipped to meet the demand. The 40,000 members commonly cited include 8,000 in various leadership and training positions within the organization. Scout leaders also suggested that the number of participants in various Tunisian Scouting activities is actually closer to 100,000. On a national level, it is estimated that 33% of Scouts are female. Both males and females participate in the same organization, thus it is incorrect to refer to the organization as the "Tunisian Boy Scouts." The number of female members is significantly lower, around 15% (estimated), in certain areas, particularly in more conservative cities and towns, such as Kairouan or Djerba.

The greatest concentration of Scouts overall is found in the eastern coastal and industrial city of Sfax, claiming some 10,000 official members. During our visit to Sfax, we found the Scouts there to be the best organized, best equipped, and most committed to Scouting in the country with strong representation among males and females. Scouting in Sfax is also reported to have a "clean image" and many local leaders were Scouts in their youth. Nevertheless, as elsewhere, problems still face the organization in this city, as they do on a national scale.

For decades, the Tunisian Scouts were one of the only civil society organizations permitted to exist in Tunisia. Under the auspices of the Ministry of Youth and Culture, the Scouts were controlled and very modestly funded by the ruling RCD party of Ben Ali. After the 2011 Revolution, non-RCD Scout leaders, previously marginalized, have emerged to have their voices heard and are actively trying to rebuild Tunisian Scouting without RCD connections and as a major force for the new Tunisia. However, the organization is now left with no primary source of funding and a stigma still exists among the populace that the Scouts are a wing of the RCD. This perception is complex and varies from place to place.

During a meeting with Scout leaders, I discussed the need to clearly formulate the core identity of the Scouts and how to reinforce that identity with particular narratives as they work to improve their public image. This argument is based on the theory articulated by scholar Walter Fisher. He argued that humans as cultural beings are best understood as *homo narrans*, or humans as storytellers (Fisher, 1984; Halverson et al., 2011, p. 16). Human beings make decisions, choices, and actions, as well as imagine identities, within a narrative framework. Theorists in religious studies may recognize Fisher's narrative paradigm in the connections evident between "myths" and religious rituals, sacred spaces, texts, and laws.

1.2 COMMUNICATION STRATEGY

Although I offered advice about how to proceed, finding an answer to important questions about communication and message crafting was ultimately the responsibility of the Tunisian Scouts. The values that the Scouts continuously spoke about were entirely in-line with those of Western liberalism. They cited human rights, pluralism, tolerance, gender equality, environmentalism, entrepreneurship, and democracy, among others. This was all music to American ears. However, a common critique of these values found within the broader Arab and Muslim world is that these concepts are part of a colonial project, or "intellectual colonialism," introduced and imposed by foreign powers, such as the United States. This is the dominant view taken by many Islamist movements. As such, if the liberal values that the Scout leaders conveyed are the basis for the identity that they seek to promote, then it is a contested identity to say the least, especially when we note the recent electoral success of Tunisia's main Islamist party, Ennahda. At the same time, there is no desire on the part of the Scout leaders to create a schism in the organization, where a pro-Islamist faction splits into a separate group. Unity remains a priority. And those attending to the business of narrative must keep that in mind.

The communication strategy that I informally recommended to Scout leaders was the dissemination of particular narratives (through a range of formats) that ground their values in the long history of Scouting and Tunisia. This is to say that the success of civil society initiatives in Tunisia rests not only on resources, but on the ability of these groups to pass the litmus test of "Tunisian authenticity." U.S. support for such initiatives must be careful not to undermine this component of any communication strategy. Furthermore, if Tunisian civil society organizations do not attend to the narratives themselves, someone else might do it for them with disastrous results.

Let us consider a brief and tangential example from another Arab state, Egypt. I lived in Cairo during the 2000-2001 academic-year and returned to the U.S. about six weeks before 9/11. During my stay, a British supermarket chain, Sainsbury's, opened throughout the city, bringing much-needed jobs to Egypt's economy and quality goods with it. One store opened near my flat in the Zamalek district and I observed that it was always busy and a good place to shop. In a matter of months, however, stories began to circulate that the British chain was owned by a wealthy British Zionist and Sainsbury's profits were going to support Israel. Despite efforts to refute these unfounded charges, that narrative framework was so powerful that Egyptians turned against the chain, a boycott (and even violence) ensured, the stores shut down, and the entire business withdrew from Egypt, all within one year. That's an example of letting someone else dictate a narrative. Tunisian civil society organizations have to avoid such serious pitfalls when seeking Western partners, so hesitation, caution, and wariness are all to be expected.

Since Scouting was one of the only licensed or legal civil society organizations during the reign of the RCD, many political forces of various trends are interested in co-opting the organization in the wake of the revolution, as one of the few established civil institutions. Tunisia's main Islamist party, Ennahda, is actively trying to do so and making gains. Ennahda now largely controls Scouting in certain cities and towns. Members of the Scout leadership informed us that "four or five" city Scout councils are now dominated by Islamists. It is further believed that the leader of Ennahda, Rached Ghannouchi, has personally directed his organization to expand into the Scouts. Since our departure from Tunisia in September, Ennahda has gone on to dominate the first post-revolution elections, taking 42% of the seats in the constituent assembly. That said, they have also since formed a ruling coalition with two secular-left parties (Congress for the Republic and Ettakatol), which may allay some fears about their attitude toward political pluralism.

This interest in Scouting may sound strange to an American audience. Scouting has deep roots in Tunisian society and functions in ways that surpass the average American perception of Scouting in the U.S. context. Tunisia is also arguably the most important center for Scouting in the Arab or North Africa and Middle East regions. Scout groups from other Maghreb and Arab countries travel to Tunisia to participate in camps and workshops. In 2005, the World Scout Conference (WSC) held its triennial meeting in Tunisia; it was the first and only time the WSC has met in an Arab or Muslim country. As such, support for Tunisian Scouting may prove to have a broader impact on the entire region. During our site visits, we were informed that the Tunisian Scouts were recently involved in cooperative efforts with Scouts from Denmark, Germany (specifically Stuttgart), and the United States. The cooperative work with Denmark created controversy however and it was opposed by some in Tunisia due to the Danish "cartoon crisis," which occurred in 2005.

During our visit to the island of Djerba in the south of the country, near the Libyan border, we met with a local Scout leader for tea. He was the most religiously conservative of all the people we met and he expressed his suspicions about our agenda. He said that American assistance must work as a partnership and that the Tunisian Scouts will not "execute a foreign project or agenda." This was coming from someone generally open to our work in Tunisia. One can imagine what someone hostile to our work might have said. If the writings of Islamist extremists (which I have studied a great deal) are any indication, an opponent might say that: "This is a plot by the global Crusader-Zionist alliance to infiltrate Tunisian society." This claim may sound ridiculous to our ears, but I assure you that such narratives do resonant in certain sectors of the Arab world. Controlling the narrative that accompanies any work with NGOs in the region is essential. Given that the Scouts are already forced to combat the view that they are an RCD-affiliated organization, attention to the dissemination and circulation of narratives has become all the more important in their case.

It is important for policymakers to understand that what is intended as benevolent in one culture can be perceived with suspicion and viewed as malevolent in another cultural context. Oftentimes, these perceptions are informed by narrative frameworks tied to historical events. From a Tunisian perspective, the U.S. government was a major supporter of Ben Ali and other dictators in the region, especially Hosni Mubarak of Egypt. U.S. Senator John McCain of Arizona early on described the pro-democracy Arab Spring revolutions, which began in Tunisia, as a "virus" spreading throughout the Arab world. The U.S. government also remains the foremost ally of Israel, the traditional foe of the Arab states for the last century, sending billions of tax dollars in military aid and weaponry to the IDF, which continues to occupy Arab territory captured in the 1967 Six-Day war. Add in the Iraq war, widespread and well-publicized post 9/11 anti-Muslim and Arab sentiment among Americans, and so on, and you can understand the reservations that Tunisians may have about working with the U.S. government. The Scouts and other civil society initiatives can ill afford the perception that they "sold-out" or switched patrons and now serve as a vehicle for U.S. interests. As such, those that are working to strengthen Tunisian civil society must be mindful of the sensitivities at play.

One strategy is working indirectly and engaging civil society organizations in the U.S. as the channel through which any such assistance and aid is given. This is hardly an infallible approach, as anything or anyone that is American can be associated with negative perceptions of U.S. foreign policy, but it still carries better narrative possibilities than direct U.S. government involvement.

Let us consider another scenario though, where too little involvement might occur. Imagine that U.S. support for the Tunisian Scouts is successful and helps to strengthen their role, values, and activities in a new democratic Tunisia. I mentioned earlier that Islamists are working to strengthen their involvement in the Scouts. So what if a newly revitalized Tunisian Scouting organization, funded by the US, becomes dominated by Islamists who use the NGO to indoctrinate Tunisian youth in Islamist ideology? In other words, if the U.S. is too distant in its supporting role, perhaps to avoid negative attention, it may inadvertently strengthen an organization that is ultimately co-opted by parties with radically divergent ideological interests.

If we can consider a wildly different but broadly relevant example, the U.S. sent aid to Afghans fighting the Soviets in the 1980s. Those U.S.-funded fighters, after the Soviet withdrawal, gave rise to arguably the most hardline Islamist militias in the world. Obviously, unintended consequences or blowback come with the territory and working with the Tunisian Scouts is a very different scenario, but Afghanistan reminds us that funds are not enough. There must be an ongoing and sustained partnership that extends beyond a particular strategic moment.

2 CONCLUSIONS

I want to conclude this paper by answering the question that I posed at the onset: Is it better for the U.S. government to do nothing at all? My answer, after all of these considerations, is: No, it is not. I asked this question to highlight the delicate nature of this sort of work. Working with civil society initiatives in other cultures will not succeed by simply directing money at the challenge. Supporting civil society initiatives in post-revolution Tunisia, or in other cultural contexts, will require a cautious and culturally informed communication strategy that offers the best narrative possibilities.

That strategy is the responsibility, first and foremost, of the local parties themselves, because they will face that narrative framework on a daily basis. For example, the Tunisian Scouts might craft a messaging campaign about working with an American civil society organization to introduce new innovative programs into the newly independent Tunisian Scouts. To support that strategy, the U.S. government would provide grant funding to the American NGO as a buffer and avoid undermining the narrative by pursuing direct funding to the Scouts. But local parties, like the Tunisian Scouts, must also work in cooperation with area specialists and social scientists that can serve as an intermediary with the original funding party, including the U.S. government. It is this latter group that can preserve a personal partnership through the duration of the work, while bureaucrats may move onto their next assignments and budgets. Without culturally-informed intermediaries and buffers, the bureaucratic management model, so common among government agencies, will likely not only fail to effectively support civil society initiatives at this critical historical moment, but could actually do the opposite and derail their development and delay democratic transition further.

ACKNOWLEDGMENTS

The author would like to acknowledge the support of Summer Allen of MITRE and grant support from the Office of Naval Research (N00014-09-1-0872).

REFERENCES

Fisher, Walter. 1984. Communication Monographs, Volume 51: 1-22.

Halverson, Jeffry R., H. L. Goodall, and Steven R. Corman. 2011. Master Narratives of Islamist Extremism. New York: Palgrave Macmillan.

Würth, Anna, and Claudia Engelmann. 2010. Governmental Human Rights Structures and National Human Rights Institutions in the Middle East and North Africa. Ed. Hatem Elliesie. Islam und Menschenrechte. Frankfurt: Peter Lang: 239-256.

Section IV

Architecture for Socio-Cultural Modeling

CHAPTER 20

Modeling and Assessing Multiple Cultural Perspectives

Dr. Lora Weiss[1], *Dr. Erica Briscoe*[1], *Dr. Elizabeth Whitaker*[1],
Ethan Trewhitt[1], *Dr. John Horgan*[2]

[1]Georgia Institute of Technology
Atlanta, GA 30332
Lora.Weiss@gtri.gatech.edu

[2]The Pennsylvania State University
University Park, PA 16802
horganjohn@psu.edu

ABSTRACT

For those seeking to understand and stop terrorism, a number of social, cultural, and behavioral perspectives are being developed by experts worldwide. These perspectives may complement each other or they may be in conflict, but they both likely contribute to a broader understanding of behavior. Our research is developing quantitative models to analyze and experiment with differing views, opinions, and perspectives of potential influences on population behavior. We have developed methods to capture multiple perspectives of behavior in a systems approach to support quantitative assessments of influences on behavior. The research includes the creation of influence diagrams to structure the interconnected and related concepts within, as well as multi-scale models of behavior across, perspectives. The approach expands initial models associated with terrorist recruitment and creates a suite of submodels that can be docked to support evaluation and assessment of differing views. Diverse types and levels of influences, ranging from individual, group, organizational, cultural, and community-based, can then be evaluated by replacing a model of one perspective with a model of an alternative perspective.

Keywords: multi-scale modeling, multiple perspectives, adversarial behavior

1 INTRODUCTION

In a 2008 presentation to the New York Police Department, Dr. John Horgan noted that in trying to understand the terrorist and in developing approaches to disengaging individuals from terrorisms, answers cannot lie in any single theory (Horgan 2008). Instead, knowledge from multiple disciplines is needed and must be aggregated within more conceptual frameworks. Furthermore, many approaches from psychology and criminology are still underdeveloped and could benefit from alternatives to evaluating their effectiveness. Quantitative models offer an approach to support development of these theories by expanding the ability to explore aspects of socio-cultural influences and to identify potential indicators that increase a person's propensity to being radicalized, recruited, deradicalized, or disengaged from adversarial behavior. Quantitative models are valuable in that they provide a means to analyze and experiment with differing views, opinions, perspectives, theories, and impacts of potential influences on population behavior. Many existing models that attempt to conceptualize these dynamics typically offer a fixed perspective rather than integrating multiple perspectives and across multiple scales, even within a single discipline (Zacharias 2008). In addition, there has been limited success in integrating qualitative, process-based perspectives into quantitative models. The research presented here seeks to capture multiple perspectives of behavior in a systems approach to support quantitative assessments of differing perspectives and to support an analytic understanding of influences on behavior.

Behavioral and cultural models are useful in assisting analysts and decision-makers to understand terrorist activities and adversarial behavior. Models are needed to analyze and experiment with the impact of potential influences on population behavior. The multiple modeling paradigms that have resulted represent different aspects of this space, and subject matter experts are often used to help couple the social behavior and theoretical aspects with modeling approaches. By integrating behavioral aspects of adversarial activities with computational methods, new approaches can be developed to support what-if experiments related to understanding these activities and to ascertain potentially effective intervention points to disrupt the process of individuals engaging in adversarial behavior (Briscoe 2010, 2011; Weiss 2010, 2011).

Our research explores techniques to identify multi-scale phenomena (macro, meso, and micro-scale) within cultural perspectives and to capture them as connecting concepts within a modeling construct. It enables communication among models that were developed by a variety of parties by federating them via model-docking that defines points of interaction and transformations for data exchange. This permits individual models to focus on their core competencies while allowing other models to act as providers of external information. This differs from the conventional approach to federated models that involve specific submodels which are typically designed to be part of a larger, federated system (Benjamin 2007).

2 MODELING MULTIPLE PERSPECTIVES

Addressing the threat of terrorism has led to multiple, diverse opinions and theories as to its causes, indicators, and influencers. This diversity arises not only from differences in opinions and exposures, but also from the originating disciplines of the analysts. Our research evaluates the process of becoming involved in adversarial behavior by agnostically modeling different, but equally valid (as determined by a subject matter expert), perspectives on the issue. The micro and meso-scale perspectives are being captured using agent-based models that are allowed to communicate with representative models of radicalization within a macro-scale system dynamics framework (see Sterman 2000, Zacharias 2008 for descriptions of these methods). A key finding of this paper is that it presents an approach that allows different models having different scales to interoperate and to be swapped out for other models of the same phenomena developed by other parties, without the requirement of a system design that incorporates the specific models from the start. Having a suite of swappable submodels allows for evaluation and assessment of different views, so that a submodel based on a particular set of assumptions about culture or behaviors can be replaced by a different submodel. For illustration, this paper presents two macro-scale system dynamics models and two meso/micro-scale agent based models. Such models describe facets of micro-scale behavior associated with individuals, as well as meso-scale behavior indicative of groups. These micro and meso-scale characteristics then interface with higher-level system dynamics models to provide an understanding at the macro-scale.

The two macro-scale models are perspectives of radicalization and illustrated in Figure 1, where the first is based on Silber & Bhatt (2007) (commonly known as the "NYPD model" of terrorist radicalization), and the second is based on Precht (2007). Although the two emerged somewhat in parallel, from a model docking approach, both can be analyzed in a computational construct that allows them to be interchanged to evaluate or understand the macro-scale behavior.

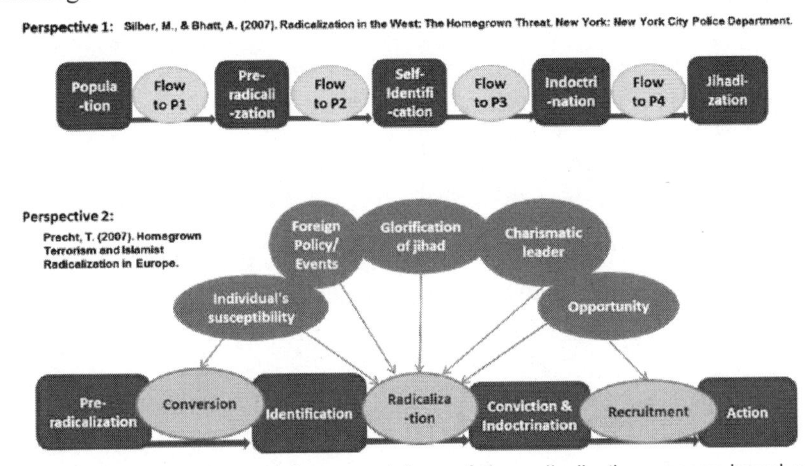

Figure 1: System dynamics model representations of the radicalization process based on (1) Silber & Bhatt (2007) and (2) Precht (2007).

The two agent-based models represent alternate 'pathways' to 'homegrown' terrorism as described by Gartenstein-Ross and Grossman (2009). These models are agnostic in that they have not been specifically developed to interoperate with each other or with either of the system dynamics models of radicalization. Instead, they are self-contained reflecting attributes and thematic focus specific to the perspectives being modeled. A summary of each model is provided to glean an appreciation of the level of multi-scale interactions involved.

The first agent-based model describes radicalization as caused by religious ideology. Though certainly not the only contributing factor and often a contentious topic, religious ideology may play an important role in the radicalization of individuals. To represent the situations in which ideology is a primary instigating factor to adopt radical beliefs (leading to terrorist acts), the agent-based model simulates radicalization as resulting from relationships with religious leaders who encourage radical ideas. Within this model, the actions of the population are captured by the formation and dissolution of friendships between individuals and the resulting changes in those individuals in terms of their (simplified) beliefs related to radicalized behavior, such as the personal attraction, persuasiveness, or charm of others (represented via other agents). A person's "fervor" represents his or her religious status in terms of his or her desire to convert others, or his or her motivation to try to change the world based on his or her own religious beliefs. Each person is associated with a neighborhood, religion, and specific church, each of which is used to characterize the likelihood of various members of the population becoming friends during a particular time step within the model. In this model, a spiritual mentor is defined by a particularly high personal relationship combined with a high "fervor." Friendships are randomly formed between members of the population at rates based on their neighborhoods, religions, and churches, and have strengths associated with them to capture the different levels of friendship between people, ranging from casual to intimate. Spiritual mentor-mentee relationships may also be formed, where for each encounter, the mentee becomes more radicalized. A mentee whose fervor reaches a certain height is then considered a radical.

A second agent-based model was developed to capture factors that contribute to the adoption of terrorism as resulting from clique formation and the feeling of isolation. The primary concept of radicalization in this model is based on the formation of small social groups or cliques. Once the clique is formed, the group can radicalize in a manner which none of the individuals would have demonstrated separately. This model focuses on the formation of the cliques which enable this form of radicalization, rather than on the radicalization process thereafter. The formation of the cliques is modeled primarily as dependent on a single variable, the feeling of isolation. The cliques are formed via the assembly of people with similar backgrounds (religion and home country) who have similar experiences in their current country. The "isolation" of a person represents their alienation from society. This feeling is increased by interactions with members of the majority who have nothing in common with that individual or by interactions with minority members who are experiencing the same isolation. For most of the population, these feelings are offset by the presence of diverse friends and family who provide adequate emotional support and context to their interactions with the majority. All of the minority population is assigned a family that can include extended relatives, in-laws, and very close friends who have been close since childhood. These ties are considered closer than the average friendship. At each time step, friendships are randomly formed between members of the population based on their personal backgrounds, and friendships are also randomly dissolved since few people who meet during random encounters rarely become close friends. Once a clique link is established, it puts a strain on the other relationships existing with those two people. Their intimates and acquaintances are effectively forced to choose to join the clique or cease having a relationship, as time passes. This represents the clique's reduced tolerance for other perspectives, whether they pertain to society, religion, politics, or something else. In this model, no distinction is made between the types of cliques formed, nor how those cliques may choose to handle their isolation from society. Since the actual rate at which people become radicalized is quite low, some of the agents' rules and probabilities can be adjusted so that the process of radicalization can be visualized in a relatively small number of time steps.

3 MODEL INTEROPERABILITY

Models developed independently are often built upon different foundations. While any two models may represent similar aspects of a culture, they may use different terminology, time scales, syntax, or grammar. A common vocabulary for models of the same system enables easy substitution of models, but such a shared vocabulary does not exist among models developed by different parties. In this research, experimentation among different models with disparate vocabularies is enabled by providing a transformation to convert concepts between models.

Tolk and Maguira (2003) provide Levels of Conceptual Interoperability to describe the degree with which models are interoperable. They are summarized in Table 1 and begin with Level-0 disconnected models. Our research enables Level 3 interoperability by creating a framework of Connecting Concepts to dock multi-scale models, and in particular, to dock agent-based micro and meso-scale models with macro-scale models. The approach uses transformation functions such as aggregation or distribution of data between pairs of models, and supplies the semantic relationships for Level 3 interoperability. The current research still requires manual connections, but it provides a library of functions to be reused and modified to enable the connections to reflect the expert view of how variables transform when they move from one model to another. As the research progresses, a more computational implementation of these relationships will be developed to move toward higher levels of interoperability without manual connections.

The implementation of model interoperability is via a blackboard that supports model docking and is depicted in Figure 2. Although the figure shows two agent-based models and two system dynamics models, they do not execute simultaneously. Rather, one agent-based model executes with one of the system dynamics model. Then, that same agent-based model executes with the other system dynamics model to expose and identify the understanding provided by each. Similarly, the second agent-based model would be paired with each of the system dynamics models for further behavioral understanding.

Table 1. Tolk and Maguira's Levels of Conceptual Interoperability (2003)

Level	Description
Level 0	No connection
Level 1	Physical connectivity allowing messages to be exchanged
Level 2	Syntactic level (protocols and formats)
Level 3	Semantic level (meaning of data is defined by common reference models)
Level 4	Pragmatic/dynamical level (information and its use and applicability)
Level 5	Conceptual level (common view of the world: knowledge and relationships among concepts)

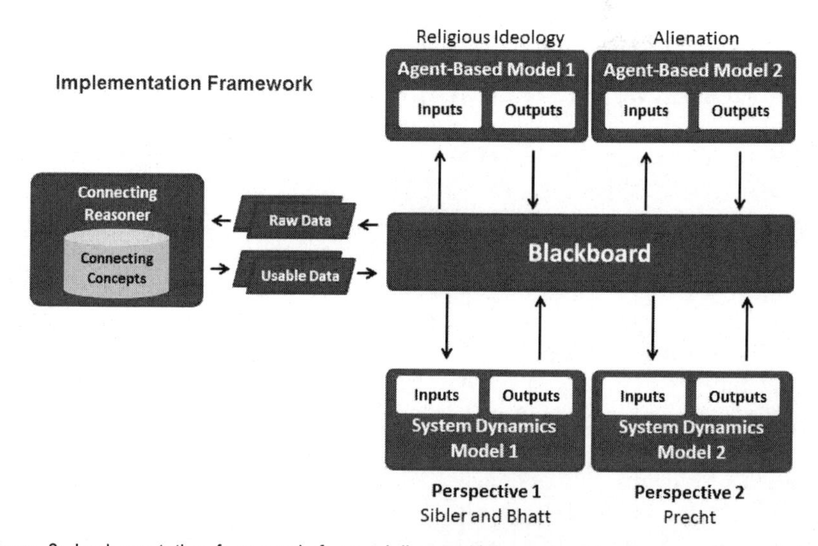

Figure 2: Implementation framework for modeling multiple socio-cultural perspectives. Not all submodels execute simultaneously. Combinations are selected to develop an understanding of behavior exposed from the different combinations of perspectives.

3.1 The Blackboard Shared Memory Area

To enable submodels to interact without requiring dedicated interfaces, a blackboard is used to dock the models. A blackboard is a shared working memory area that enables collaboration among a set of models that write information to and read information from the blackboard. Blackboard frameworks were developed to address complex and ill-defined problems for which existing techniques were inadequate (Whitaker 1992, Engelmore 1998). They can post partial data, e.g., information that they have learned or generated, that may be of use to other models. In this way, the blackboard supports asynchronous, opportunistic modeling by a group of submodels that collaborate indirectly. The asynchronicity enables models to take information when they need it based on their individual time-scales.

The blackboard approach inherently enforces little structure upon the models that make use of the shared memory area. This makes it an attractive method for communicating and sharing data in an experimental system where concepts can be exchanged and where the models can be substituted with other submodels representing alternate perspectives.

3.2 Connecting Concepts

In a blackboard system, any data written by any model is available to all other models within the system. In federated systems containing models that were developed specifically to interoperate, data written by one submodel is specified in a way for it to be read specifically by another submodel as opposed to all other submodels. Since the interfaces are specific, there is no need to translate data between models. However, when experimenting with multiple models that were not designed to interact, the data written by one model will not necessarily exist in a format usable by another model. This necessitates the use of Connecting Concepts.

Connecting Concepts define the types of inputs and outputs of models along with transform functions that convert the outputs of various models to inputs of other models. A Connecting Concepts Reasoner is used to perform variable transformation during federated model execution. Figure 3 depicts how data flows from the models through the blackboard to the Connecting Concepts Reasoner.

The Connecting Concepts Reasoner acts as an outside intermediary that monitors the blackboard for particular types of data, creating a transformed version of that data, and writing the transformed data back to the blackboard. Figure 3 shows an agent-based model and a system dynamics model exchanging data via the Connecting Concepts Reasoner. The Connecting Concepts Reasoner monitors the output of both models and transforms the data for use as input to each other.

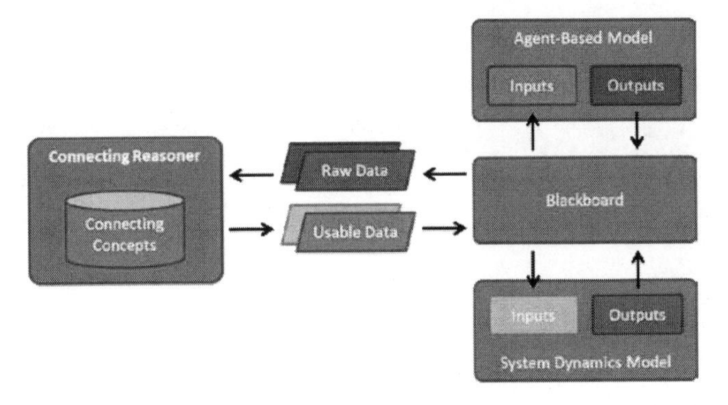

Figure 3. The Connecting Concepts Reasoner within a blackboard system

The simplest transformation has a one-to-one relationship between the input and output data structures. This type of transformation includes renaming variables, converting units, or changes in which one piece of input data yields one piece of output data. However, the true benefit results from more complex transformations.

Transformations are typically more complex when docking models of different scales. We developed two such transformations that are used regularly: Distribution and Aggregation. Distribution transformations (Figure 4) support one-to-many relationships, typically based on probability distributions to convert a single output into multiple inputs. Aggregation transformations (Figure 5) support many-to-one relationships, combining many outputs into a single input.

Transformations also may specify conditionals that instruct the receiving model(s) to accept an input only under certain conditions. For example, a distribution function may specify that a subpopulation of models is to receive the output of a higher-level model, as shown in Figure 6. By allowing transformations to be specified via functions, more complex transformations can be attained.

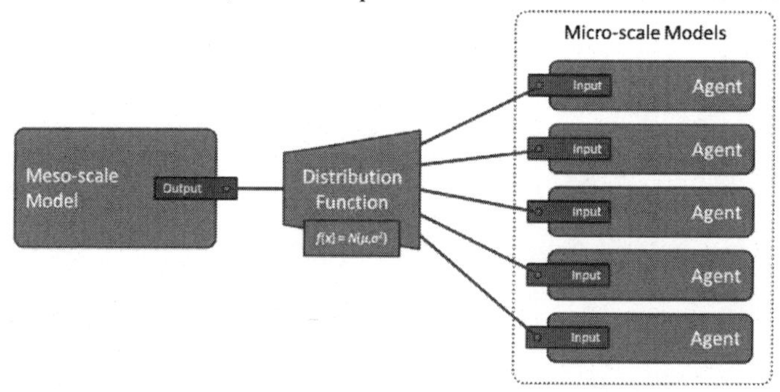

Figure 4. Distribution transformation converting single output to many inputs from a meso-scale model to several micro-scale models.

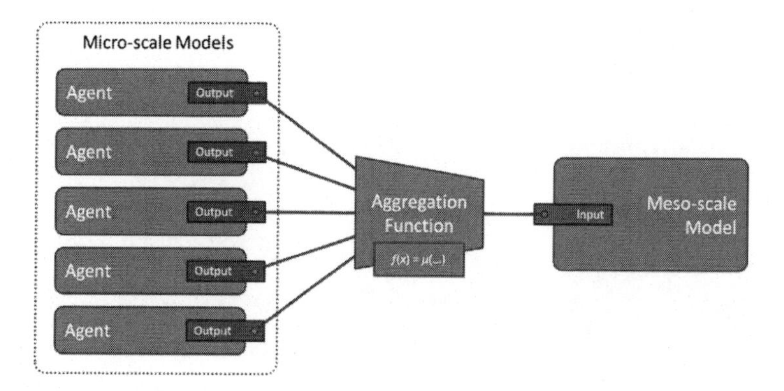

Figure 5. Aggregation transformation converting multiple outputs to a single input from several micro-scale models to a meso-scale model.

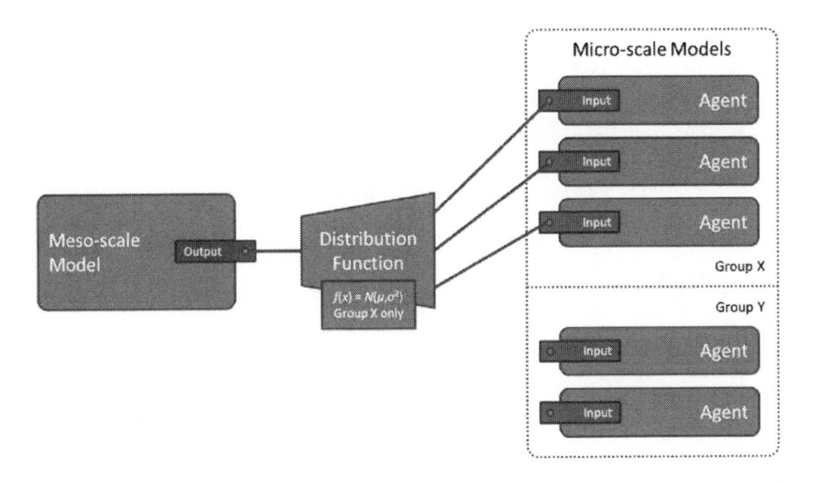

Figure 6. Distribution transformation function selectively applied to a subgroup of models.

4 TRANSFORMATIONS & INTERCHANGEABLE MODELS

As an example of how transformations support interchanging submodels, consider the models in §2, which presented two agent-based models that were independently developed representing different 'pathways' through which a person may become radicalized. In the Silber and Precht system dynamics model of Figure 1, the second stage of radicalization is conversion and identification. The manifestation of identification is quite different in the two agent-based models. In the first, identification results from an agent building a close relationship with a mentor; in the second, identification occurs when an agent finds and begins to relate to a terrorist group. Both conceptualizations require that the agent-based models provide information, in terms of an aggregation of those agents who exhibit strong mentor relationships (in the first perspective) or negative relationships with culturally-opposite groups (in the second perspective). By using aggregation, this information can be communicated to the macro-scale model. Similarly, the agent-based models may provide individuals who identify with a particular culture. This information passes through the Connecting Concepts Reasoner to compute, via an aggregation, the percentage of the population that identifies with each particular culture. This is then used within the macro-scale system dynamics model as a computation measure representing, e.g., the degree of isolation. The result is that the construct of model docking with transformations and connecting concepts enables interoperability among independently developed models.

5 CONCLUSIONS

Quantitative models allow exploration of many aspects of socio-cultural influences and may provide methods to identify indicators that increase one's propensity to being radicalized, recruited, disengaged, or deradicalized from the adverse behavior. At present, many models attempting to conceptualize these dynamics tend to be limited with respect to integrating and assessing multiple qualitative perspectives into quantitative models. If properly developed, rigorous quantitative models may be valuable in that they provide a means to analyze and experiment with differing views, opinions, and perspectives of potential influences on population behavior. This research expands a modeling construct to support quantitative analysis of diverse perspectives and views associated with influences on those who may become involved with or who are already engaged in adversarial behavior.

REFERENCES

Benjamin, P., Akella, K., &Verma, A. (2007). Using Ontologies for Simulation Integration. in S. G. Henderson, B. Biller, M.-H. Hsieh, J. Shortle, J. D. Tew, and R. R. Barton(Eds.) , Proceedings of the 2007 Winter Simulation Conference (pp.1081-1089). Pscataway: IEEE Press.

Briscoe, E., Trewhitt, E., Whitaker, E., and Weiss, L. (2010), "Mitigating Issues Related to the Modeling of Insurgent Recruitment," in Advances in Social Computing, S. Chai, J. Salerno, P. Mabry (eds.), Springer, Berlin, pp. 375-381.

Briscoe, E., Weiss, L., Whitaker, E., Trewhitt, E. "A Systems-level Understanding of Insurgent Involvement in IED Activities," Systems Research and Behavioral Science, 28(4), 391-400, 2011.

Engelmore, R. S., & Morgan, A. (Eds.): Blackboard Systems. Addison-Wesley, Reading (1988)

Gartenstein-Ross, D. & Grossman, L. (2009). Homegrown Terrorists in the U.S. and U.K.: An Empirical Examination of the Radicalization Process. FDD Press, Washington, DC.

Horgan, J. Disengagement & Deradicalization from Terrorism, Presentation to NYPD, Dec 2008

Precht, T. (2007). Homegrown Terrorism and Islamist Radicalization in Europe.

Sterman, J. *Business Dynamics: Systems Thinking and Modeling for a Complex World,* Boston: Irwin McGraw-Hill, 2000, p.42.

Silber, Mitchell D., and Arvin Blatt, *Radicalization in the West: The Homegrown Threat,* The New York City Police Department, 2007. As of March 25, 2008: http://sethgodin.typepad.com/seths_blog/files/NYPD_Report-Radicalization_in_the_West.pdf

Tolk, A. and Muguira, J.A. (2003). The Levels of Conceptual Interoperability Model (LCIM). Proceedings IEEE Fall Simulation Interoperability Workshop, IEEE CS Press

Weiss, L., Whitaker, E., Briscoe, E., and Trewhitt, E., "A Model for Evaluating counter-IED Strategies," Defense and Security Analysis, 27(2), 135-147, 2011.

Weiss, L., Whitaker, E., Briscoe, E., Trewhitt, E., and Gonzalez, R., "Cultural Influences Associated with Adversarial Recruitment," in Advances in Cross-Cultural Decision Making, D. Schmorrow, D. Nicholson (eds.), CRC Press, 2010.

Whitaker, E., & Bonnell, R: A Blackboard Model for Adaptive and Self-Improving Intelligent Tutoring Systems. Journal of Artificial Intelligence in Education (1992).

Zacharias, G., MacMillan, J., & Van Hemel, S. (Eds.). (2008). Behavioral Modeling and Simulation: from Individuals to Societies. Washington, D.C.: National Research Council.

CHAPTER 21

Cultural Engagements: Leveraging a Modeling Framework for Cultural Interaction Simulations

Nick Drucker, Phil Jones

MYMIC
Portsmouth, VA
Nick.drucker@mymic.net

ABSTRACT

Today, most cultural training is based on or built around "cultural engagements" or discrete interactions between the individual learner and one or more cultural "others". Often, success in the engagement is the end or the objective. In reality, these interactions usually involve secondary and tertiary effects with potentially wide ranging consequences that may result in changes to the attitudes of a populace. In these instances, attitude changes may travel, impacting both the immediate human terrain and that of neighbors located in other regions. The extent to which these cultural interactions branch across societies necessitates an approach to model this phenomenon. The concern is that learning culture within a strict engagement context, not accounting for this information transfer, might lead to "checklist" cultural thinking that will not empower learners to understand the full consequence of their actions. In this paper, we explore methods to create simulations, which more accurately relay the complex nature of a cross-cultural interaction. Understanding that accurately modeling human cultural and societal behaviors is extremely difficult we propose several methods to construct simulations that may serve as a first effort. The models described in this paper represent a broad effort to model information transfer across a large population and their reactionary behaviors. However, these models represent a blue print that we feel is repeatable and, over time, refine-able. Ultimately, we propose that the models we explore here serve as a baseline set for simulations of populace behavior which would not need to completely capture accurate outcomes, but which would demonstrate to users the complexity of the cultural landscape. Thus, they would not necessarily support learning of what to do and think, but how to do and think. We begin by proposing the use of agent based modeling (ABM) to collect, store, and, simulating the effects of social networks and promulgate engagement effects over time, distance, and consequence. ABM is one method that allows us to capture individual behaviors, which we then aggregate to larger sections of a society. The information transfer, and subsequently storage of information, by agents in the model provides us the detail at the individual level to transfer information, and the ability at the macro level to demonstrate how it impacts populace behaviors. The ABM development allows for rapid adaptation to model any number of population types, extending the applicability of the model to any requirement for social modeling. From this point, we move to exploring other methods that may augment an ABM and create a more robust simulation tool. The results of our study demonstrate that creation of simulations to capture the effects cultural interactions over time and space is more ideal and relatively easy to implement. The results also highlight that cultural training programs need to capture and relay this sort of information to users in order to fully prepare them to function in cross-cultural settings. The framework we outline here would serve to augment a cultural training program which is focused on the cultural encounter and reinforce this process by adding the complex nature of a human network.

Keywords: Cultural Simulation, Training, Modeling Architecture

1 INTRODUCTION

In the emerging global village, many thinkers agree that cultural competency is and will continue to be an essential skill. The military has found this to be true for our service men and women in the ongoing operations in Iraq, Afghanistan, and other places. The United States and its partners have worked hard to create training capabilities to provide these skills. For many reasons, most of these current capabilities use training conditions defined by a "cultural engagement" or a discrete interaction between the individual learner and the cultural "other" or others. The engagements are discrete because all trainee attention is focused on a determinate outcome tied to particular training objectives. In reality, however, cross-cultural interactions in the current operational environment invoke secondary and tertiary effects with potentially wide ranging consequences. The concern is that learning culture within a strict engagement context leads to "checklist" cultural thinking that will not empower learners to understand the full consequence of their actions nor tie their cross-cultural success to larger, more strategic aims.

The authors of this paper propose an evolved approach to culture training, one that is "campaign" based, with the potential to be an individual or collective training tool, and a method for its execution founded on agent based modeling (ABM). Such an approach would include multiple, cross-cultural engagements and extend their significances across distance, time, and consequence. Such an approach would develop a broader, more strategic perspective on cross-cultural conditions while learning to achieve mission success in cross-cultural conditions.

As the authors have experimented with this new approach within serious games, the paper will use that modality for discussion, though the approach is not limited to game-based solutions.

2 ENGAGEMENT VS. CAMPAIGN

Today, military operational success must be achieved "amongst the people" (Patraeus, 2010), and cultural competency is essential to achieving that success (Gates, 2011). One way of delivering culture training is through game-based training systems. Such systems provide an efficient and engaging learning environment and outcomes. Game-based training systems promote learning retention through engagement and require minimal overhead compared with systems dependent on instructor time, materials, and brick and mortar facilities.

There are already several efforts leveraging games for culture training. Some include BiLAT™, Tactical Iraqi™, and Virtual Cultural Awareness Trainer (VCAT™). These efforts share characteristics, a primary being that they are focused on or built around a "cultural engagement" or discrete interaction between the individual learner and one or more cultural "others". Further, because simulated engagements tend to serve as merely a platform to demonstrate and teach aspects of cultural knowledge, they tend to be highly focused on a single objective and with extremely limited consequences. In reality, however, cultural interactions are not ends by themselves but rather means to achieve military or other objectives. Additionally, these interactions cause secondary and tertiary effects with potentially wide ranging consequences. Therefore, individuals and groups who are effective in achieving goals under cross-cultural conditions are the focus of true operational objectives. The concern is that learning culture within a strict engagement context might lead to "checklist" cultural thinking that will not empower learners to understand the full consequence of their actions nor foster the initiative sought by the military (Shamir, 2011).

It is well understood that training is best when it matches actual application: the "train as you fight" paradigm. Learning culture and language in a manner that enables one to apply what has been learned to a broad set of possible situations requires a new approach. Attempting to learn culture through discrete engagements will tend to limit training to heuristics that will be difficult to apply in the complexity of actual operations. Furthermore, it will miss the real target—achieving mission success under cross-cultural conditions.

Current virtual environment-based culture trainers tend to employ a single tier of consequence when interacting with cultural-other characters. This is not representative of the real world; in reality, consequences of one's actions will propagate along social networks throughout and beyond communities, having effects across a depth of time, distance, and consequence. These actions will affect future interactions with individuals a trainee has not yet met as well as affecting the interactions of "flanking" individuals and organizations. The negative conditioning or training of this limited approach can be easily seen.

The documentary Restrepo (Sebastian & Hetherington, 2010), illustrates the risks of training cultural interaction strictly within the "engagement" context. In the documentary, the American company commander conducts what appears to be a well-coordinated Shura (gathering of village elder's in Afghanistan). However, in the Shura, the commander makes a point of negatively comparing his predecessor to himself. As the senior representative on the ground this could be perceived as disparaging the International Security Assistance Force (ISAF) or America in general. Had the company commander been trained to see cultural interaction as a campaign versus a series of discrete engagements, he would have better understood the linkages between his actions and those of his predecessor and possibly have had a more effective strategy.

The 2007 Marine Corps Operating Concepts for a Changing Security Environment identifies that our likely military opponents, unable to achieve decisive results, are expected to leverage duration as a means to achieve their aims (USMC, 2007). This will result in "prolonged operations." Further, the Operating Concept discusses the predominance of insurgencies in the past half century and into the future. These are wars among the people and for the support of the people. These same concepts are seen as a basis for current operational theater requirements, as the Commander of ISAF spelled out in his Counterinsurgency Guidance: secure and serve the population, live among the people, hold what is secured, foster lasting solutions, be a good guest, consult and build relationships, fight the information war, manage expectations, and maintain continuity through unit transitions (Patraeus, 2010). These are requirements to engage in a prolonged interaction with and for the people where cross-culture is a condition for long-term success.

The Depart of Defense defines engagement as "A tactical conflict, usually between opposing lower echelon maneuver forces" (Joint Chiefs of Staff 1-02, 2010). Within the taxonomy of operations are, in order, engagements, battles, campaigns, and wars. The DoD defines a campaign as "A series of related major operations aimed at achieving strategic and operational objectives within a given time and space" (Joint Chiefs of Staff 1-02, 2010). Because of the nature of current operations, the campaign is a better context for envisaging cross-cultural performance than the engagement.

What would be required to create a campaign-based approach to cross-cultural training? The authors propose changes to how we model individuals, the cultural others, and new methods to modeling the collective community.

3 MODELING INDIVIDUALS

Humans are inherently complex and difficult to accurately model. To deal with this complexity, or possibly to side-step it, many traditional training solutions have used intricate branching logic. This solution, unfortunately, leads to stilted effects and is unable to reflect the dynamics of the human condition.

At the individual level, a thorough framework must include the three domains of human existence: physical, cognitive, and emotional. Physical and cognitive modeling is relatively simple and exists in most first-person games. The development of emotional structures, such as the Global Structure of Emotion Types (Ortony, Clore & Collins, 1988), represent tools for developing computer frameworks that will allow response by AI-driven NPCs.

Sociology and philosophy are also underused disciplines for providing a framework that allows modeling of cross-cultural dynamics. Frame analysis, as introduced largely by Erving Goffman (1974) has provided a basis for decoding how people understand and react to situations. The Social Theory of Practices is another approach for appreciating the natural behaviors of individuals (Bourdieu, 1972; Bourdieu, 1980; Turner, 1994; Schatzki, 2002). These theories, taken together, postulate the presence of context or environment to which individuals bring differing frames of reference and personal and cultural practices. The theories illuminate why an office worker acts differently in an office setting than in a social setting or why a local in a foreign community would respond differently than an American based. These theories provide an approach for modeling a much richer and more complete cultural-other, one who responds more complexly and realistically to changed context or predominant frame.

The next logical step is to tie these more complete individual models to an accurate model of the collective community.

4 MODELING THE COLLECTIVE

The human community is comprised of complex systems. These include networks of people whose interactions result in observable and noteworthy behaviors. People live their lives based on rules (norms), interacting with one-another. Combining groups of people into observable groupings (i.e. a neighborhood, towns, provinces, etc.) provides insight into people composed systems-of-systems. Scaling the systems up creates larger cross sections of society that in turn provides insight on how that society functions. Observing massively complex human systems requires a set of tools suited for recreating individual behavior based on a known set of rules. One such tool is Agent Based Modeling (ABM). ABM is a method whereby experimenters construct complex systems consisting of numerous individual entities called agents (Epstein, 2006). These agents follow a set of rules and interact with each other as they mimic the inner workings of complex systems (Miller & Page, 2007). Using ABM, researchers re-create extremely complex social systems for study.

Researchers traditionally use ABM as a tool to examine and analyze complex societies. However, the authors propose using ABM as a societal model to capture, store, and propagate the results of cross-cultural engagements, or as the bond that links these engagements across space and time to enable a cross-cultural campaign.

One of ABMs primary uses is for the observation of emergent behaviors in groups of individual entities. Emergent behaviors are the aggregate of the behavior of all the agents in the ABM system; behaviors which are unforeseen because we cannot normally observe them or notice them in the real world (i.e. one person dropping a rock in a lake will result in a small ripple, but 500 people performing the same act would result in a significant sized wave). Researchers model a targeted population to include means for individual agents to interact, create an initial state, and then run the system, observing the emergent behaviors of the modeled agents. Researchers can subsequently modify agent states and run additional tests, building a set of relevant results to gain insight on the targeted population.

In the approach offered by the authors, the targeted, cultural-other community is modeled within the ABM with several designated agents representing specific characters from the game environment. These characters design are such that they possess the same initial, attitudinal states as their game environment characters. The results of the trainee's engagement will modify the characters' attitudinal states and the system will relay this change to their ABM counterparts. The system cycles, allowing the initial changes to propagate across the ABM population, changing the states of other individual agents. The system then transfers these subsequent changes back to the virtual environment in terms of changes to the attitudes, beliefs, and behaviors of the characters that trainees will subsequently engage. In a training application, these behaviors become the source of new information to the trainee.

These changes will modify future interactions at the local community level, thus creating a dynamic, consequence-included environment for trainees; a trainee's actions will have future ramifications for themselves but also, because ABM social networks will span separate local communities, those actions will have ramifications for neighbors. This unique expansion of impact over time, space, and consequence will create the conditions for a collective, campaign-based approach to culture training versus an engagement-based approach.

The ABM will also allow a management system that will capture trainee actions and resultant system reactions in a manner that will enable the learner to view the impacts of each interaction as it propagates through the broader social network. This will enable runtime or after action review of the impacts of actions serving as a feedback mechanism.

While the initial prototype of a virtual community employs ABM, the authors envision additional solutions. While ABM is well suited for capturing individual behaviors and aggregating them into a population level view of a society, it does not function as well when faced with capturing the behaviors of much larger groups of individuals. In order to augment the ability of ABM to capture finite individual level details, the authors envision being able to hook the society/village level ABM to a larger system dynamics model which would then represent a country or multi-country level view. During a long-term campaign, it would be important for a model of cultural behavior to capture larger influencing factors, such as the effect of neighboring communities or countries' ability to influence the target population's behaviors. Conversely, it could be beneficial for the trainee to observe how their actions, and the subsequent populace reactions, may result in such far ranging consequences as a change in the trainee's "home" countries opinion of the mission. Providing them with this level of scope, in terms of the reach of their actions, will further reinforce the importance of appropriate cultural interactions.

In addition to utilizing a system dynamics model to propagate information from an ABM to a multi-country level, the authors envision being able to incorporate elements of Game Theory within each agent in the model. During initial development of such a set of models the Game Theoretic models would be simplistic, serving to allow ABM populace members to solve basic decision processes. Introducing Game Theory components within the ABM will allow agents to determine if passing on information about an interaction is beneficial to them. Along the same lines Game Theory within the ABM would allow some agents to determine if they want to pass on false information in order to further their own gains.

5 AGENT-BASED MODELING REQUIREMENTS

In order to create a system where cultural-other agents in the ABM can pass changes to the rest of the population, it is important to model accurate social networks. Social networks are complex groups of individuals socially connected through some form of communication (Barabási, 2002). For example, a group of people who attend a church, temple, or mosque together are independent of one another, yet they routinely gather to worship together, and in this way share information. This social network is, in turn, part of a larger network of social networks. ABM is highly suited to modeling these networks to duplicate information transfer, or to accurately replicate the mechanisms of information transfer.

ABM, in this context, represents a tool with which a training system can set and modify interactional / relationship parameters in real time to affect both immediate and extended dynamics between the learner and the virtual, cultural-other community. In addition, game-based systems, which require the presentation of learning experiences tailored to the trainee, are more efficient if those experiences update in real or near-real time through assessment of the trainee's responses.

During the initial prototype, the authors were able to construct a successful ABM that reached both up to a high level System Dynamics model and incorporated Game Theoretic components to provide more detail to agent behavior. While the prototype did not include either a System Dynamics model or Game Theory elements, the authors were able to construct both systems external to the ABM. The next step in the process for this architecture would be to further refine the System Dynamics model to incorporate the complexity of capturing multi-country level behaviors, and further refine the Game Theory models to allow individuals to model out the potential result of their behaviors. Once both models are more mature, the authors propose developing a prototype system that incorporates the added models, resulting in more accurate behavior of the NPCs in a training environment.

2 CONCLUSIONS

Cultural competency consists of prolonged interactions within a community. The training individuals receive may prepare them to understand some components of a foreign culture, but may not adequately prepare them and their organizations for mission success under cross-cultural conditions. The combination of ABM as a back end to game based cultural trainers presents a solution to the problem of encompassing the complex nature of societies within game-based solutions. The further development of additional modeling paradigms will facilitate a finer level of detail for NPC behavior in a cultural model. With an ABM incorporating Game Theory capturing the behavior of individuals, and this system connected to a System Dynamics model of a much larger population the authors expect to more accurately model the far reaching implications of individual cultural interactions. Until cultural trainers are able to address the prolonged nature of interactions, there is a risk that trainees will not fully grasp the scope of their actions. The risk is Soldier's who may understand the basic components of a foreign culture but do not fully grasp how mistakes or success will affect future interactions. Additionally, if Soldier's do not realize that their actions will impact other units located in the vicinity, and those that follow on; they run the risk of creating cultural issues which they were not trained to handle.

REFERENCES

Barabási, A. 2002. *Linked: The New Science of Social Networks*. Cambridge, MA: Perseus Publishing.

Bourdieu, P. 1972. *Outline of a Theory of Practice*. Translated by Richard Nice. New York, NY: Cambridge University Press.

Bourdieu, P. 1980. *The Logic of Practice*. Translated by Richard Nice: Stanford, CA: Stanford University Press.

Epstein, J. 2006. *Generative Social Science Studies in Agent-Based Computational Modeling.* Princeton, NJ: Princeton University Press.

Gates, R. 2011. "Speech Delivered to the United States Military Academy, 25 Feb 2011". Accessed April 8, 2011, http://www.defense.gov/speeches/speech.aspx?speechid=1539.

Goffman, E. 1974. *Frame Analysis: An Essay on the Organization of Experience*, Boston, MA, Northeaster University Press.

Joint Chiefs of Staff 1-02. 2010. *Joint Publication 1-02, Department of Defense Dictionary of Military and Associated Terms*. Washington, D.C.: Joint Chiefs of Staff.

Miller, J. H., & S. E. Page. 2007. *Complex Adaptive Systems: An Introduction to Computational Models of Social Life*. Princeton, NJ: Princeton University Press.

Ortony, A., G. L. Clore, & A. Collins. 1988. *The Cognitive Structure of Emotions*, New York: Cambridge University Press.

Patraeus, D. 2010. *COMISAF's Counterinsurgency Guidance*. HQ, International Security Assistance Force, Kabul, Afghanistan.

Schatzki, T. 2002. *The Site of the Social: A Philosophical Account of the Constitution of Social Life and Change*. University Park, PA: Pennsylvania State University Press.

Sebastian J., and T. Hetherington, (Co-Filmmakers). 2010. *Restrepo: One Platoon, One Valley, One Year* [Documentary], Location: Outpost Films and National Geographic Channel.

Shamir, E. 2011. *Transforming Command: The Pursuit of Mission Command in the U.S., British, and Israeli Armies*. Stanford Security Studies, Stanford, CA.

Turner, S. 1994. *The Social Theory of Practices: Tradition, Tacit Knowledge, and Presumptions*. Chicago, Il: University of Chicago Press.

USMC. 2007. *Marine Corps Operating Concepts for a Changing Security Environment, Commanding General.* Marine Corps Combat Development Command Deputy Commandant for Combat Development and Integration, June 2007.

CHAPTER 22

Three-dimensional Immersive Environments to Enable Research in Cross Cultural Decision Making

Amy A. Kruse, John R. Lowell

Intific, Inc.
Alexandria, VA, USA
akruse@intific.com

ABSTRACT

Cross-cultural investigations are ideally carried out in the most realistic environments suitable for the type and complexity of the research being conducted. However, it is not always possible to collect the volume and variety of data needed for in-depth analysis in this arena. Three-dimensional immersive simulations offer an opportunity for researchers to create complex environments and realistic user interactions in a manner that reveals culturally relevant information. These simulations allow for the efficient collection of quantitative data and replication of experimental conditions necessary for statistical comparisons to be conducted. We describe a 3D immersive game engine and scenario creation toolkit, RealWorld, that can be used to create realistic and varied cultural interactions as stimuli for research studies. Work to date has demonstrated the use of this toolkit for construction of both training simulations and serious games, each with sufficient cross-cultural depth to serve as a test bed for investigators in this area. Additionally, because of the toolkit features in the immersive system, researchers can themselves change aspects of the scenarios and games to answer specific cultural decision-making research questions. This is a flexibility not found in other cultural simulation software. We have also created the ability to embed decision-making tasks inside the gaming environment, thus allowing for additional scoring and data collection opportunities based on the scenes and interactions encountered. Through our current efforts, we have enabled this immersive toolkit to serve as an experimental platform for neuroscience and psychophysiology researchers through construction of our "NeuroBridge" data interface. This interface enables the software to log relevant in-game events, user interactions and other salient data in real-time with millisecond resolution, and stream the data for correlation with neurophysiological and psychophysiological data. Finally, the "NeuroBridge" is two-way and has been designed to facilitate adaptive feedback to influence in game or scenario events based on the real-time state of the user or player. Taken together, all of these features allow for a realistic, compelling and immersive cross-cultural experience for the player that still addresses the needs and requirements of the researcher.

Keywords: Cross-Cultural, Decision Making, NeuroCognitive Testbed, 3D immersive environment, NeuroBridge, RealWorld

1 UTILITY OF SIMULATION ENVIRONMENTS FOR CROSS CULTURAL INVESTIGATIONS

Researchers spend significant time and resources preparing and conducting investigations into cultural differences in behavior, decision-making and communication. These investigations require access to realistic, detailed environments that support, rather than undermine, their research goals. Funding constraints can compromise access to authentic environments, or restrict the volume and variety of the data acquired. These compromises often weaken the conclusions of a given investigation by limiting statistically relevant supporting evidence.

One potential solution to this challenge is to utilize researcher-controlled, instrumented three-dimensional immersive simulations. Advances in computer graphics have enabled the construction of virtual environments with sufficient visual realism to meet many objectives of cross-cultural investigations. However, to provide repeatability and sufficient performance at a reduced overall cost, many additional features are needed, as highlighted in Table 1 below. Properly constructed simulation scenarios show enormous promise as research tools in the cross-cultural environment.

Table 1 Virtual Simulation Features Supporting Cross-Cultural Investigations

Research Need	Required Virtual Simulation Supporting Features
3D immersion	First-person perspective 3D visual simulation with culturally relevant locations, characters and objects.
Realistic cultural interactions	Flexible and powerful environment and scenario construction system.
	Character animation system able to match culturally specific non-verbal cues.

Experimental controls	Control of presented content and tools to manage content provisioning for specific experiments.
	Repeatable scenarios to reduce unnecessary variability.
Data acquisition	Data logging and data connectivity support, including data management tools.

2 USER CONTROL OF SIMULATION CONTENT

Intific began development of the RealWorld® PC-based rapid mission rehearsal and training system under DARPA sponsorship in 2005. This software enables rapid generation of simulated geo-specific locations and multiplayer, multi-role first- or third-person perspective simulations of proposed missions on the actual mission terrain with actual mission equipment. DARPA was prompted to create this system because there existed an opportunity to save both lives and money by leveraging state-of-the-art computer game technology to create a PC-based mission rehearsal system. DARPA recognized they could reduce or possibly eliminate an opponent's "home-field" advantage if an easy-to-use, widely distributed, realistic mission rehearsal system could be created that would allow war fighters to see the actual terrain and rehearse virtually "in situ" before mission execution. RealWorld version 2.0 was released to the government in August 2011, and will be sold commercially beginning Spring 2012. Although initially conceived as a mission rehearsal platform, users quickly realized that the platform had substantial utility as a research tool. From a research perspective, RealWorld gives non-programmers the ability to construct and execute a wide range of multi-player virtual scenarios in a repeatable manner via a high-fidelity display system.

Figure 1 Screen shot from the RealWorld rapid mission rehearsal and training system.

2.1 Environment and Scenario Construction – Creating Culturally Relevant Locations

The RealWorld platform is a collection of software, including the RealWorld Object Creator, RealWorld Builder and the RealWorld Ground Synthetic Environment, which is the "engine" that executes the created levels. RealWorld allows for the construction of environments and scenarios through the RealWorld Builder software and the mission scripting system. Using geo-specific data and RealWorld scenario authoring tools, Intific recently completed a project modeling the U.S. Army's National Training Center at Fort Irwin, CA to ground level accuracy within the RealWorld simulation environment that illustrates the power of these authoring tools. During this project, government and commercial Geographic Information System (GIS) data and imagery were used to produce accurate 3D environments within Intific's mission rehearsal simulation system. These efforts are relevant, since they demonstrate the ability to accurately recreate specific complex geographic locations that may serve as important context cues for culturally-specific research investigations.

Figure 2 Virtual environment of Fort Irwin, California.

Of particular interest to most cross-cultural investigators is the ability to produce smaller, detailed areas for their experiments. The same tools used to construct the virtual environment above were used to construct a vigilance trainer for soldiers deploying overseas. The Eagle-Eye IED Detection game used the RealWorld engine and RealWorld Builder to create a stand-alone software application that has the potential to increase the ability of soldiers and Marines to detect IEDs in the current operational environment through an engaging gaming experience. As shown in the screen shot below, the game includes an extremely detailed, culturally specific backdrop upon which the training activity takes place, including the sounds and background activity of typical locations where soldiers and Marines deploy. Although the Eagle-Eye environment is a specific example of a polished, art intensive scene, it serves as a guide for the types of environments that can be created with these tools. Typically researchers have the time resources needed to produce environments of high quality – as contrasted with the pressures and schedule of the commercial environment. As another benefit to researchers, RealWorld scenarios, data files and assets can be exchanged between anyone with the RealWorld software platform, substantially increasing collaboration and sharing between laboratories.

Figure 3 Screen shot from the Eagle Eye IED Detection Trainer showing culturally specific locations used as backdrops for vigilance training for deploying soldiers and Marines.

2.2 Character Animation and Behavior - Making Culturally Relevant Virtual Characters

To visualize all aspects of a culturally relevant presentation to a participant, the RealWorld platform utilizes character animations rigged for vocal performances, an appropriate regional environment, facial and body animations that vary with context and scale in intensity and other presentation aspects as required. Character rigs can be re-skinned as necessary to create a geographically typical look and feel.

Figure 4 High resolution Middle Eastern avatar from RealWorld immersive simulation system.

People express almost a million nonverbal cues, a quarter of a million facial expressions and over 70,000 unique cultural gestures. Since human behaviors are exceedingly complex when engaged in conversation, the character animation system must be flexible so that it can expand to include specific behaviors needed for a particular cross-cultural study. The RealWorld character animation system was designed to deal with physical, emotional and cultural behaviors of face and body, and thus able to support a wide array of experimental variety. These capabilities were used to construct the High Altitude Trainer, which augments classroom activities for Navy medical corpsman by training diagnosis of High Altitude Pulmonary Edema (HAPE) and Acute Mountain Sickness (AMS). The software uses the RealWorld engine for both the facial avatar and full body motion across the mountain terrain. The software generates symptom presentation (face color, coughing, breathing) via observation and free-form interview capability that can be leveraged across a variety of interpersonal training needs. Profiles can be created for a wide variety of physical types and health states with environmental effects such as altitude and temperature applied to high fidelity avatars in interview and observation scenarios. Although used for a different purpose, the tools created for this effort provide cultural researchers the capabilities needed for detailed character animation to fuel their studies.

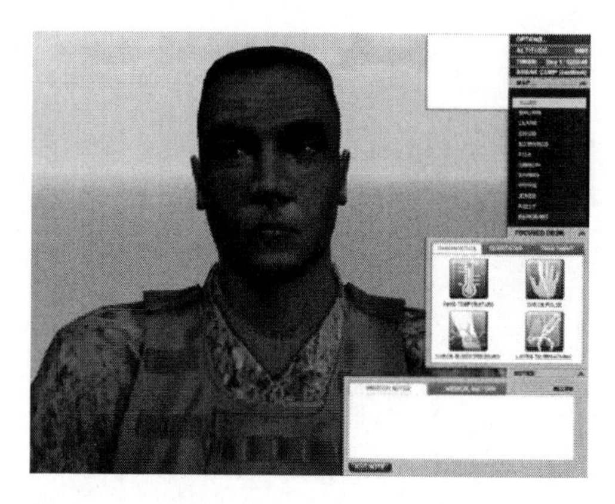

Figure 5 High Resolution character taken from the High Altitude Trainer. Note the unusual coloring of the face indicative of illness; similar parametrically-driven alterations can be used to populate and efficiently repurpose material for cross-cultural studies.

To drive moment-by-moment changes in actor animation states, the RealWorld platform utilizes scripting and a Goal-Oriented Behavior system that includes behavior authoring tools for non-programmers. The underlying drivers for each behavior can be prioritized, character-specific implementations can share common scripts, and complex physical actions can be combined to produce extremely complicated character behaviors. Involuntary emotional or idle symptoms can be injected into facial and body motion. As an example, behavior scripting tools can be used to produce highly realistic interactive conversations by linking segments of branching conversations that are be triggered while a scenario simulation is underway. The behavior scripts and other scenario content can be shared among researchers at different locations, so that configuration control of experimental protocols can be maintained with relative ease.

3 VIRTUAL ENVIRONMENT DATA ACQUISITION

Inspired by the potential for use of RealWorld in research environments, for both non-defense and defense purposes, we have created a NeuroCognitive Testbed for neurophysiological and behavioral studies. The goal of this testbed, consisting of RealWorld 2.2 and the NeuroBridge plug-in, is to support research data acquisition with simultaneous use of a virtual environment. This combines the controlled environment of immersive content delivery (RealWorld) with a comprehensive neurophysiological testing suite for the brain indicators of cultural influence (NeuroBridge plug-in). The NeuroBridge allows for millisecond resolution time-stamping of scenario events and user actions within RealWorld to be logged and communicated to outside applications. This occurs in real-time, and provides the opportunity for measured neurophysiological outputs from the user to directly drive aspects of the simulation. Using this software, virtual activity from pre-constructed scenarios, whether scene based or avatar acted, can be accurately aligned with the neurophysiological measurements being acquired in real-time. Since this feature was created with the researcher in mind, the key feature of RealWorld with NeuroBridge is its flexibility. Neuroscience researchers can develop any desired communication protocol between their applications and RealWorld. Current RealWorld communication and logging support is available for sophisticated neurophysiological measurement tools such as EEG and eye tracking. To date, Intific has integrated RealWorld with NeuroBridge with the Advanced Brain Monitoring B-Alert X10/X24 EEG system.

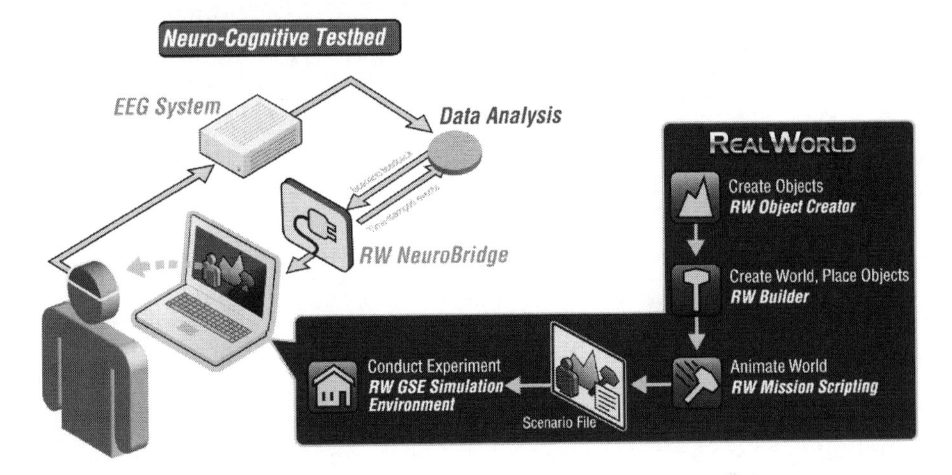

Figure 6 Schematic of RealWorld and RealWorld NeuroBridge for linking of EEG system to authored scenario for cultural experimentation.

The current commercial release of RealWorld with the NeuroBridge plug-in includes RealWorld 2.2, which supports multi-player scenarios, triggering of scripted scenario events based on real-time brain activity and a host of other features. In addition to providing sophisticated data logging, a real-time data presentation dashboard was built into the NeuroBridge plug-in. By presenting raw and conditioned data streams in real-time researchers have the opportunity to watch multiple subjects in one window. They can watch the scenario unfold, annotate data from within the Dashboard and use the RealWorld Builder software to make rapid adjustments to scenarios. In addition to the current supported hardware, both NeuroBridge and the dashboard can be configured for a wide range of external adaptors, hardware, algorithms or other applications.

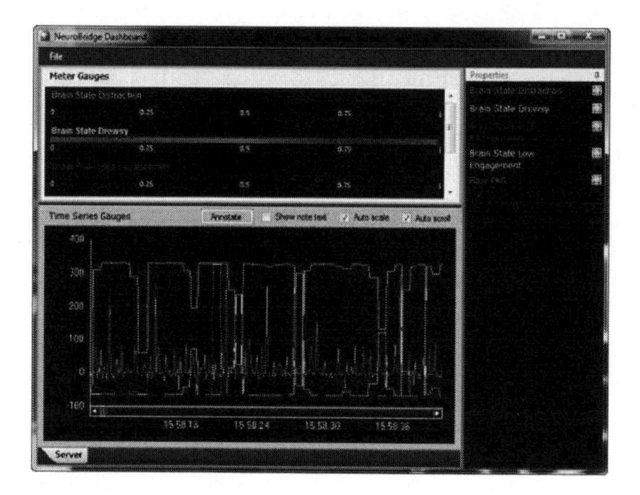

Figure 7 NeuroBridge Dashboard displaying real-time signals and metrics generated by the B-Alert X-10 EEG Analysis Application.

4 LIVE SCENARIO GAME PLAY

To support researchers, we have developed a number of other useful gameplay features. The first is the ability to create mini-games, or quiz-like activities, and embed them within the scenario. This was initially implemented in Eagle-Eye and is now utilized on several demonstration levels in the RealWorld with NeuroBridge software release. This gives the researcher flexibility in probing for responses to specific stimuli within the scenario without interrupting the game play or scenario flow. These responses can generate scoring implications or even trigger scripted events within the scenario.

The second feature is the ability to "close the loop" and cause scripted events to occur within the scenario based on the real-time neurophysiological activity of the participant. This is critical for augmented cognition research, and perhaps one of the first software platforms explicitly designed with this in mind. This has been implemented in one particular demonstration level within the current software release, the 'Vigilance Driving Scenario.' In this scenario, an extended drowsy or distracted state, as calculated in real-time by the ABM B-Alert cognitive state measures, feeds into the scenario and causes scripted events – like pedestrians entering the roadway – to occur.

Figure 8 Image from Vigilance Driving Scenario in RealWorld 2.2. .

5 CONCLUSIONS

The capabilities offered by the RealWorld with NeuroBridge software platform are currently unmatched in any existing system. Other immersive software products are available with culturally relevant simulations and content. However, there are distinguishing features that set the RealWorld with NeuroBridge platform apart from any of these existing products. None of them offer the flexibility and user programmability of the RealWorld platform. Unlike simulations that require constant intervention and modifications from the software provider, RealWorld is designed with the operator, and researcher, in mind. The entire platform has been constructed with a toolkit approach that exposes every aspect of world, environment and scenario creation to the user. Thus the user is able to go as deep into the creation process as they like – from simple building and terrain placement to mission scripting through a LUA programming interface – thereby designing the precise environment needed for their experiment.

The emerging capabilities outlined in this paper point to rapid future advances in cultural experimentation. One can imagine future multi-subject experiments performed using virtual environments and high-speed network connections with real-time data collection and analysis. Experiments could be configured to switch the presented culture mid-experiment based on collected data, or even to construct poly-cultural environments not available through traditional experimental methods. As both software and hardware tools mature, these approaches hold enormous promise for quantitative data collection in the cross-cultural research environment.

ACKNOWLEDGMENTS

The authors would like to thank their colleagues on the NeuroCommercial development team, particularly Caleb Crawford, Jose Garcia, Belinda Heywood, Raul Iglesias, Chuck Lupher, Eugene Moore, Tony Morone, Brendan Seagraves, and Justin Williams. We would also like to thank Pete Bonanni and John Main for their support of the product. Finally, we would like to thank our colleagues at Advanced Brain Monitoring and the Wadsworth Center, who assisted on early experiments with external devices.

REFERENCES

Forsythe, C., Kruse, A., Schmorrow, D., 2004, Augmented Cognition. In C. Forsythe, M.L. Bernard & T.E. Goldsmith (Ed.) Cognitive Systems: Human Cognitive Models in Systems Design. Mawah, NJ: Lawrence Erlbaum Associates.

Johnson, R.R., Popovic, D.P., Olmstead, R.E., Stikic M., Levendowski, D.J., and Berka, C. 2011. Drowsiness/alertness algorithm development and validation using synchronized EEG and cognitive performance to individualize a generalized model. Biological Psychology 87(2): p. 241-50.

Kruse AA. 2007. Operational Neuroscience: Neurophysiological measures in applied environments. Aviation, Space and Environmental Medicine 78(5, Suppl.):B191-B194.

Stikic M, Johnson RR, Levendowski DJ, Popovic DP, Olmstead RE and Berka C. 2011. EEG-derived estimators of present and future cognitive performance. Frontiers in Human Neuroscience. 5: 70.

CHAPTER 23

Enhancing Cultural Training Platforms with Integrated Psychophysiological Metrics of Cognitive and Affective States

Chris Berka[1], Veasna Tan[1], Cali Fidoppiastis[2], Anna Skinner[3], Dean Martinez[1], Robin R. Johnson[1]

[1]Advanced Brain Monitoring, Inc.
Carlsbad, CA, USA
chris@b-alert.com
[2]University of Alabama Birmingham
Birmingham, AL, USA
cfidopia@uab.edu
[3]AnthroTronix, Inc.
Silver Spring, MD, USA
askinner@atinc.com

ABSTRACT

Serious gaming platforms incorporating immersive video gaming technology or real-time collaboration within Massive Multiplayer Online Games (MMOG) are increasingly replacing more conventional classroom training in industry, academia and the military. Simulations and MMOG environments are accessible, easy to use and low cost when compared to "live" training. These gaming platforms offer many advantages including reality-based geo-spatial maps, seemingly infinite options for data presentation, and unprecedented opportunities for team collaboration. In addition, the training offers opportunities for understanding and improving subtle communications and interactions that are only beginning to be understood. Training or mission rehearsals can be recorded and analyzed later to explore patterns or identify trends that may be missed in a live training. The proposed work is designed to further enrich MMOG and other simulation training by adding another layer: analysis of the subjective states of the gamers.

Keywords: EEG, training, gaming, neurophysiology

1 INTRODUCTION

Serious gaming platforms are replacing more conventional classroom training in industry, academia and the military. Simulation environments offer many advantages for cultural training including the potential to create multiple social and cultural scenarios and to record and playback sessions for after action review. Virtual worlds and Massive Multiplayer Online Games (MMOG) such as Second Life and Big World are accessible, easy to use and low cost when compared to "live" training and support opportunities for team collaboration. Trainees not co-located can meet team members for the first time in the virtual world. This new training reality offers opportunities for understanding cross-cultural differences and improving social interactions that are only beginning to be understood.

Our research is designed to further enrich simulation-based training by adding another layer: analysis of the subjective states of the gamers. The convergence of advances in ultra-low power consumer electronics, ubiquitous computing and wearable sensor technologies enables real-time monitoring of cognitive and emotional states providing *objective, timely, and ecologically valid* assessments of psychophysiological states such as attention, alertness, mental workload and arousal. These metrics provide a window on the internal states of the trainees with the goal of clarifying *why a training exercise is a success or failure.* In addition to accessing the states of the trainee in real-time, neurosensing outputs can be used to drive changes in serious game scenarios based on real-time physiological metrics.

Real-time acquisition of electroencephalography (EEG) with ECG and GSR offers access to cognitive and affective states while participants are engaged in simulation environments. Preliminary work suggests psychophysiological metrics can track, in real-time, participants' levels of engagement, boredom, mental workload and fatigue during training. Data is currently being acquired to create a taxonomy of emotion characterized on two dimensions: valence (i.e., positive or negative) and level of intensity. From these and other psychophysiological metrics a matrix will be derived that will be useful in characterizing experiential states currently only described subjectively as "presence, immersion and flow". "Presence" is understood in terms of a participant's sense of "being in" a virtual world (Lim and Reeves 2007); several scales and metrics have been designed to measure the degree to which participants perceive themselves to be in a different place, the reality of that place and the reality of other people and characters in that place. "Immersion" is the experience, obtained through the senses of being surrounded by and engulfed in the content of a virtual world (Steuer 1993). Virtual interactions have been shown to increase engagement levels with associated patterns of heightened physiological responses and brain activation (Chaminade, Meltzoff et al. 2005; Ravaja, Saari et al. 2006; Lim and Reeves 2007; Nacke and Lindley 2008; Ravaja, Turpeinen et al. 2008; Salminen and Ravaja 2008).

The "flow" state is often described by gamers as the quintessential experience with complete involvement in the selected activity, a suspension of time and a sense of pure fulfillment and joy with no dependence on goal or anticipated future achievement (Chaminade, Meltzoff et al. 2005; Quinn 2005; Ravaja, Saari et al. 2006; Nacke and Lindley 2008; Salminen and Ravaja 2008). An early model of the flow state was introduced by Csikszentmihalyi in 1990 (Csikszentmihalyi 1990) based on research on the subjective experience of artists, musicians, chess players, and athletes. Game designers view the flow state as a highly desirable outcome of the game experience, and significant effort is invested in game mechanics with the goal of creating environments that enhance immersive and flow states. There are few validated measurement tools for assessing flow; most common are self-reports, which are inherently subjective and must be completed by the user following a gaming session so as not to interrupt the game or the flow state, thus relying on accurate recollection of states experienced during game play.

In a series of studies conducted by Advanced Brian Monitoring, Inc. (ABM), a psychophysiological profile was identified as characteristic of peak performance in archery, golf and marksmanship. Expert marksman, golfers and archers exhibited increases in EEG alpha and theta synchronization with heart rate deceleration during the pre-shot period (3-8 seconds preceding a shot), and this profile was associated with optimal performance (Berka, Behneman et al. 2010). Experts acknowledged internal perception of the state and demonstrated the ability to reproduce it even under stressful and challenging conditions. This state shares some phenomenology with the concept of flow and supports the utility of psychophysiological metrics in characterizing trainee state changes in simulated environments. Our team is constructing multiple psychophysiological metrics for accessing internal states to optimize simulation design and to ensure participants are fully engaged in the scenarios.

2 Methods

Preliminary data acquired at ABM using the International Affective Picture System (IAPS) stimulus set to elicit a range of emotional experiences suggests that physiological variables can be used to create a taxonomy of emotional valence (positive-negative) and to estimate the level of intensity of the emotion.

Participants

Twenty-six subjects (12 male, 14 female) participated in this study. The age range was 19-45 with 73% between the ages of 19-29 and 27% older than 30.

Materials

The testbed consisted of the IAPS stimulus set to elicit a range of emotional experiences. This study focused on two primary dimensions: affective valence (pleasant and unpleasant) and arousal (calm and excited) with the goal of deriving physiological variables to create a taxonomy of emotional valences for positive-negative and to estimate the level of intensity of the emotion. The IAPS is a set of over 1000 images (e.g puppies, loaded gun) selected and calibrated for emotion elicitation by the NIMH Center of Emotion & Attention (CSEA). Normative ratings obtained from large and diverse samples show mean ratings for valence and arousal levels that are highly consistent across populations; therefore the IAPS is ideal for this application. Physiological response was assessed via a multi-sensor suite combining the ABM B-Alert® easy-to-apply EEG and EKG wireless sensor headset with the Bodymedia Sensewear armband to support development of neural and physiological markers of emotion on two orthogonal dimensions: level of arousal and emotional valence (positive or negative) to provide signatures that detect multiple cognitive and affective states.

Procedure

Subjects viewed a total of 180 images in 3 block increments. Using E-Prime, each block contained 60 images (20-Neutral, 20-Negative, and 20-Positive) that were randomly interspersed and appeared for a total of 5 seconds each followed by a valence rating screen. Subjects were instructed to rate these images using the keyboard's numerical pad as either positive (1), neutral (2), or negative (3).

Electroencephalographic Measures

The B-Alert X10 system was used to acquire EEG data. Channels according to the international 10-20 system include Fz, F3, F4, Cz, C3, C4, POz, P3, and P4 referenced to linked mastoids. Standard impedances values for all subjects were maintained below 40kΩ. Automatic artifact decontamination identified 5 artifact types: EMG, eye blinks, excursions, saturations and spikes. Event Related Potentials (ERPs) were processed for two seconds from the onset of the image presentation. Trial by trial maximum and minimum difference greater than 100μV were manually removed. Artifact identified as saturation, EMG, and eye blinks were also manually removed.

3 RESULTS

Building a preliminary psychophysiological classifier for positive and negative emotions The kNN (k-nearest neighbor algorithm) is a non parametric method for classifying instances based on the closest training examples in the feature space. The algorithm finds k nearest observations in the training set, and classifies the instance based on the majority vote of its neighbors (i.e. according to the most frequent class among the found neighbors). When the referential channel's (Fz, Cz, POz, F3, F4, C3, C4, P3, P4) wavelet variables (from 0-2Hz, 2-4Hz, 4-8

Hz, 8-16 Hz, 16-32 Hz, 32-64 Hz, and 64-128 Hz) were submitted to a k-NN analysis using k=2, an effective classifier was identified with *100% sensitivity and specificity*. This non-parametric classifier approach was the only effective classifier identified, with parametric approaches (including those utilizing PSDs, heart rate, ERP features, actigraphy and GSR variables) resulting in classification error rates ranging from 27%-86%. Additional variables (heart rate, GSR and actigraphy) will be added once the sample size is increased and are available to integrate. While k-NN algorithms are computationally inefficient, there are standard approaches to improve the utility of such an approach. The primary drawback of the current data is inadequate sample size (n=26) to develop a more effective classifier at this time, and as additional data becomes available, it is likely that more efficient, parametric approaches will become more effective.

Event-Related Potential correlates of emotion The graphs in Figure 1 show the stimulus locked grand mean event-related potential (ERP) waveform analysis for all 26 subjects across 9 channels (Fz, F3, F4, Cz, C3, C4, POz, P3, and P4) plotted for Negative, Neutral and Positive valences. The ERPs are shown across a span of 2 seconds (2000ms) from stimulus onset. Images were classified by IAPS as either negative, neutral or positive, and when subjects responded, only answers that matched the IAPS response were included. For example, an image that is classified by the IAPS system as Negative with a subject response of Negative (i.e. IAPS Valence/Subject Response : Neg/Neg) were used in the analysis. Any image classified by IAPS as Negative with any other subject response was excluded from this analysis (e.g. Neg/Pos). In other words, the subject's valence response had to match the designated IAPS valence.

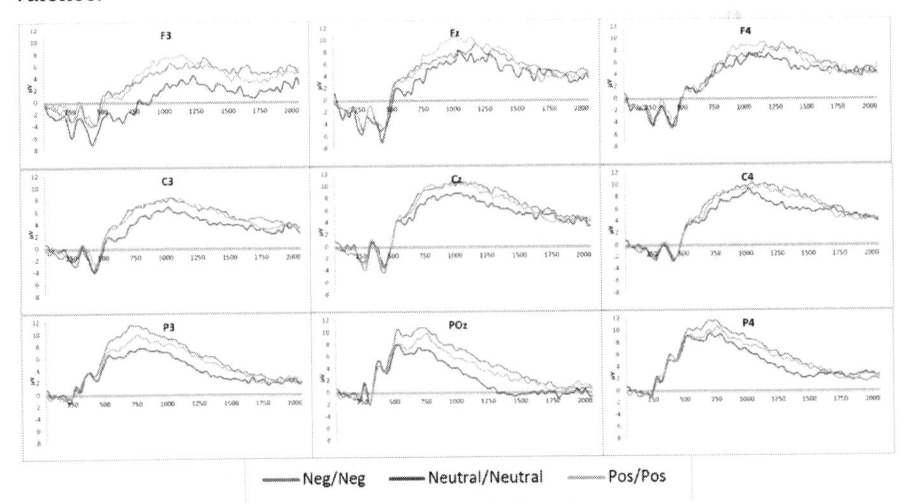

Figure 1 Grand mean analysis for the ERP waveform of 26 subjects across all 9 EEG channels. The ERPs are stimulus locked from the onset of the image and plotted to show the relationship of the valences for each sensor site location.

Figure 2 collapses the Frontal, Central and Parietal sites (collapsed across left, midline and right). Both sets of graphs in Figures 1 & 2 show a change between frontal and parietal sites where the positive images have a higher late positive amplitude in the frontal region but the negative images have a greater late positive amplitude along the parietal regions.

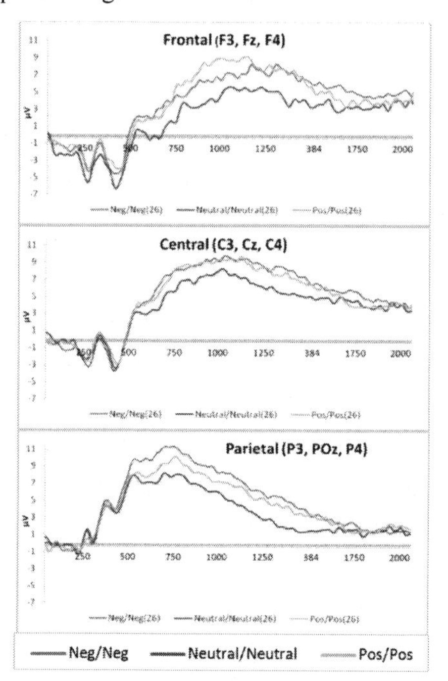

Figure 2 Grand mean ERP waveforms collapsed across frontal, central, and parietal.

The ERP waveforms in the next two figures (Figures 3 & 4) collapses the analysis by laterality (left, midline, right). Figure 3 illustrates the left, midline, and right hemispheres plotted by each valence of negative, neutral, and positive. The laterality of valences indicate that the right and midline areas generally have a higher late positive amplitude compared to that of the left.

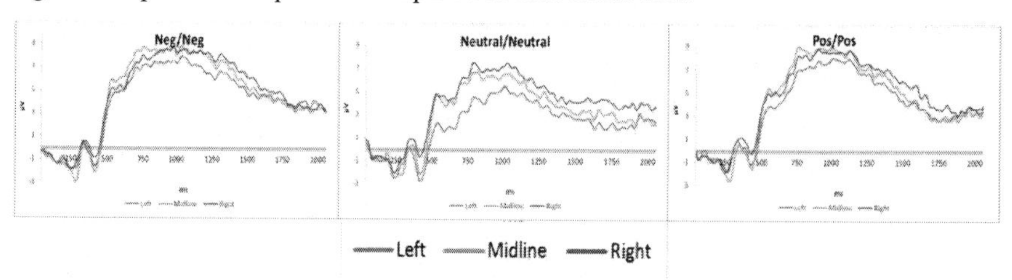

Figure 3 Grand mean of the ERP waveforms collapsed by laterality and plotted by left, midline, and right across the valences.

However, when the ERP waveform is plotted by valence across the left, midline and right, there doesn't seem to be any differences between the negative or positive images on the left or right sides.

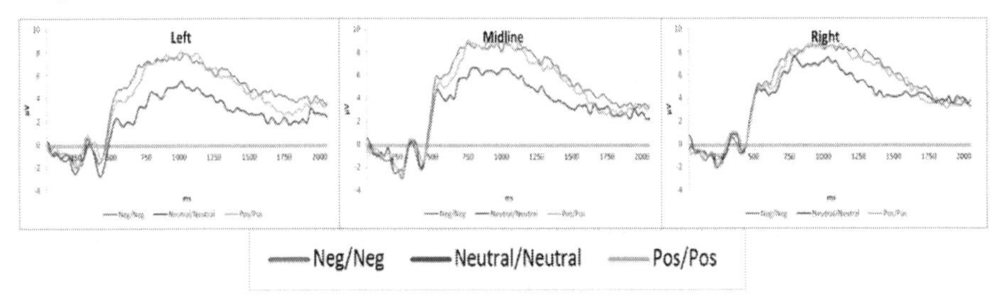

Figure 4 Grand mean of the ERP waveforms collapsed across laterality plotted by negative, neutral, and positive valences across laterality.

For further analysis, the maximum amplitude of the late positive component for each individual subject was measured using a range from stimulus onset of 750-1250ms for frontal sites and 500-750ms for the parietal sites. A 2(Valence: Neg, Pos) x 2(Position: Frontal, Parietal) ANOVA was performed. Results show a significant 2-way interaction for **channel*valence**, $\underline{F}(3,100) = 7.41$, $P < .001$, with a main effect of **channel**, $\underline{F}(1,102) = 21.47$, $P < .0001$. There was no main effect for valence.

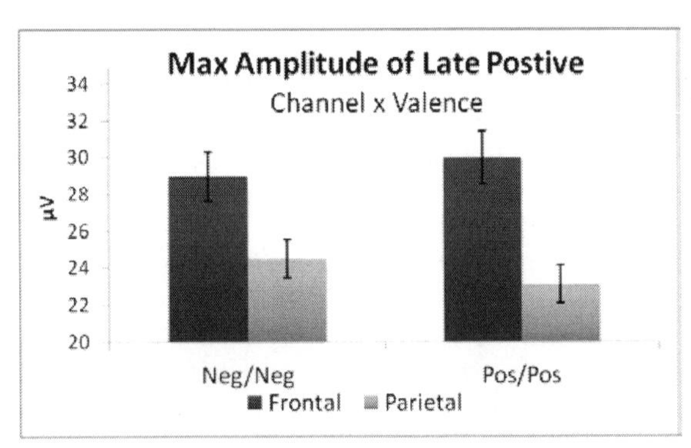

Figure 5 Graphs the averaged max amplitude of the late positive component of the ERP waveform. A significant 2-way interaction was found for channel x valence.

These results primarily show differences in the frontal and parietal sights for negative and positive valences. Many studies have found that emotional valences associated with feelings of joy, happiness and positivity (such as those elicited in response to images or music) tend to show dominant right hemispheric ERP results (Kayser, Tenke et al. 1997; Cunningham, Espinet et al. 2005) specifically within the frontal regions (Schmidt and Trainor 2001; Zhang, Zhou et al. 2011). The study conducted by Zhang et. al 2011 used the IAPS testbed for both variables of valence (negative positive) and arousal (high, low) to assess emotional hemispheric brain asymmetry. The current study, however, only measured valence, but 13 of the 26 subjects were prompted to record the strength (or arousal level) for their chosen valence. Based on a scale of 1-9, the subjects chose how strongly they felt about the negative, neutral or positive image. This scale will provide the categorization for the level of arousal in future analyses.

Brief videos elicit EEG indices of positive and negative emotional states A separate ongoing study conducted at Azusa Pacific and Loma Linda Universities involves scenarios of eustress (Laugh) and distress (Stress) (Berk, Cavalcanti et al. 2012). Preliminary results showed significant differences between the eustress and distress conditions for the EEG power spectral density (PSD). The analysis calculated the difference between the right and left hemispheres, (log[right] - log[left]). The scale can then be interpreted with higher scores indicating relatively greater left activity and lower scores relatively greater right activity (Coan and allen 2004). The difference between the log right and left alpha was calculated and evaluated using a one-way ANOVA (state). The ANOVA revealed significant right-left differences occurring across state for alpha, $F(1,62) = 4.83$, $p < .05$, but not for theta or beta. Similar effects were found for the parietal region. The Laugh scenario exhibited greater left frontal alpha activity and the Stress scenario had a slightly greater right frontal alpha activity (Berk, Cavalcanti et al. 2012).

4 DISCUSSION

The team plans to leverage and extend this prior work to integrate multiple psychophysiological measures and use those measures to assess the impact of various scenario designs on trainees though measurement of engagement, boredom and mental workload. These cognitive state metrics have been developed and validated in prior work (Berka, Levendowski et al. 2004; Berka, Levendowski et al. 2007; Stevens, Galloway et al. 2007; Stevens, Galloway et al. 2008; Berka, Behneman et al. 2010; Berka, Pojman et al. 2010; Johnson, Popovic et al. 2011). Preliminary data suggest that psychophysiological parameters including EEG, ECG and GSR can be derived to create a taxonomy of emotion characterized on two dimensions: positive-negative and level of intensity. An approach similar to the cognitive state implementation will be used to translate the emotion metrics to run in real-time. The combined cognitive and emotional states can then be mapped to a matrix that will be useful in characterizing experiential states that are currently only described subjectively such as “presence, immersion and flow”.

The long-term goal of this work is to translate the real-time cognitive and emotion metrics into either easily interpreted visuals (e.g. colors, gauges) within the MMOG or other simulation platform or used to directly control natural actions of the avatars within the MMOG (changes in facial expressions or gestures). An integrated platform will be designed to allow rapid prototyping and evaluation of a variety of linkages between the physiological metrics and avatar behavior.

Real-time integration and display of the users cognitive and affective states within MMOG or other training and simulation scenarios should increase engagement and immersion and facilitate student-teacher or team interactions.

Identification of cognitive and emotional states within the MMOG should also improve communication and lead to enhanced training and mission rehearsal strategies and tactics, particularly in cross-cultural scenarios where subtle movements and body position may play a critical role in communication.

REFERENCES

Berk, L., P. Cavalcanti, et al. (2012). EEG Brain Wave Band Differentiation During a Eustress State of Humor Associated Mirthful Laughter Compared to a Distress State. Experimental Biology, San Diego, CA.

Berka, C., A. Behneman, et al. (2010). "Accelerating Training Using Interactive Neuro-Educational Technologies: Aplications to Archery, Golf, and Rifle Marksmanship." International Journal of Sports and Society **1**(4): 87-104.

Berka, C., D. Levendowski, et al. (2004). "Real-time Analysis of EEG Indices of Alertness, Cognition, and Memory Acquired with a Wireless EEG Headset." Int J Hum Comput Interact **17**(2): 151-170.

Berka, C., D. Levendowski, et al. (2007). "EEG Correlates of Task Engagement and Mental Workload in Vigilance, Learning and Memory Tasks." Aviation Space and Environmental Medicine **78**(5): B231-B244.

Berka, C., N. Pojman, et al. (2010). NeuroGaming: Merging Cognitive Neuroscience & Virtual Simulation in an Interactive Training Platform. Applied Human Factors and Ergonomics. Miami, FL, CRC Press / Taylor & Francis, Ltd.

Chaminade, T., A. N. Meltzoff, et al. (2005). "An fMRI study of imitation: action representation and body schema." Neuropsychologia **43**: 115-127.

Coan, J. A. and J. allen, J.B. (2004). "Frontal EEG asymmetry as a moderator and mediator of emotion." Biological Psychology **67**: 7-49.

Csikszentmihalyi, M. (1990). The Psychology of Optimal Experience. New York, Harper & Row.

Cunningham, W. D., S. D. Espinet, et al. (2005). "Attitudes to the right- and left: Frontal ERP asymmetries associated with stimulus valence and processing goals." NeuroImage **28**: 827-834.

Johnson, R. R., D. Popovic, et al. (2011). "Drowsiness determination through EEG: development and validation." Biological Psychology **87**(2): 241-250.

Kayser, J., C. Tenke, et al. (1997). "Event-related potential (ERP) asymmetries to emotional stimuli in a visual half-field paradigm." Psychophysiology **34**: 414-426.

Lim, S. and B. Reeves (2007). Responses to Interactive Game Characters Controlled by a Computer Versus Other Players. annual meeting of the International Communication Association. San Francisco, CA.

Nacke, L. and C. A. Lindley (2008). Flow and immersion in first-person shooters: measuring the player's gameplay experience. Proceedings of the 2008 Conference on Future Play: Research, Play, Share, Toronto, Ontario, Canada.

Quinn, R. W. (2005). "Flow in knowledge work: High performance of experience in the design of national security technology." Administrative Science Quarterly **50**: 610-641.

Ravaja, N., T. Saari, et al. (2006). "Spatial prescence and emotions during video game playing: Does it matter with whom you play?" Presence: Teleoperators and Virtual Environments **15**: 381-392.

Ravaja, N., M. Turpeinen, et al. (2008). "The psychophysiology of James Bond: Phasic emotional responses to violent video game events." Emotion **8**(1): 114-120.

Salminen, M. and N. Ravaja (2008). "Increased oscillatory theta activation evoked by violent digital game events." Neuroscience Letters **435**(1): 69-72.

Schmidt, L. A. and L. J. Trainor (2001). "Frontal brain electrical activity (EEG) distinguishes valence and intensity of musical emotions." Cognition and Emotion **15**(4): 487-500.

Steuer, J. (1993). "Defining Virtual Reality: Dimensions Determining Telepresence." Journal of Communication **42**(4): 73-93.

Stevens, R., T. Galloway, et al. (2007). Allocation of Time, EEG-Engagement and EEG-Workload Resources as Scientific Problem Solving Skills are Acquired in the Classroom. Augmented Cognition: Past, Present, & Future. D. Schmorrow, K. Stanney and L. Reeves. Arlington, VA, Strategic Analysis, Inc.

Stevens, R., T. Galloway, et al. (2008). Assessing Student's Mental Representations of Complex Problem Spaces with EEG Technolgies. Human Factors and Ergonomics Society 52nd Annual Meeting, New York, NY.

Zhang, J., R. Zhou, et al. (2011). "The effects of valence and arousal on hemispheric asymmetry of emotion." Psychophysiology **25**(2): 95-103.

CHAPTER 24

Use of Psychophysiological Measures and Interactive Environments for Understanding and Training Warfighter Cross-Cultural Decision Making

Anna Skinner[1], *Jack Vice*[1], *Chris Berka*[2], *and Veasna Tan*[2]

[1] AnthroTronix, Inc.
Silver Spring, MD
{askinner, jvice}@atinc.com

[2] Advanced Brain Monitoring, Inc.
Carlsbad, CA
{chris, vtan}@b-alert.com

ABSTRACT

Cross-cultural decision-making involves perspective-taking and an understanding of inherent biases in attention and interpretation of operationally and emotionally relevant cues within the environment. Training warfighters to make operational decisions within culturally unfamiliar environments requires trainees to not only understand cultural norms and customs in order to interpret intent and predict actions, but to also recognize and account for personal biases. A theoretical framework is proposed for using psychophysiological measures to develop a greater understanding of the processes underlying cross-cultural decision-making, and to provide targeted feedback within interactive virtual environments for training these critical skills.

Keywords: cross-cultural, decision-making, EEG, training, virtual environments

1 BACKGROUND

Effective decision-making relies on accurate perception of the environment and interpretation of environmental cues. US Marines are trained to use a decision process referred to as the OODA loop: Observe, Orient, Decide, Act; in order to successfully execute appropriate actions, Marines are trained to observe the environment, using appropriate cues to develop accurate situational awareness and orient to contextual and circumstantial factors, before a decision can be made and acted upon. "Intuitive" decision-making relies on this process, but at a pace that is too rapid to be decomposed and assessed effectively using standard methods of during- and after-action review (Vice et al., 2011). Additionally, attention and perception are strongly influenced by attitudes, biases, motives, interests, past experience, and expectations. Human decision-makers rely on heuristic principles and pattern matching in order to reduce the complexity of the problem space, particularly when assessing probabilities and predicting outcomes under conditions of uncertainty (Tversky & Kahneman, 1974). Such heuristics are subject to cognitive bias, attribution errors, halo effects, contrast effects, projection, and stereotyping. Thus, while heuristics provide useful mechanisms for reducing cognitive load and facilitating expedited information interpretation and decision-making under conditions of complexity and ambiguity, these cognitive short-cuts often result in systematic errors or deviations in judgment and decision-making.

Research has specifically identified fundamental differences in information processing and reasoning across cultures. For example, it has been shown that East Asians and Westerners utilize qualitatively different cognitive reasoning approaches to problem solving (holistic versus analytic) (Nisbett, Peng, Choi, & Norenzayan, 2001).

Decision-making within unfamiliar foreign cultures and environments presents a particularly complex series of issues as biases in attention, perception, and decision criteria are often introduced. In unfamiliar cultural environments, decision-makers may misinterpret perceptual cues, or apply inappropriate or contextually irrelevant heuristics and decision criteria. Effective cross-cultural decision-making involves perspective-taking and an understanding of inherent biases in attention and interpretation of operationally and emotionally relevant or salient cues within the decision-making environment.

Training warfighters to make operational decisions within culturally unfamiliar environments requires trainees to not only understand cultural norms and customs, but to account for underlying differences in information processing and cognitive biases of others. Trainees must develop the ability to take on the perspectives of both friendly and enemy personnel in foreign cultures in order to accurately interpret intent and predict actions.

Currently, culturally-driven subtle differences in attention, perception, and decision-making, are assessed via indirect measures such as task performance and implicit association tests, or subjective assessments. Sensitive and objective metrics are needed to decompose these processes within the context of immersive and interactive virtual training environments or real world environments.

Psychophysiological measures have the potential to provide important information regarding the interactions between culturally-mediated perception, attention, and decision-making within such contexts, as well as objective metrics of cognitive and emotional states throughout an interaction or training experience, including those associated with perceptual biases.

Additionally, visualization of one's own scan patterns in response to static, dynamic, or interactive stimuli, particularly preceding a decision, can be used to provide trainees with insights into their own thought processes and attentional biases. Research has also demonstrated the effectiveness of feed-forward training of expert eye scan patterns (Nalanagula et al., 2006; Sadasivan, Greenstein, Gramopadhye & Duchowski, 2005). By using feedforward techniques to display the eye-scan patterns of persons from other cultures, trainees can experience immersive perspective-taking and obtain a better understanding of the thought processes that lead to differential decision-making across cultures, enabling U.S. warfighters to more effectively team with indigenous allied forces and to predict the decisions and actions of enemy forces.

As is the case when learning a foreign language, acquiring an understanding of foreign cultural norms and attitudes is done most effectively using techniques of immersion and interaction. Thus, an immersive and interactive testbed platform may provide the ability to collect data identifying differences in information processing and decision-making across cultures and training backgrounds, as well as providing a means for training improved cross-cultural decision-making skills.

2 PROPOSED FRAMEWORK

A framework is proposed for using psychophysiological measures to develop a greater understanding of the processes underlying cross-cultural decision-making, as well as targeted feedback within interactive virtual environments for training these critical skills. This framework involves a series of basic experiments aimed at identification of psychophysiological indices related to cultural bias and decision-making, and integration of the resulting metrics within interactive virtual training environments to enable real-time and post-hoc assessment of perceptual biases throughout training scenarios, as well as customized feedback relevant to the specific biases detected by the training system.

2.1 Psychophysiological Indices of Cultural Bias and Decision Making

Eyetracking and EEG have been shown to provide reliable measures of cognitive workload, engagement (e.g., Berka et al., 2004), and attention allocation (Carroll, Fuchs, Hale, Dargue, & Buck, 2010), as well as cognitive processing changes due to fidelity and stimulus variations within virtual training environments (Crosby & Ikehara, 2006; Skinner, Vice, Lathan, Fidopiastis, Berka, & Sebrechts,

2009; Skinner, Sebrechts, Fidopiastis, Berka, Vice, & Lathan, 2010; Skinner, Berka, O'Hara-Long, & Sebrechts, 2010). Research also indicates that EEG indices are sensitive to emotional valence (positive or negative) and intensity in response to visual imagery (Bradley & Lang, 2007; Berka et al., in press). Additionally, Berka et al., (in press) have identified EEG-based indices of emotional valence in response to emotionally provocative videos.

Decomposing perceptual and decision-making characteristics based on EEG is complex, as multiple factors may elicit similar responses (Jacobs, Hwang, Curran, & Kahana, 2006), particularly when averaged across stimuli and task parameters; however, Ratcliff, Philiastides, and Sajda (2009) demonstrated that a single-trial EEG neurophysiological measure for nearly identical stimuli can be used to differentiate behavioral response times and selections related to the quality of decision-relevant evidence. A recent study conducted by Behneman, Berka, et al., (2012) explored the difference between experts and novices within the context of deadly force judgment and decision-making (DFJDM). The physiological process of making a dire decision showed changes in heart rate variability (HRV) and EEG alpha power spectral density (PSD) suppression. Experts in particular had a significantly lower low-frequency heart rate variability (LF HRV) profile during DFJDM scenarios as compared to rest periods, indicating less influence by the sympathetic nervous system. Additionally experts showed greater suppression and control of alpha power compared to that of the novices and a significantly higher pass rate than novices. These results imply that the experts have a better grasp on the OODA loop, and that the autonomic nervous system and PSDs may help to characterize the neurophysiology of intuitive behavior. The proposed framework seeks to leverage such empirical studies to identify culturally-dependent information processing variations using similar metrics.

The first step is to conduct a series of experiments aimed at differentiating between various types of culturally-dependent biases affecting information processing and decision-making. Both past experiences and formal training introduce significant biases in information processing and decision-making. Past experiences generate the perceptual models and ideals that influence thought processes. Experience separates the familiar from the unfamiliar and contextualizes information, influencing the development of mental models, schema, and scripts. Formal training may influence the formation of perception, attention, and decision-making criteria. For example a Marine sniper scanning an urban scene looks for and detects are very different cues from those of a helicopter mechanic. The mechanic may see the scene and look for possible landing zones as field resources, whereas the sniper looks for possible enemy sniper positions and other specific target indicators within the terrain. The two individuals are processing the scene almost as if they are from two different cultures within a given branch of service. Similarly an Afghan civilian and American civilian would process images of Los Angeles and Jalalabad very differently. Understanding and quantifying these differences would enable more effective cultural training.

The USMC Combat Hunter program of instruction (POI) provides an ideal context in which to examine the proposed framework. This specialized training program was developed in response to increased sniper attacks in Iraq in late 2006. The goal of this program was to train Marines to become hunters, rather than the hunter, emphasizing proactive observation and preemptive action within the context of irregular warfare. Core tenets of the training were developed based on hunting and tracking skills as well as urban streets smarts. Marine commanders observed that both Marines from rural areas who hunted while growing up and Marines from inner-city areas were particularly adept at identifying threats based on observation of environmental cues. The findings of a preliminary research effort completed with support from the I Marine Expeditionary Force (I MEF) and the Marine Corps Warfighting Lab (MCWL), suggested that observational skills are critical to both situational awareness (SA) and tactical decision-making (Carroll, Milham, Champney, Eitelman, & Lockerd, 2007). The Combat Hunter POI emphasizes observation skills in order to increase battlefield SA and produce proactive small-unit leaders with a proclivity for preemptive action. World-class experts in hunting and tracking, as well as seasoned law enforcement personnel from urban areas aided in the development of the course curricula, including practical application exercises. Included in this training is an emphasis on combat profiling, which involves interpretation of complex environmental cues and human behaviors to establish a baseline and identify anomalies or deviations from that baseline within both urban and rural environments.

Within the proposed theoretical framework, a series of controlled empirical studies would be used to identify and diagnose various types of perceptual and decision-making biases. Building on the Combat Hunter conceptual framework, a sample experimental scenario might examine terrain biases (e.g., perceptual biases within urban versus rural terrains) within the context of threat detection. The goal of this experiment would be to identify psychophysiological measures associated with a bias against unfamiliar environments, and to use that metric to differentiate culturally-mediated perceptual and decision-making biases from performance errors related to training such as improper scanning techniques. For example, trainees from rural areas with experience hunting could be compared to trainees from inner-city urban areas. By presenting both sets of trainees with imagery depicting scenes from both urban and rural areas in which potential threats are present and decisions must be made, psychophysiological measures could be used to categorize the physiological response, in addition to behavioral performance, of each cultural group to both familiar and unfamiliar task environments. Eyetracking could be used to assess their scan patterns within both the culturally familiar and unfamiliar terrains, while measures such as EEG power spectral density (PSD), heart rate variability (HRV), and galvanic skin response (GSR) could be used to assess variations in cognitive and emotional state in response to the various task conditions. EEG event-related potentials (ERPs) could be used to further decompose the responses to the various stimulus types (urban versus rural, and threat versus no threat present) in order to identify potential differentiations in the early stages of perceptual processing; later stages of feature extraction, pattern matching, and object recognition; and finally, decision-making and response preparation. The next step in the framework would then be to collect data for trained observers from a foreign culture, as well as unskilled and skilled American warfighters, applying the metrics identified in the initial experiment in order to identify cross-cultural differences in perception and decision-making that can be translated to an interactive training environment.

2.2 Interactive Virtual Environments

A variety of paper-based and computer-based tests have been shown to elicit and assess cognitive biases; however, existing tests may not provide the ecological validity to support transfer of training to dynamically changing real-world events that include rich sources of information and contextual influences. For example, the Implicit Association Test (IAT) provides a means for assessing implicit attitudes related to a wide variety of social issues (Greenwald, McGhee, & Schwartz, 1998).

Additionally, in learning and memory, it is well understood that most learning requires multiple trials under similar, but not identical, conditions in order to ensure generalizability of the learned process. Virtual environments (VEs) provide a means for exposing trainees to multiple training scenarios with subtle or significant variations in a short period of time with little overhead cost or risks.

Perceptual biases can be induced by culturally and emotionally relevant factors, as well as experiential factors, including those encountered in a virtual training environment. A common example of visually-induced bias is that estimation of the distance between an observer and an object is often based on clarity of the object; objects that are seen less sharply are judged to be further away, and thus under conditions of low visibility (i.e., fog, low lighting), distances are often overestimated (Tversky & Kahneman, 1974). This is a very simple example of bias; however, it demonstrates the importance of visual cues in bias and decision-making. Within VEs, high visual fidelity is often costly in terms of development time and resources. Thus, it is necessary to determine which aspects of VE fidelity are critical to the specific VE tasks to be trained and completed. In order to elicit bias responses within the VE, environmental cues represented within the VE may require a certain level of fidelity.

Although extensive theoretical and empirical research has been conducted examining the transfer of training from VEs to real world tasks (e.g., Lathan, Tracey, Sebrechts, Clawson, & Higgins 2002; Sebrechts, Lathan, Clawson, Miller, & Trepagnier, 2003), objective metrics of transfer are limited and there is currently a lack of understanding of the scientific principles underlying the optimal interaction requirements a synthetic environment should satisfy to ensure effective training. Previous research has demonstrated that ERPs are sensitive to even slight variations in visual fidelity within simulated task environments requiring visual discrimination, even in cases in which task performance does not significantly differ across fidelity conditions (Skinner, Vice, Lathan, Fidopiastis, Berka, & Sebrechts, 2009).

Training in VEs also provides novices with experiences from which to draw when making decisions in the real world. By shaping and monitoring these experiences and providing appropriate feedback, it is possible to prevent inappropriate biases from forming or to enable the trainee to recognize such biases and correct for them, either in real-time or during scenario after-after review.

Thus, the proposed framework seeks to incorporate validated psychophysiological metrics within the context of an interactive VE. As described by Carroll, Milham, and Champney (2009) feedforward techniques can be used to enhance perceptual skills training by demonstrating the scan patterns and observation techniques of an expert and highlighting differences between the expert and novice scanning techniques. Additionally, by displaying not only the eye-scan patterns of foreign personnel, but also the environmental interactions of users from the target culture within an explorable virtual task environment (e.g., paths selected, and decisions made), trainees can experience immersive perspective-taking and obtain a better understanding of the thought processes that lead to differential decision-making across cultures, enabling U.S. warfighters to more effectively team with indigenous allied forces and to predict the decisions of enemy forces.

3 CONCLUSIONS

The proposed framework seeks to leverage cutting edge research and technology capabilities in psychophysiological measurement and assessment, as well as interactive virtual training environments and training methdologies. With a method for determining the type and extent of perceptual bias, virtual training environments can dynamically tailor the instruction to each individual by specific biases. The virtual training environment could then guide the trainee through decision-making scenarios most relevant to the targeted biases using feedforward and feedback to train the proper decisions based on specific culturally-relevant criteria. This same framework could also be used to predict an individual's decisions based on real-time bias assessment, providing real-time error mitigation.

REFERENCES

Berka, C., Levendowski, D.J., Cvetinovic, M.M., Davis, G.F., Lumicao, M.N., Popovic, M.V., Zivkovic, V.T., & Olmstead, R.E. (2004). Real-Time Analysis of EEG Indices of Alertness, Cognition and Memory Acquired with a Wireless EEG Headset. *International Journal of Human-Computer Interaction 17(2),* 151–170.

Bradley, M. M., & Lang, P. J. (2007). The international affective picture system (IAPS) in the study of emotion and attention. In J. A. Coan and J. J. B. Allen (Eds.). Handbook of Emotion Elicitation and Assessment. Oxford: Oxford University Press.

Carroll, M., Fuchs, S., Hale, K., Dargue, B., Buck, B. (2010). Advanced Training Evaluation System (ATES): Leveraging Neuro-physiological Measurement to Individualize Training. *Proceedings of I/ITSEC 2010.*

Carroll, M. B., Milham, L., Champney, R., Eitelman, S., & Lockerd, A. (2007). ObSERVE initial training strategies report (Program Interim Rep., Contract No. W911QY-07-C-0084). Arlington, VA: Office of Naval Research.

Carroll, M. B., Milham, L., Champney, R. (2009). Military Observation: Perceptual Skills Training Strategies. Proceedings of the Interservice/Industry Training, Simulation & Education Conference (I/ITSEC).

Greenwald, A. G., McGhee, D. E., & Schwartz, J. L. K. (1998). *Measuring individual differences in implicit cognition: The implicit association test. Journal of Personality and Social Psychology, 74,* 1464-1480.

Jacobs J, Hwang G, Curran T, Kahana MJ (2006) EEG oscillations and recognition memory: theta correlates of memory retrieval and decision making. NeuroImage 15:978–987.

Nisbett RE, Peng K, Choi I, & Norenzayan A. (2001). Culture and systems of thought: holistic versus analytic cognition. *Psychol. Rev., 108,* 291–31.

Nalanagula, D. Greenstein, J. S., & Gramopadhye, A. K (2006). Evaluation of the effect of feedforward training displays of search strategy on visual search performance. *International Journal of Industrial Ergonomics, 36(4),* 289-300.

Ratcliff, R., Philiastides, M. G., & Sajda, P. (2009). Quality of evidence for perceptual decision making is indexed by trial-to-trial variability of the EEG. *Proc Natl Acad Sci U S A 106,* 6539–6544.

Sadasivan, S., Greenstein, J.S., Gramopadhye, A.K., & Duchowski, A.T. (2005). Use of eye movements as feedforward training for a synthetic aircraft inspection task. In *Proceedings CHI '05*, April 2-7, 2005, Portland, OR, ACM.

Skinner, A., Sebrechts, M., Fidopiastis, C. M., Berka, C., Vice, J., & Lathan, C. (2010). Psychophysiological measures of virtual environment training. In P. E. O'Connor & J. V. Cohn (Eds.), Human performance enhancement in high risk environments: Insights, Developments, & Future Directions from Military Research (pp. 129-149). Santa Barbara, CA: Paeger.

Skinner, A., Vice, J., Lathan, C., Fidopiastis, C. M., Berka, C., and Sebrechts, M. (2009). Perceptually-Informed Virtual Environment (PerceiVE) Design Tool In D. D. Schmorrow, I. V. Estabrooke, and M. Grootjen (Eds.), Augmented Cognition, HCII 2009, LNAI 5638 (pp. 650–657). Berlin: Springer-Verlag.

Stanovich K.E. & West, R.F. (2000). Individual differences in reasoning: Implications for the rationality debate? *Behav Brain Sci., 23(5):*645-65.

Tversky A. & Kahneman, D. (1974). Judgment Under Uncertainty: Heuristics and Biases. *Science, 185*:1124-30.

Vice, J.M., Skinner, A., Berka, C., Reinerman-Jones, L., Barber, D., Pojman, N., Tan, V., Sebrechts, M.M., Lathan, C.E.: Use of Neurophysiological Metrics within a Real and Virtual Perceptual Skills Task to Determine Optimal Simulation Fidelity Requirements. *In HCI (13)*(2011) 387-39.

CHAPTER 25

Cross Cultural Training through Digital Puppetry

Angel L. Lopez, Charles E. Hughes, Daniel P. Mapes, Lisa A. Dieker

University of Central Florida
Orlando, FL, USA
ceh@cs.ucf.edu

ABSTRACT

Digital puppetry refers to the control of virtual characters by human puppeteers. The intent of such an approach is to provide mediation for human-to-human communication. The intricacies of such human interaction include non-verbal as well as verbal communication, the latter being necessary but not sufficient to create an aura of believability. In fact, nonverbal cues are essential to the message a person conveys, and are as much nuanced by culture, as are language and dress.

Characters that exhibit believable, natural behaviors are key elements in an experience designed to trigger emotional/empathic responses. Although expressive behaviors can be scripted for many circumstances, direct interactivity with the user is difficult for computational approaches, when conversation is open-ended. In fact, this complex level of communication seems presently doable only by humans.

The focus of this paper is on the use of such puppetry to assist individuals and organizations to better understand people of other cultures and to discover recondite biases that may be negatively affecting their abilities to assess and communicate with diverse individuals. As such, our employment of digital puppetry is primarily centered on using it to place people in unfamiliar and often uncomfortable situations. The purpose of these scenarios is to achieve positive outcomes, even when the short-term goal is to bring about discomfort. Obvious uses of puppetry in cross-cultural scenarios include preparation for deployment into a foreign country as a soldier or businessperson. Less obvious is preparation for a career as a police officer, counselor, or teacher, the latter being the central application discussed here.

Keywords: avatar, cross cultural, digital puppetry, teacher preparation, virtual reality

1 GEPPETTO AND THE TEACHLIVE SYSTEM

1.1 Digital Puppetry

Characters that exhibit believable, natural behaviors are key elements in virtual experiences designed to trigger emotional/empathic responses. Although it is possible to script expressive behaviors for many circumstances, direct interactivity with the user is a problem for computational approaches, when conversation is open-ended. This complex level of communication seems presently doable only by humans. The intricacies of human-to-human interaction include non-verbal as well as verbal communication, the latter being necessary but not sufficient to create an aura of believability in virtual environments. In fact, nonverbal cues are essential to the message a person conveys, and are nuanced as much by culture, as are language and dress (Ekman, 1971; Ekman, 1993). Preparing a person for another culture, understanding one's recondite biases in unfamiliar contexts, or bringing back memories that are couched in another culture requires more than scenery, clothing or automated dialogue. Thus, a major challenge in creating cross-cultural training and uncovering unrecognized cultural biases is getting the affective component right in any type of computer-based environment. In environments that are "non-human" such as game-based environments or even immersive simulated environments finding a way to represent cultural aspects and to use those attributes to discover unintended side effects related to bias form the basis of this paper.

1.2 Geppetto

The Geppetto system (Mapes, Tonner & Hughes, 2011) provides an infrastructure for digital puppetry, supporting multiple forms of user interfaces, ranging from motion capture (literal control) to marionette-style manipulation (figurative or gestural control). The system provides means for human-to-human communication that include non-verbal, as well as verbal communication. Geppetto's support for nonverbal cues that are triggered by but are not mimics of the puppeteer's gestures is essential to the messages being conveyed. This system allows simulated environment to be used to prepare people for cultures with which they are not familiar, as well as allows for the testing of hypotheses about the effects of these perceived and actual cultural differences on user behaviors.

The Geppetto user interaction paradigm (Figure 1) (gestural control) places low cognitive and physical loads on the puppeteer. It employs a communication protocol that results in low network bandwidth requirements, supporting the transmission of complex behaviors between geographically remote sites. Geppetto incorporates a process for the iterative design and development of the appearance and behaviors of virtual characters. This process involves artists, puppeteers/interactors (Wirth, 1994), programmers and, most importantly, the requirements of users of the resulting system. Components of this include model design and creation, and verbal and non-verbal behavior selection and implementation (puppeteered and automated). Geppetto also provides support for session recordings that can be used in after-action review and to seed new automated behaviors.

The Geppetto system uses the software concept of a "sandbox", which allows for a flexible and ever-changing environment that can shift and change depending upon the attributes wanted and the characteristics of the avatars placed within the sandbox. This approach isolates core elements of the system from experimental components. The core elements are well-tested and provide a stable set of common services. The experimental extensions can implement any type of environment that involves human-to-human interaction, mediated by digital puppetry. This plasticity allows a wide range of applications as evidenced by existing projects involving teacher education (the focus of the TLE TeachLivE™ discussed in the next section), cross cultural communication (Barber et al., 2010), peer pressure resist/avoid strategies (Wirth et al., 2011), interviewer training and communication skills development for young adults with autism.

Figure 1 Puppeteer controlling students in virtual classroom. Uses head tracking and Hydra. Note Skype window that allows puppeteer to observe teacher's non-verbal behaviors.

1.3 TLE TeachLivE™

Imagine walking into a room that looks like a middle-school classroom, with props, whiteboards and, of course, children. This room is not a traditional classroom; it is a virtual setting and the students in the classroom are avatars. The virtual students may act like typically developing or not-typically developing students, depending on the objectives of the experience. Participants can interact with students and review previous work, present new content, provide scaffolding or guided practice in a variety of content areas, and monitor students while they work independently. In an environment like this, prospective teachers can learn the instruction and pedagogical skills needed to become effective teachers, and practicing teachers can hone and refine their skills. Using the Geppetto system teachers can do all of this without putting actual children in harm's way and without facing consequences of failure in a safe, controlled environment. Issues of remediation and open reflective discourse can occur in real time, allowing participants to step away and then immediately return to the classroom after receiving some coaching with the virtual students having no memory of their missteps. Consequently, successful and learned virtual rehearsal experiences for each participant that instills both confidence and builds reflective teaching practices are created.

The above scenario details what we provide with the TLE TeachLivE™ Lab, which comprises the TLE TeachLivE™ system (built on top of Geppetto), and a set of pedagogies, content and processes, created as a structure for universities to positively impact teacher preparation (Figure 2). The environment delivers an avatar-based simulation intended to enhance teacher development in targeted content and/or pedagogical skills. Teachers have the opportunity to experiment with new teaching ideas in the TLE TeachLivE™ Lab without presenting any danger to the learning of "real" students in a classroom. If a teacher, novice or experienced, has a bad session, he or she can reenter the virtual classroom to teach the same students the same concept or skill. In real classrooms, students might get bored and become difficult to manage when an instruction routine is repeated. Moreover, in real classrooms only one or two teachers can practice an instructional routine with a group of students. In the TLE TeachLivE™ Lab, instructional routines may be repeated with an individual teacher or across several teachers using the same instructional context until the routine is mastered. The instruction context may then be changed systematically to examine how participants respond to a changing classroom environment.

Figure 2 Virtual class of five students. Left screen shows virtual students on task; right shows them distracted, except for student in front row left who is presently controlled by puppeteer.

2 APPLICATION OVERVIEW

Our employment of digital puppetry is primarily centered on using it to place people in unfamiliar and often uncomfortable situations. The purpose of these scenarios is to achieve positive outcomes, even when the short-term goal is to bring about discomfort. Obvious uses of puppetry in cross-cultural scenarios include preparation for deployment into a foreign country as a soldier or businessperson. Less obvious is preparation for a career as a police officer, counselor, or teacher. Perhaps even less obvious is employment for prolonged exposure therapy after traumatic events, whether in war or in civilian life.

This paper will present details on the use of the TLE TeachLivE™ Lab in an experiment to discover unrecognized biases that negatively affect the performance of teachers and their engagement with students. Viewing teaching as a cross-cultural experience may at first seem a stretch of the imagination, but further introspection makes it clear that teachers, especially of middle school students, have vast cultural differences with their students (Ferri & Conner, 2005). The first and most noticeable difference is ethnicity, with over 87% of today's classroom teachers being middle-class white females, while classrooms each year become increasingly more diverse; the second is geographical with the majority of schoolteachers living in rural areas or the suburbs versus the inner city; and the last is generational. Digital puppetry-based simulations that account for this range of differences provide a safe place for self-discovery and rehearsal – safe for the teachers and for the real children they will work with upon graduation.

3 CULTURAL BIAS AND HIGH SCHOOL DROPOUT RATE

More than one in every three Americans is a minority or something other than a Non-Hispanic single-race white. The current number of minorities in the United States now exceeds one hundred two million. While the Hispanic population continues to climb steadily, the dropout rate among Hispanic youth is alarmingly high. A report from the U.S. Census Bureau (2008) indicated that, of youth between ages sixteen and twenty-four, Hispanics accounted for forty-one percent of all current high school dropouts. Hispanics are listed as having the highest dropout rate of any cultural group and yet represent the largest minority population in the U.S.

Rodriguez (2008) asserts that the reason Hispanic youth are dropping out of high school is because the system is discriminatorily pitted against them and there is a scarcity of social policies in place. This discrimination is the foundation for this study related to potential bias against Hispanic males by both placing the label of emotional disturbance upon this population with the resulting lowered academic expectations and outcomes due to this label. Nelson (2000) reported that fifty to sixty percent of students labeled ED drop out of high school. Data retrieved from ideadata.org reported that from 2002-2007 the Hispanic population in the United States saw a two percentage point increase in the number of students labeled ED. The identification of Caucasian students fell by almost three percentage points while the increase in all other demographic groups was negligible moving anywhere from a hundredth to a tenth of a percentage point.

The purpose of this study is to determine the potential for educator bias of pre-service teachers (predominantly white and female) on their interactions with virtual secondary male Hispanic students identified with ED compared to those without an identified label of ED. In this true-experimental mixed methods design with a weightless control study, data were collected on two non-equivalent groups of six participants for a total of twelve participants. All participants were undergraduate students pursuing degrees in secondary education with only field or student teaching experience. All participants were enrolled in an exceptional student education college course and were randomly assigned to either a control or experimental group. Each participant took two online implicit association tests via Harvard University's Project Implicit®, one on race and the other on disability. In addition, participants took a baseline survey via the Understanding Prejudice webpage.

After these initial tasks were completed, participants interacted with virtual students within the TLE TeachLivE™ Laboratory four times over a period of two weeks. All participants rated how they expected individual students to perform within each simulated classroom experience based solely on a brief description of that student. After each experience, participants completed a brief reflection of their experience that was video-taped – that process is hereafter referred to as an After Action Review (AAR). After the second and fourth visits in the TLE TeachLivE™ Lab the experimental and control groups respectively viewed and discussed modules on emotional disturbances and cultural competence produced by Vanderbilt University's IRIS Center and housed online on the Department of Education's IDEA Partnerships webpage called the Learning Port. During their final week of the study AAR questions were given in a written format to both the control and experimental groups and participants once again completed the baseline survey given in week one. The control group completed the same modules and discussions conducted by the experimental group.

3.1 Design of Experiment

The first goal (Question One) was to determine, within a simulated classroom environment, to what extent the identification and swapping of the label emotionally disturbed on two virtual secondary Hispanic male students increase, decrease or maintain the teacher's frequency of:

a) Positive comments
b) Negative comments
c) Proximity
d) Cultural statements

The second goal (Question Two) was to determine to what extent providing a pre-service teacher with a module on Cultural Linguistic Diversity and a module on Emotional Disturbance would influence the frequency of each of the above five factors with secondary Hispanic male students identified with and without emotional disturbances.

We used multiple measures to evaluate the influence of bias on randomly selected secondary pre-service teachers' interactions with two virtual male secondary Hispanic students. The quantitative data gathered included a pre-test that allowed for participants to self-report bias, a Likert scale ranking of perceived virtual student performance, frequency counts of the dependent variables collected through video recordings and a post-test.

3.2 Quantitative Analysis

Our team used video recordings to observe participant interactions with virtual students within the TLE TeachLivE™ Lab. For each session, one alternating student was identified as having an emotional disturbance; however, behaviors for the alternating Hispanic male secondary student remained consistent throughout all experiences. In research question two, each dependent variable was analyzed pre-post with the independent variable being time.

Inter-rater reliability on all video recordings for both live interactions and the intervention were set at 80% or higher agreement. The observed videos were viewed and frequency counts for each dependent variable were compiled then compared with two research assistants. Agreement between coded video for all parties was at 100% on each recording. Discrepancies in agreement occurred for multiple videos, due to a hard drive crash and audio distortions causing the agreement to fall to less than the 80%. Raw footage from days and/or sessions where discrepancies occurred was given to the research assistants and a 100% agreement between the primary researcher and the research assistants was established. Video recordings that required additional viewing included both control and experimental participants. Overall fidelity of treatment was established at or above 80% and 100% of the videos were viewed. Research questions one and two segment d, which looked for cultural statements, were eliminated from analysis; as no such statements were observed. Frequency counts for each listed dependent variable were tallied and entered into SPSS for statistical analysis as detailed below.

Research Question One

A multivariate analysis of variance was used to compare differences between interactions of pre-service teachers and students with and without the ED label and between the other individual characteristics of the identified students on positive comments, negative comments, and proximity.

Hotelling's Trace was not statistically significant for interactions between the ED label and Student (F (1, 11) = 2.322, p=0.144) or ED Label (F (1, 11) = .688, p=0.582). However there was a statistically significant effect for Student, (F (1, 11) = 4.838, p=0.028).

Although there was no effect for ED label participants gave more positive comments, negative comments, and used proximity more often with Marcus than with Vince suggesting that there was a difference between the individual characteristics of the students and participant interactions in the TLE TeachLivE™ Lab

Research Question Two

A multivariate analysis of variance was used to compare differences between a) students with and without the ED label, b) other individual characteristics of the identified students, and c) pre and posttest measures on positive comments, negative comments, and proximity.

Hotelling's Trace was not statistically significant for interactions between the ED label, Student, and Time ($F (1, 11) = 1.098$, $p=0.399$); ED label and Student ($F (1, 11) = 1.952$, $p=0.192$); ED label and Time ($F (1, 11) = 0.263$, $p=0.850$); Student and Time ($F (1, 11) = 1.577$, $p=0.262$). Moreover, there were no significant differences for main effect of ED Label ($F (1, 11) = 1.966$, $p=0.190$); or time ($F (1, 11) = 0.754$, $p=0.547$). However there was a statistically significant effect for Student, ($F (1, 11) = 4.037$, $p=0.045$).

Although there was no effect for multiple dependent variables, participants gave more negative comments, and used proximity more often with Marcus than with Vince, again suggesting that there was a difference between the individual characteristics of the students and participant interactions in the TLE TeachLivE™ Lab. Interactions were significant for Negative Comments and Proximity when looking at Student.

4 CONCLUSIONS

Human-to-human communication is complex and includes both verbal and non-verbal communication. The Geppetto system allows for a training and virtual rehearsal environment whereby characters/avatars that exhibit believable, natural behaviors interact with participants to create virtual experiences designed to trigger emotional/empathic responses. This augmented reality environment can be used for multiple purposes including to prepare people for cultures with which they are not familiar, as well as for the testing of theories about the effects of perceived and actual cultural differences on user behaviors.

Understanding culture and access to cross cultural training are important in many areas of commerce and the military. Training customer service professionals to successfully navigate a foreign market or military personnel to operate in countries they are unfamiliar with are just a couple of applications the plasticity of this system allows. An additional culture that this research team identified as consistently divergent is the culture within the educational institutions that serve to educate the youth of our country. The cultural bridge as stated by McKown and Weinstein (2008) that exists between the learner and the teacher is a critical topic to consider. McKown and Weinstein see the relationship between student and teacher as one of contention. These authors further state that the divide perseverates within the educational institutions that produce teachers. Looking at the demographic make-up of professional educators Picower (2009) reports the majority of teachers are female and near ninety percent are Caucasian. According to a report compiled for the National Center for Education Statistics (NCES) in 2009, about forty-two percent of all students in K-12 schools come from a minority background.

The majority of the children in school today are from minority populations; with minority student populations exceeding seventy-five percent in high poverty areas. Consequently, many educators who are primarily Caucasian, female and come from middle-class backgrounds have very little in common with those they teach. McKown and Weinstein (2008) share that as a result of this cultural divide and the propensity for dominance of both being white and female in the field of education increases the potential for racism, classism and unjust mistreatment of students classified within any minority rank.

The results of this study, which showed statistically significant interactions between the identified virtual students and the pre-service teachers participating in this study, clearly suggest that additional research and inquiry are needed. The cultural divide between teachers and students is leading to disparities in educational attainment. This educational gap ties directly to social mobility, as educational achievement is a major predictor for future life successes including job opportunities (Stamps and Bohan, 2006).

ACKNOWLEDGMENTS

The material presented in this publication is based on work supported by the National Institutes of Health (R15NR012189), the National Science Foundation (CNS1051067) and the U.S. Department of Education Office of Special Education Leadership Program. Any opinions, findings, and conclusions or recommendations expressed in this material are those of the authors and do not necessarily reflect the views of the National Institutes of Health, the National Science Foundation or the Gates Foundation.

The authors would like to thank their colleagues at the UCF College of Education and the Institute for Simulation & Training's Synthetic Reality Laboratory. We are especially indebted to the support and inspiration provided by Mike Hynes, Jacqueline Rodriguez, Eric Imperiale and Edward (Chip) Lundell.

REFERENCES

Barber, D., S. Schatz and D. Nicholson. 2010. AVATAR: Developing a military cultural role-play trainer. *Proceedings of 1st International Conference on Cross Cultural Decision Making(CCDM).* July 17-20, CRC Press.

Dieker, L., M. Hynes, C. E. Hughes and E. Smith. 2008. Implications of Mixed Reality and Simulation Technologies on Special Education and Teacher Preparation. *Focus on Exceptional Children*: 1-20.

Ekman, P. 1971. Universals and cultural differences in facial expressions of emotion. *Nebarska. Symposium on Motiation* 19:207-283.

Ekman, P. 1993. Facial expression and emotion. *American Psychologist, 48,* 384–392.

Ferri, B. A. and D. J. Connor. 2005. Tools of exclusion: Race, disability, and (Re)segregated education. *Teachers College Record,* 107(3), 453-474.

Mapes, D. P., P. Tonner and C. E. Hughes. 2011. Geppetto: An Environment for the Efficient Control and Transmission of Digital Puppetry. *Lecture Notes in Computer Science, Volume 6774,* Springer-Verlag, Heidelberg: 270-278.

McKown, C., & Weinstein, R. S. (2008). Teacher expectations, classroom context, and the achievement gap. *Journal of School Psychology,* 46(3), 235-261.

National Center for Education Statistics. (2009). The condition of education 2009 (NCES Publication No. 2009-042). Washington, DC: U.S. Government Printing Office.

Nelson, J. R., and M. L. Roberts. 2000. Ongoing reciprocal teacher-student interactions involving disruptive behaviors in general education classrooms. *Journal of Emotional & Behavioral Disorders*, 8(1): 27-37, 48.

Picower, B. (2009). The unexamined whiteness of teaching: How white teachers maintain and enact dominant racial ideologies. Race, *Ethnicity & Education*, 12(2), 197-215.

Rodriguez, L. F. (2008). Latino school dropout and popular culture: Envisioning solutions to a pervasive problem. *Journal of Latinos & Education, 7*(3), 258-264.

Stamps, K., & Bohon, S. A. (2006). Educational attainment in new and established latino metropolitan destinations. *Social Science Quarterly (Blackwell Publishing Limited), 87,* 1225-1240.

U.S Department of Commerce Bureau of the Census. 2008. *Current population survey, various years, Dropout rates.*

Wirth, J.., A. E. Norris, D. Mapes, K. E. Ingraham and J. M. Moshell. 2011. Interactive Performance: Dramatic Improvisation in a Mixed Reality Environment for Learning. *Lecture Notes in Computer Science, Volume 6774,* Springer-Verlag, Heidelberg: 110-118.

Wirth, J. 1994. *Interactive Acting: Improvisation, and Interacting for Audience Participatory Theatre*. Fall Creek Press, Fall Creek Oregon.

CHAPTER 26

Adding GeoNames to BEN: Event Extraction in Real-Time and Space

Kacia Strous

Social Science Automation
Hilliard Ohio, USA
Kacia.strous@socialscience.net

ABSTRACT

A human understands that Susan Rice in the sentence "Susan Rice opposed the UN veto to condemn Syria" is a person. However, an automated coding scheme using a location database would tag Susan as a reference to As Susan, an area within the Aleppo Governorate of Syria. When developers at Social Science Automation (SSA) added the open source location database GeoNames to the automated coding scheme Behavior and Events from News (BEN) human names doubling as locations and several other issues required resolutions before BEN events could receive location identification.

Keywords: location identification, event data, data mining

1. BEHAVIOR AND EVENTS FROM NEWS (BEN)

BEN is a user-built coding scheme that was developed by SSA and is run on the text analysis tool Profiler Plus to monitor news feeds for events of interest. BEN evolved from the CAMEO codebook developed by Dr. Phil Schrodt (Penn State Event Data Project), and was developed to use the full text of a document to extract event information. Each extracted event is assigned a 4-digit number that indicates the corresponding BEN category. Categories are separated into overarching groups of conflict and cooperation, and further specified within each. BEN began in an attempt to answer the question "who did what, to whom?" That question was expanded to "when", and most recently, "where".

Location identification emerged as an issue for BEN developers during the DARPA/ICEWS project, which sought to monitor and forecast international crises. Although it was simple to identify countries and large cities, BEN developers quickly discovered a large portion of events reported in the news occurred in small neighborhoods, with an endless variety of possible spellings. Once the task of a user adding each location to a coding scheme was deemed insurmountable, the BEN team turned to the possibility of using the GeoNames location database.

2. GEONAMES ADDITION TO PROFILER PLUS

GeoNames was chosen because it is an open source and well-maintained database of locations. GeoNames contains millions of locations, and is frequently revised. Attempting to convert the database to a coding scheme would be slow and also require the user to update the coding scheme with each revision to GeoNames. For efficiency, referral to the database is instead included as a prebuilt component of Profiler Plus. Location information for any country is downloaded from the GeoNames website as a zipped file. The user selects the country files they want to include, and Profiler Plus calls on the GeoNames module to match each token in a document against the country files to identify potential locations. Users may search a single country, multiple countries, or every country available in the GeoNames database.

Once GeoNames locations were added several issues were discovered, namely:

1. Common words that are locations
2. Differentiating between humans and locations
3. Multiple GeoNames results
4. Secondary location references
5. Locations not in GeoNames
6. No location provided

2.1 Common words that are locations

The most visible and immediate problem was that nearly every word was labeled a location. For instance, the text "as" is labeled a location because of a small beach in Egypt named As. Similarly, "and", "of", and "the" all receive location labels in certain countries. This was solved by creating rules in a coding scheme to limit potential GeoNames locations to only words that begin with an uppercase letter. This depends on proper capitalization of locations, which for the news articles used to gather BEN data has not proven an issue, but if location identification were desired for less formal texts, it could become a concern.

2.2 Differentiating between humans and locations

Another common occurrence was human names that double as location names. Bashar al-Assad receives a location code because Bashar is a city within the Idib governorate in Syria. If Bashar al-Assad has sent troops to Homs, it is important that there is not a false positive in this sentence, so that Homs can be correctly identified as the location of the event, not Bashar. This issue was addressed by using the HumanMaleFemale scheme developed by SSA to identify humans. If the text is labeled human, or if surrounding text has contextual clues like a title, it is not labeled a location. Correctly identifying humans and locations is a tricky task for developers. To limit false locations, coders erred on the side of over-identifying humans. Because of this, there are locations that are not captured because they have an incorrect human label. This will therefore be a focus of future development.

2.3 Multiple GeoNames results

A frequently-mentioned problem of automated location identification is that many locations throughout the world share the same name. For example, Homs is a governorate in Syria. But Homs is also the name of a district and city in Syria, as well as the name of a city in Libya. GeoNames results in Profiler Plus were built to include every potential location to which text may refer. Figure 1 shows a sentence prior to GeoNames, and Figure 2 demonstrates how GeoNames adds tokens to provide the user with all possible locations. The highlighted text in Figure 2 is every potential location the text of Homs could be.

Users choose which countries to include in a search, so if the user only selects Syria, the Libyan result will not be included. However, most users desire identification in multiple countries, and as demonstrated in the Homs example, even within a single country one word can refer to any number of possible locations. To solve the problem of multiple locations, BEN uses the SSA-developed Location Identification Disambiguation (LID) coding scheme.

Ind...	Class	Lemma	Pos	Tense	Text	Trut...	Slot3	Slot7	Slot8	Slot9	Slot13
1	BOD	(bod)									
2	BOP	(bop)									
3	BOS	(bos)					en				
4		5	number	1	5						number
5	actor	protestor	noun		protestors	subject		ProtestActor	common		Not-GeoNames
6		be	be	past	were						Not-GeoNames
7		kill	verb	past	killed						Not-GeoNames
8		in	particle		in		244			prep-loc	
9		homs	noun	default	Homs	subject				location	
10		today	adv		today						

Figure 1: A sentence viewed in Profiler Plus prior to GeoNames

Ind...	Class	Lemma	Pos	Tense	Text	Trut...	Slot1	Slot2	Slot3	Slot4	Slot5	Slot6	Slot7	Slot8
1	BOD	{bod}												
2	BOP	{bop}												
3	BOS	{bos}							en					
4		5	number	1	5									
5	actor	protestor	noun		protestors	subject							ProtestActor	common
6		be	be	past	were									
7		kill	verb	past	killed									
8		in	particle		in				244					
9		homs	noun	default	Homs	subject							1	
10		homs	noun	default	Al Khums	subject	P.PPLA	LY	82	seat of a first-order administrative division	Libya		4	2219905
11		homs	noun	default	Muhafazat Hims	subject	A.ADM1	SY	11	first-order administrative division	Syria	Homs	4	169575
12		homs	noun	default	Mintaqat Hims	subject	A.ADM2	SY	11	second-order administrative division	Syria	Homs	4	169576
13		homs	noun	default	Hums	subject	P.PPLA	SY	11	seat of a first-order administrative division	Syria	Homs	4	169577
14		today	adv		today									

Figure 2: A sentence viewed in Profiler Plus after GeoNames

If there are multiple location candidates from the same country, then LID uses category information provided by GeoNames. For example, one result for Homs is in the A category and receives a specific code of ADM1 as a first-order administrative division. Another result for Homs is in the P category and receives a specific code of PPLA as the seat of a first-order administrative division. Rules were made within the LID coding scheme to establish a hierarchy of these categories. If any surrounding text indicates which category is correct by using words like district, province, or city, that information is used to choose the correct location. If there are no surrounding indicators, the hierarchy is used to pick the largest location. For the Homs example, the A category is larger than the P category, so the last result in Figure 2 is eliminated. For the remaining two results, ADMI1 is a larger location than ADMI2, so the result chosen in this case is Muhafazat Hims.

If there are multiple locations from different countries, then the LID coding scheme is built to search surrounding text for contextual clues to determine the correct country. If the surrounding text does not offer useful information, the larger category of the locations is selected. For Figure 2, the Syrian result would be chosen over the Libyan location as it is a member of the larger category. Finally, if the locations are the same category the first location is chosen.

2.4 Secondary location references

After locations are initially identified in news articles, the following sentence will often use a secondary location term such as city or town, rather than identify the location by name again. To capture these references rules were created in the LID coding scheme to use the GeoNames location hierarchy to correspond with the generic references to city, town, etc. Therefore, if the city of Homs is identified it receives a code of PPLA. When the text "city" next appears LID will identify the city as the last mentioned PPLA location.

2.5 Locations not in GeoNames

Although GeoNames includes millions of locations, there are invariably still locations not within the GeoNames database. There is no perfect way to capture these locations, but we attempted to do so in the LID coding scheme by using contextual information in the sentence. If the city of Circleville appears and receives no location code from GeoNames, then based on the provided information Circleville is identified as a city, and the last mentioned country code is used to identify the country. Rules within the LID coding scheme contain specific requirements that the first letter of the tagged location must be uppercase to prevent capturing non-specific references such as city of dreams. This does lead to some falsely identified locations, however, so it was built as a component of the coding scheme that can be excluded if desired.

2.6 No location provided

If an event occurs in the lead sentence of an article it sometimes does not contain a location. For example, if the lead sentence is "Syrian troops fired on protestors today" the reader may infer that this event happened in Syria, but the same should not be inferred by BEN, because Syrian troops could just as easily be acting in Lebanon. In this case the event is captured, but cannot be tied to a location.

3. CASE STUDY: 2011 SYRIAN UPRISING

The implementation of GeoNames within BEN is illustrated with data from the 2011 Syrian protests. Documents pertaining to the Syrian conflict from January 2011 through June 2011 were downloaded from BBC News, Reuters, and the New York Times. Due to time constraints, only three BEN categories were used for this project: 6000 Social Conflict, 9000 Law Enforcement Conflict, and 9500 Military Conflict. The measured accuracy of these categories was not blind, and at the time of writing, the three categories had a combined accuracy of 72%.

The Syrian protests began in mid-March 2011 following the arrest of youths in Daraa for anti-regime graffiti. Their arrest caused an eruption of protests and events that led to the on-going Syrian uprising. Figure 3 displays the number of mentions of specific events throughout the measured 6-month period. The event data is consistent with reports of the Syrian Revolution. According to timelines provided by Wikipedia (Timeline of the 2011-2012 Syrian Uprising), in this time period the largest protests took place in April. Protests dropped off in May when the Syrian government increased crackdowns against protests, and then picked back up in June. Three sources were used because of the assumption that news outlets may report different details of events, or report events in a different manner. However, because multiple sources were used single events are likely reported several times, and the results are not discrete events, but a collection of mentions in media.

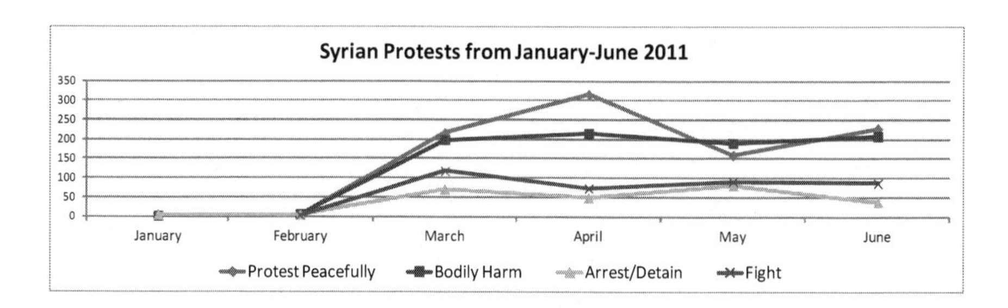

Figure 3: Syrian protest data from a six month period; data from BBC News, Reuters, and the New York Times

To attempt to measure discrete events, one news source, Reuters, was chosen to measure a single month. Figure 4 shows the results of this month. The March Reuters data looks at the first month of the protests by week and, according to the Wikipedia timeline of events (Timeline of the 2011-2012 Syrian Uprising), corresponds with the major events that took place. Two upticks in the category of Protest Peacefully match the dates of the two largest protests of the month, known respectively as the Day of Rage protests, and the Friday of Glory protests. The category of Bodily Harm shows the first date Syrian troops clashed with protestors and, in combination with the Fight with Small Arms category, shows the large amount of violence that took place during the Friday of Glory protests.

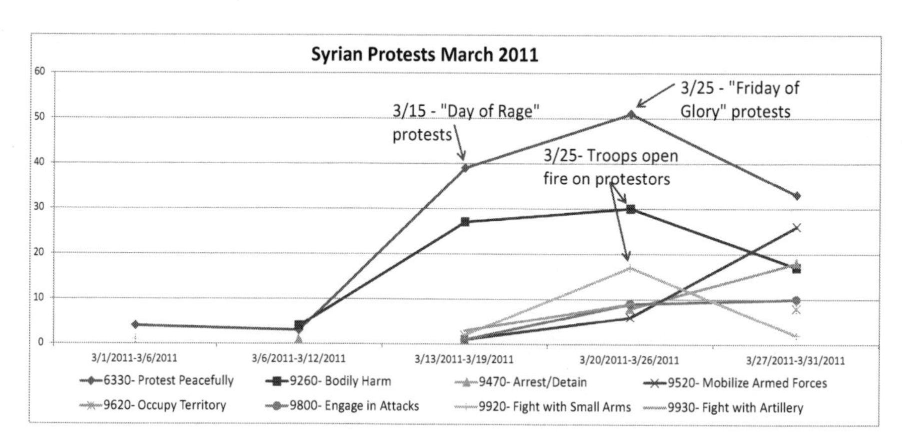

Figure 4: March results from Reuters

To test the usefulness of location identification, the Syrian protest results were broken down by location. Figures 5-8 show each month of data during the conflict by specific location. Unsurprisingly, in the month of March Daraa is the location of the most protests and violence, as it was the starting point of the uprising. In the month of April, Damascus, Homs, and Aleppo emerge as locations with a large number of protests. Interestingly, in April the Mobilize Armed Forces category shows that Syrian troops moved to Homs in April. In the following month, Homs shows the largest amount of Bodily Harm and Fighting, corresponding to the "Siege of Homs" that took place in May. In the month of June, after being the subject of a crackdown by the Syrian military, the number of protests in Homs drops off sharply. By viewing the results by location, we are able to track this series of events in Homs.

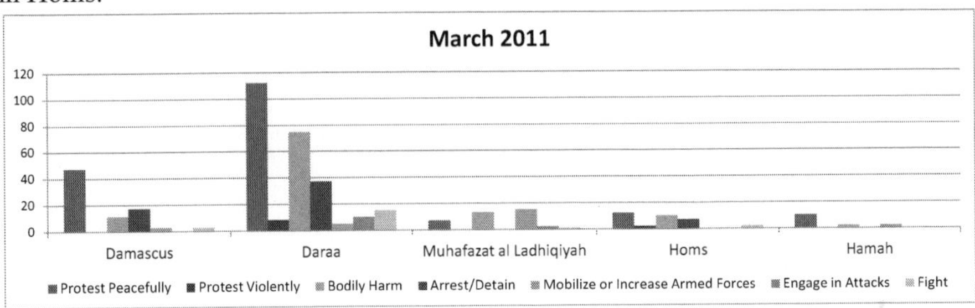

Figure 5: Syrian protest data from BBC News, Reuters, and the New York Times in March 2011 by location

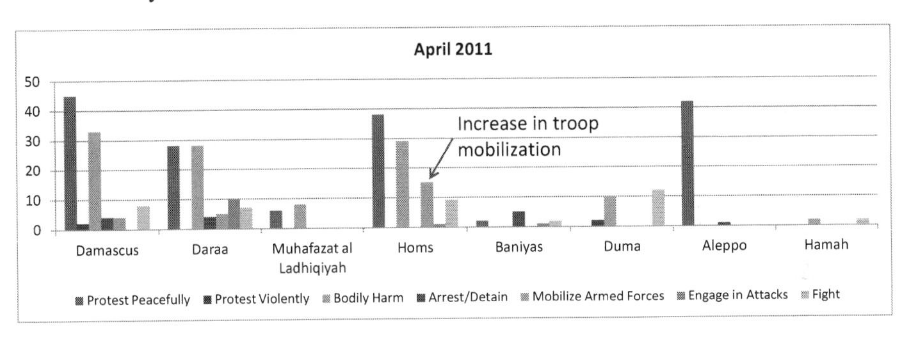

Figure 6: Syrian protest data from BBC News, Reuters, and the New York Times in April 2011 by location

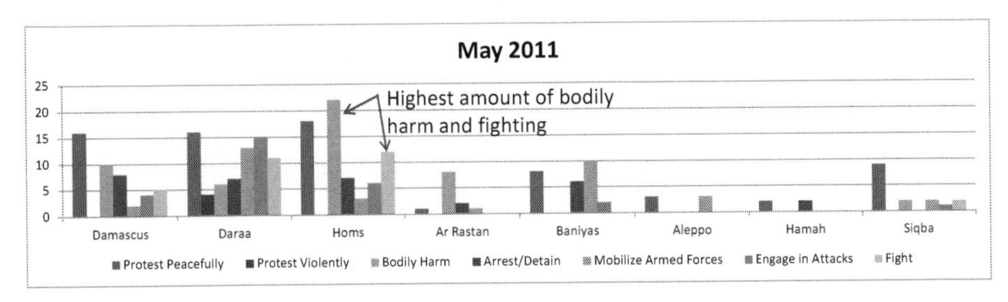

Figure 7: Syrian protest data from BBC News, Reuters, and the New York Times in May 2011 by location

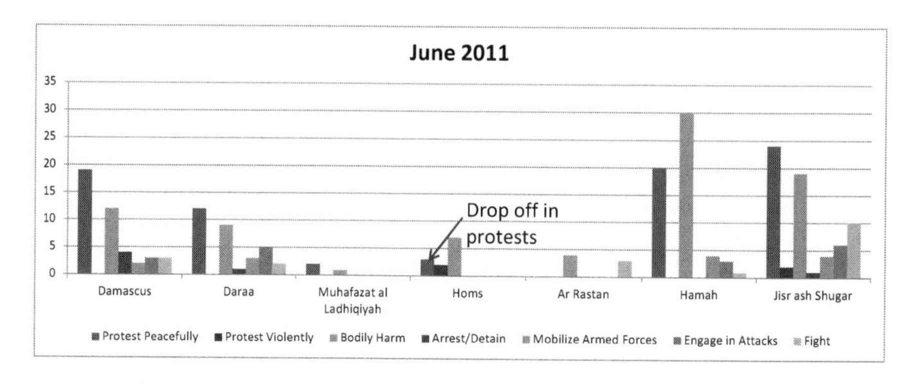

Figure 8: Syrian protest data from BBC News, Reuters, and the New York Times in June 2011 by location

4. FUTURE DEVELOPMENT

The Syrian study successfully demonstrated the usefulness of location identification by allowing the user to delve deeper into the conflict. To further explore connecting locations to BEN events, SSA will use the longitude and latitude of GeoNames-identified locations to connect BEN results to Google Earth, to demonstrate distance between event happenings. Future work on BEN will focus on capturing discrete events, so that results will show true data of events rather than falling subject to how many media reports exist of a particular event.

REFERENCES

"The Penn State Event Data Project," Accessed February 28, 2012, http://eventdata.psu.edu/index.html

"Timeline of the 2011-2012 Syrian Uprising (January-April 2011)," Accessed February 24, 2012. http://en.wikipedia.org/wiki/Timeline_of_the_2011%E2%80%932012_Syrian_uprising_(January%E2%80%93April_2011)

"Timeline of the 2011-2012 Syrian Uprising (May-August 2011)," Accessed February 24, 2012. http://en.wikipedia.org/wiki/Timeline_of_the_2011%E2%80%932012_Syrian_uprising_(May%E2%80%93August_2011)

CHAPTER 27

Dynamic Network Analysis (DNA) and ORA

Kathleen M. Carley, Jürgen Pfeffer

Carnegie Mellon University
Pittsburgh PA, USA
kathleen.carley@cs.cmu.edu, jpfeffer@cs.cmu.edu

ABSTRACT

Dynamic network analysis can be used to assess complex socio-cultural systems from a network perspective. Key elements of this approach include: 1) dynamic meta-network representation of the who, what, where, how, why; representation of data from both a trail and network perspective where the nodes are attributed and the links probabilistic; extension of social networks analytics to the geo-spatial network analytics; techniques for assessing and forecasting change in networks; infrastructure tools to support data extraction, analysis and forecasting ranging from machine-learning models for network extraction to agent-based models for assessing the impact of the co-evolution of networks in various domains on human socio-cultural behavior.

Keywords: social networks, dynamic networks, geo-spatial networks

1 INTRODUCTION

Reasoning about human socio-cultural behavior requires reasoning about individuals in context. In other words, it requires understanding not just the people (who), but answering the classic journalistic queries of who, what, why, where, how and when. While social network analysis supports identification of critical actors and so answering who is important and to an extent why, it does not support the full assessment. For that, dynamic network analysis is needed. That is, dynamic network analysis places the social network within context.

Two factors underlie dynamic network analysis. First the use of both meta-network and trail representation. Second the use of an interoperable tool suite that supports moving from text to network and visual analytics to simulation and forecasting. At the heart, is the network and so the techniques for assessing and visualizing this network – and for this ORA can be used.

2 META-NETWORKS AND TRAILS

One of the key insights underlying dynamic network analysis is that networks are samples of relations formed through the intersection of individual trails. A trail is the path that an individual follows over their life course expressed in terms of who was where when doing what with whom, why and how. Within any trail there are sets of interaction events, such as Joe sends email to Helen. By specifying a slice through defined in terms of a range on a time window, a geo-spatial window, and a set of actors a network can be identified; e.g., the sets of relations that can be extracted by considering all employees of the World Bank, in Afghanistan, during 2011. Dynamic network analysis uses the duality of trails and networks to generate novel grouping algorithms (FOG), trails assessments of changes in networks, and algorithms for visualizing and grouping trails based on meta-network node linkages.

Most socio-cultural systems can be represented as meta-networks linkage Agents and Organizations (Who), Resources and Knowledge/Expertise (How), Tasks/Activities and Events (What), Beliefs (Why), Locations (Where) through time (When). These ontological classes provide a way of categorizing and segmenting nodes. Nodes in a network are instances of an ontological class; e.g., Joe and man are nodes in the Agent ontology class. Most analysts find it beneficial to have a second level in this ontology segmenting agents, organizations, locations and events into specific entities and generic entities. Trails weave patterns between any three entity classes; however, the typical trails involve who, how, or what by where and when.

Given two ontology classes there exists one ore more networks composed of the relations of this type between the nodes in this ontology class. Some of these networks are uni-modal such as social networks connecting agents to agents. Whereas, other networks are bi-partite connecting nodes in one ontology class to nodes in another ontology class such as an attendance network indicating who attended what event. For dynamic network analysis the power comes from being able to extract, assess and forecast change in both types of networks at the same time, and in exploiting the constraints implied by the entire meta-network.

3 METRICS

ORA contains over 150 metrics for assessing various aspects of networks and several dozen algorithm based tools to support various activities such as finding local patterns of interest, comparing networks, and characterizing groups. The metrics include both the standard social network metrics, bi-partite and multi-network metrics. Standard network metrics such as degree centrality, betweenness centrality, closeness and eigenvector centrality are useful for identifying nodes with undue structural influence. Bi-partite metrics, particularly those assessing specialization and redundancy are useful for characterizing the roles of agents or organizations relative to their environment. For example, imagine a people by knowledge network. In this cases agents who are specialists have exclusive or near exclusive access to knowledge other agents do not; whereas, redundancy captures the extent to which there are many individuals with access to the same areas of expertise.

Multi-network metrics such as congruencies, dependencies and load are useful for assessing key performance related aspects of the overarching system. Congruency metrics are measures of fit. They assess the extent to which there is a fit between who or what is needed and who or what is assigned or available. In general, the higher the congruence the better the overall performance of the system. Cognitive demand measures, from a meta-network perspective, the relative "business" of the agent by assessing the extent to which the agent has many others they interact with, many tasks, complex tasks, many resource to juggle, many others that they need to coordinate with and so on. In general, cognitive demand is a reasonable indicator of emergent leadership; not who will become a formal leader, but who is likely to be directing traffic behind the scenes. Finally, the various loads, such as work-load are alternative measures of overall system performance – as assessed from a network perspective.

4 CORE TECHNOLOGIES

As noted, dynamic network analysis is supported through a suite of interoperable tools. This is shown in Figure 1. Additional tools can be added and old ones dropped as technologies evolve. Each of the core technologies can be used alone or as part of an overarching process. Data extraction and cleaning can be done for raw text by using a combination of AutoMap and ORA. Analysis and forecasting of change in networks can be accomplished by using ORA and Construct. ORA, the network analysis engine is at the heart of the dynamic network analysis process. The tools described here support analysis of networks ranging in size and scope from a few nodes to 10^6 nodes per ontology class. Many of the same metrics work on much larger data sets; however, that is not the focus of these technologies.

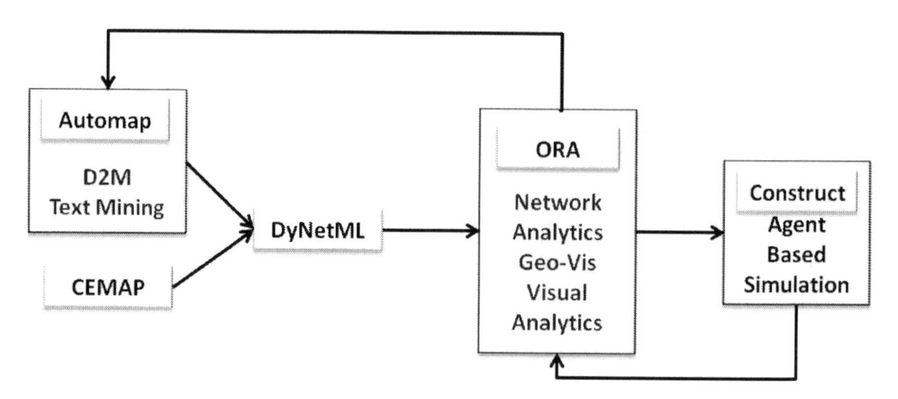

Figure 1 Illustrative interoperability of Core Technologies Supporting Dynamic Network Analysis

AutoMap (CMU: http://casos.cs.cmu.edu/projects/automap/, Carley et al, 2011a) is a mixed-initiative system for the extraction of nodes and relations from raw unformatted texts. Using advanced machine learning techniques and basic thesauri construction techniques AutoMap can be used to support content analysis (extraction of concepts and frequencies), semantic network analysis (extraction of network of concepts), dynamic-network analysis (extraction of ontologically cross-classified nodes and relations), and aspects of sentiment analysis. AutoMap uses a windowing process for link identification (Danowski, 1993) in conjunction with conditional random fields (Diesner and Carley, 2004) for concept classification and thesauri application. Specialized features include sub-tools for extraction of email and phone numbers and post-processers to augment the extracted data with the latitude and longitude of key locations. AutoMap can be used in batch or GUI mode; however, for those unfamiliar with text processing, the best mode is through the data-to-model (D2M) process in which AutoMap and ORA are used together in a predefined and optimized sequence to clean and structure the extracted meta-network data. In general, the techniques for extracting and classifying agents, organizations, and locations are more accurate than those for knowledge, resources, tasks and beliefs.

CEMAP is a mixed-initiative system for the extraction of nodes and relations from semi-structured texts, such as blogs or email. Specialized parsing tools admit the extraction of meta-networks from header data. For most semi-structured data only the agents, organizations, and tasks are extracted.

ORA (CMU: http://casos.cs.cmu.edu/projects/ora/, Carley et al, 2011b) is a powerful network analysis tool, capable of handling large 10^6 networks, and supporting meta-network data, geo-spatial network data, and dynamic network data. ORA is capable of generating over 150 metrics including all standard social network metrics for uni-modal and bi-partite networks and specialized metrics for multi-mode data including measures of loads and demands such as cognitive demand.. Relatively unique features include trail, network, and geo-network visualization, classical and fuzzy grouping algorithms, multi-mode network assessment, built in network simulators, and special reports for semantic network data. User guides, tool tips and integrated help support the user. ORA can import and export data in a large number of formats including direct imports for CSV and UCINET and export of images in png, jpg, pdf, and svg.

Construct (CMU: http://casos.cs.cmu.edu/projects/construct/, Carley, 1990, 1991; Carley, Martin and Hirshman 2009) is an agent-based dynamic-network model for assessing the co-evolution of social and knowledge networks through fundamental learning, information diffusion and belief dispersion processes. Using Construct the impact of various interventions can be assessed at the individual, group or network level under alternative communication media environments. Interaction logics based on homophily and expertise seeking govern individual choice of interaction partners, and knowledge masks accounting for general and work related knowledge and beliefs impact message construction, while forgetting, attention limits and social interaction spheres are used to characterize the boundedly rational agent. Construct gains its power for evolving change in the networks by accounting for the influence of bi-partite networks in constraining the development of the uni-modal networks.

DyNetML is an XML based language for the interchange of relational data (CMU: http://casos.cs.cmu.edu/projects/dynetml/ , Tsvetovat, Reminga and Carley, 2004). Developed as a way to extend graphml, DyNetML supports the representation of geo-temporal meta-network data, meta-data, and attributes on nodes and relations. DyNetML is used to exchange data between AutoMap and ORA with AutoMap exporting DyNetML and ORA importing and exporting DyNetML.

5 ILLUSTRATION OF CRITICAL NETWORK ANALYTIC CAPABILITIES

Critical network analytic capabilities include key entity identification, visual analytics, group identification, path finding, dynamic analytics, geo-network analytics, and forecasting. Aspects of these techniques are illustrated using data gathered on Afghanistan. This is not a complete cataloguing of all capabilities, but a taste of the overall system.

Open source data on Afghanistan were extracted from Lexis-Nexis and others provided as part of an overarching HSCB SNARC effort. The resulting 282,000 texts covered the period December 1999 to August 2011 were processed by AutoMap and a full meta-network extracted. Cleaning resulted in a merging of alternate spellings of the same words, and collapsing. The resulting data is a set of meta-networks one per year for multiple years. Over time, the networks became increasingly dense. Americans and friendly foreign leaders were removed.

Key entity identification was used to find those agents, organizations, locations, and so on that stand out on one or more network dimensions. For agents, for example, metrics indicating the criticality of the individual in the social network are used such as degree, betweenness, eigenvector and so on. A composite look across these metrics suggests that those individuals who are important on many dimensions include key leaders such as Hamid Karzai – see Figure 2. Top ranked key agents are often such leaders. This assessment is done using the key entity report in ORA.

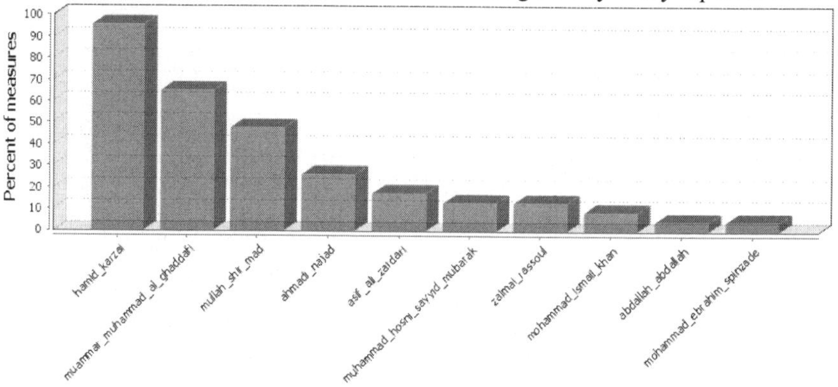

Figure 2. Most Recurring Agents as Key on Numerous Network Metrics

In large networks, these top ranked individuals may be less interesting than the "seconds", those individuals who are not top but still critical. Such individuals might be "the power behind the throne" or at least "actors of interest" to watch. In Afghanistan two of these "seconds" are Abdul Rasid Dostum and Mohammad Ismail Khan. For such individuals, the analyst might want to explore their sphere of influence. The sphere of influence is an extension of the of the ego-net concept to the meta-network. It is the set of entities regardless of ontology class that are directly connected to ego and the connections among them. Khan has a more elaborated sphere of influence in this data than does Dostum indicating a higher level of integration and ability to control more resources. These spheres of influence can be found directly in the visualizer, or through using the sphere of influence report in ORA.

Over time, the power of various agents can increase and decrease. In Figure 3 the over time profile of Dostum and Khan is shown for the network metric cognitive demand. Individuals who are high in betweenness are thought to be more influential. In this case, the analyst has selected to use ORA's over time charting capability to examine the over time changes in the betweenness of Dostum and Khan. After their appointment to the ministry their influence dropped, which was the intent of the appointment. Dostum's further dropped as he left Afghanistan after the Akbar Bai incident, rising only briefly when he returns to support Karzai's re-election. In contrast, Khan's influence is seen to be rising steeply in recent years.

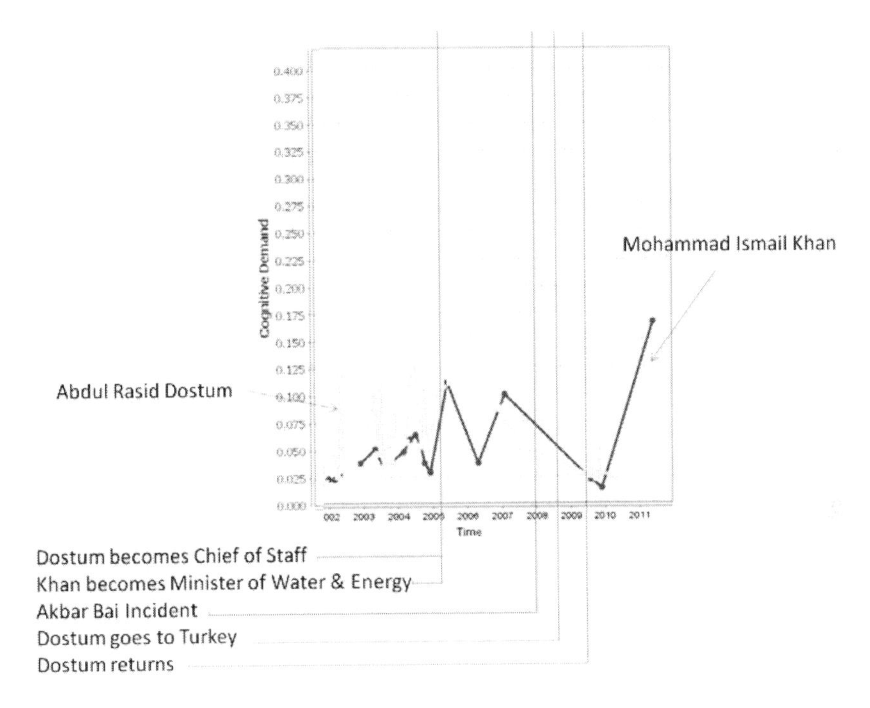

Figure 3. Dynamics of the Betweenness of Dostum and Khan

Spatially, a key issue is what is the region where an actor has influence. For this, we can examine the polygon within Afghanistan that contains the activities of the agents of interest. ORA supports geo-spatial visualization and exports using any of NASA World Wind, ArcGIS, or Google Earth. Figure 4 shows the region of influence calculated for Dostum and Khan. As can be seen, both cover most of Afghanistan, which as former warlords and then ministers in the Karzai government is not too surprising.

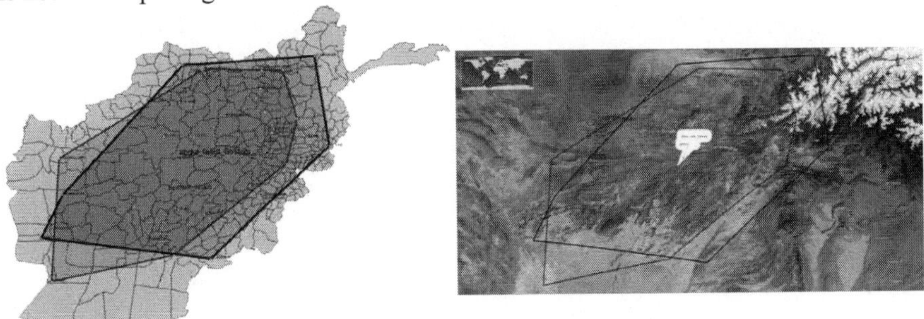

Figure 4. Regions of Influence for Dostum and Khan

Moving on to the network, we might ask where are the agents? Using the agent by location subnetwork ORA can color the regions by frequency and overlay the network. Here we see in Figure 5, that the density of agents is highest near the areas of Kunduz and Kandahar.

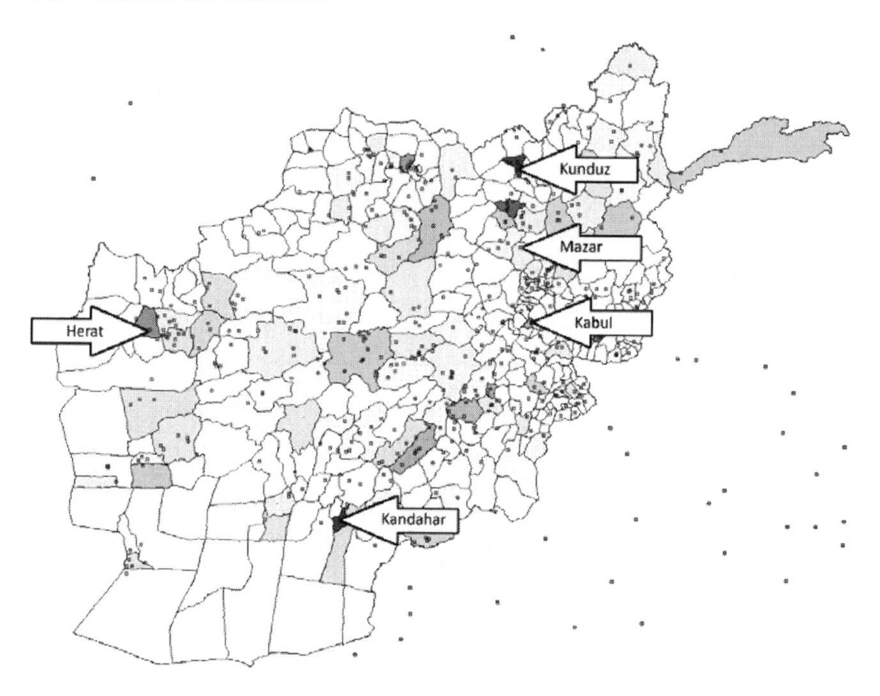

Figure 5. Regions of Concentration of Activities by Political Elite

Analyses such as those just described characterize and provide an in-situ understanding of the existing socio-cultural environment. To go beyond this is to engage in forecasting. ORA provides two distinct types of forecasting technology. The first, immediate impact, enables the analyst to examine the immediate impact on a network or meta-network of removing one or more nodes or links. Figure 6 shows the impact on Karzai's power when Dostum and Khan are removed. As can be seen, these agents are to an extent holding Karzai in check and the removal of either agent strengthens Karzai's power base. However, networks can heal themselves. That is, individuals interact and in doing so build new connections and maintain existing one. Agent-based dynamic-network modeling, via Construct, is used to assess the changes due to this "healing." Through ORA Construct can be called by using the Near Term Impact report.

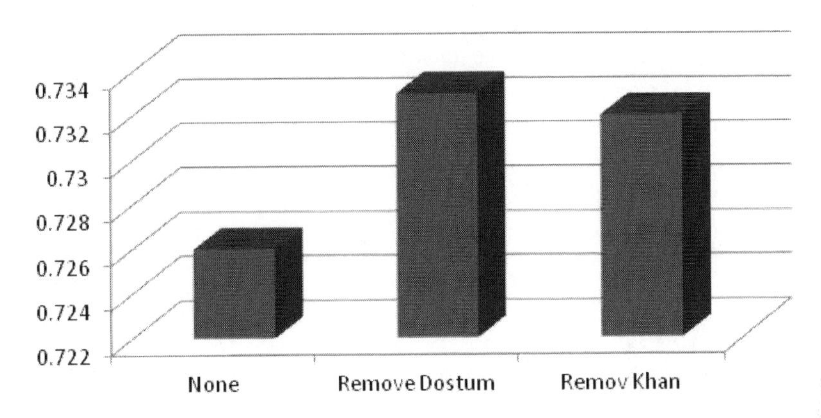

Figure 6. Impact on Karzai's Power if Course of Action Removing Either Dostum or Khan is Followed

6 CONCLUSIONS

Dynamic network analysis is a valuable approach for understanding human socio-cultural behavior and for forecasting the space of future possibilities. Importantly, these techniques can be used to reason about both own and adversarial groups. These techniques can be used to identify weaknesses and strengths in any group, and to suggest ways of overcoming those. The breadth of application is one of the core strengths of this approach.

Core challenges, that if solved would further increase the utility of these techniques include automatic interpretation of metrics; auto-identification and visualization of critical features, and improved data extraction and fusion techniques. Key advances supporting improved dynamic and spatial analysis are likely to occur in the near future. Even without meeting these challenges, ORA and Dynamic Network Analysis can support overall reasoning about individuals, groups and their activities, course of action analysis, and improve overall situation awareness. Simply moving to meta-network reasoning and the ability to focus on geo-spatially embedded networks and dynamic networks improves the analysis and enables a more contextual based understanding.

ACKNOWLEDGMENTS

This work was supported in part by the Office of Naval Research – ONR N000140811223 (SORASCS), N000140811186 (Ethnographic), the AFOSR FA9550-05-1-0388. Additional support was provided by the center for

Computational Analysis of Social and Organizational Systems (CASOS). The views and conclusions contained in this document are those of the authors and should not be interpreted as representing the official policies, either expressed or implied, of the Office of Naval Research, the Air Force Office of Scientific Research, the Department of Defense or the U.S. government.

REFERENCES

Carley, K. M. J. Reminga, J. Storrick and D. Columbus, 2011b, "ORA User's Guide 2011," Carnegie Mellon University, School of Computer Science, Institute for Software Research, *Technical Report,* CMU-ISR-11-107.

Carley, K. M., 2002, Smart agents and organizations of the future. In L. Lievrouw & S. Livingstone (Eds.), *The Handbook of New Media: Social Shaping and Consequences of ICTs* (pp. 206–220). Thousand Oaks, CA: Sage.

Carley, K. M., D. Columbus, M. Bigrigg and F. Kunkel, 2011a, "AutoMap User's Guide 2011," Carnegie Mellon University, School of Computer Science, Institute for Software Research, *Technical Report*, CMU-ISR-11-108.

Carley, K.M. 2011. Dynamic network analysis and modeling for CANS. In Belinda Bragg and George Popp (eds.), A Guide to Analytic Techniques for Nuclear Strategy Analysis. NSI, inc.

Carley, K.M., M. K. Martin and B. Hirshman, 2009, "The Etiology of Social Change," Topics in Cognitive Science, 1.4:621-650.

Carley, Kathleen M., 1990, "Group Stability: A Socio-Cognitive Approach." Pp. 1-44 in Lawler E., Markovsky B., Ridgeway C. and Walker H. (Eds.) Advances in Group Processes: Theory and Research . Vol. VII. Greenwhich, CN: JAI Press.

Carley, Kathleen M.1991. "A Theory of Group Stability." *American Sociological Review*, 56(3): 331-354.

Diesner, J. and K. M. Carley, 2008, "Conditional Random Fields for Entity Extraction and Ontological Text Coding," *Computational and Mathematical Organization Theory*, 13.3: 248-262.

Danowski, J., 1982, A network-based content analysis methodology for computer-mediated communication: An illustration with a computer bulletin board. In R. Bostrom (Ed.), Communication Yearbook, (pp. 904-925). New Brunswick, NJ: Ransaction Books.

Pfeffer, J. & K. M. Carley, Forthcoming, Rapid Modeling and Analyzing Networks Extracted from Pre-Structured News Articles. Computational and Mathematical Organization Theory.

Tsvetovat, M., Reminga, J., & Carley, K.M., 2004, DyNetML: Interchange Format for Rich Social Network Data. Carnegie Mellon University, School of Computer Science, Institute for Software Research International, Technical Report CMU-ISRI-04-105.

Wasserman, S., & Faust, K., 1994, *Social Network Analysis: Methods and Applications*: Cambridge University Press.

Section V

Strategic and Tactical Considerations for Cross-Cultural Competency

CHAPTER 28

Cultural Awareness Training for Marine Corps Operations: The CAMO Project

Jean MacMillan, Alexander Walker, Ellen Clarke, Yale Marc

Aptima, Inc.
Woburn, MA
macmillj@aptima.com

ABSTRACT

As the Marines Corps engages in diverse operations that range from peacekeeping to disaster relief in diverse locations around the world, the quality of the relationships that Marines build with people, both in and out of uniform, is critical to mission success. Marines need to be able to work effectively with people from cultures that may be significantly different from their own. The Marine Corps has developed the concept of "Operational Culture" to define and describe the cross-cultural knowledge and skills that are needed for Marines to operate effectively anywhere in the world. The goal is for Marines to develop the concepts and skills with which to "read" any culture.

Because time for pre-deployment training is a scarce resource for the Marines, they need portable anytime anywhere training in Operational Culture. To meet this need, the Cultural Awareness for Marine Operations (CAMO) project is developing an interactive on-line training system for Operational Culture. CAMO uses images and stories to provide realistic experiential training without the training time and expense of a fully immersive game or simulator.

Keywords: instructional design, cultural training, computer based training (CBT)

1. THE MARINE CORPS NEED

In 2008 the Marine Corps published a textbook, *Operational Culture for the Warfighter: Principles and Applications* (Salmoni and Holmes-Eber, 2008) to help Marines link concepts of culture and cultural differences to the planning and execution of military missions. The goals of the book were to provide a sound theoretical foundation for understanding different cultures based on five cultural dimensions: (1) the physical environment; (2) the economic dimension; (3) social structures; (4) political structures; and (5) beliefs and symbols, and to show how to apply these basic principles in actual environments in which Marines and other military personnel have deployed or may deploy. The ultimate goal is to develop the capacity of Marines to think systematically about culture as it affects their ability to perform their missions. Table 1 summarizes the goals of Operational Culture training as detailed in the textbook.

Table 1. Summary of Marine Corps' Goals for Operational Culture Training: Experiential Situation-based Training

Goals of Operational Culture Training	Salmoni and Holmes-Eber, 2008
Train cross-cultural skills, not just knowledge about one area of responsibility	"Need for this kind of learning to be non-region focused" (p. 255)
Teach "how to think," not rules and procedures	"Focus on the concepts and skills with which to 'read' any culture" (p. 238)
Use actual situations and real experiences for immediate relevance	"Marines must be willing to pay attention and 'take it on board'" (p.257)
Use situations set in multiple specific cultures	"Without a concrete regional context...the Dimensions remain sterile and without meaning" (p. 268)
Focus is on cognitive skills, using realism to add an affective dimension	"Receiving mode (Affective) precedes Comprehension (Cognitive)" (p. 253)
All situations should involve tactical considerations	"Speak in an idiom and use examples that are relevant to activities and practical concerns of Marines." (p. 269)

Based on material in the *Operational Culture* textbook, the Marine Corps' Center for Advanced Operational Culture Learning (CAOCL) has developed and implemented a Program of Instruction (POI) to teach Marines how to apply the basic elements of operational culture before they deploy, but classroom-based training is far from ideal for many Marines. It is time intensive, and pre-deployment time is limited. Also, experienced Marines may not need the same level of training as Marines who are deploying for the first time. Ideally, Marines need Operational

Culture training appropriate to their skill level that they can access at any time, from any location. This seems to indicate computer-based instruction. However, as summarized in Table 1, this training needs to be interactive and engaging, and anchored in realistic experiences that will be relevant to Marines. It must teach "how to think," not just convey textual material for rote memorization. In addition, the training is needed immediately, and must be funded from limited budgets. Can such training be delivered effectively using interactive computer-based instruction? This was the challenge confronting the CAMO project.

2. THEORETICAL LEARNING STRATEGY

Finding a way to teach Marines complex cogntive skills through a computer-based trainer, to be utilized in a distance education format, was the key goal of CAMO. Educational research has shown time and again that the traditional lecture approach to teaching is not the best way to get students to learn complex cognitive and reasoning skills (Adams, Wieman, & Schwartz; 2008; Schwartz, Lindgren, & Lewis, 2008), particularly when in a non-traditional learning environment, such as computer-based instruction (Chin et al., 2010). Instead, research has shown that a *constructivist* approach, allowing students to discover answers for themselves and explore material rather than listen to a lecture, is much more effective (Schwartz, Lindgren & Lewis, 2008). Additionally, effective training of cognitive skills should develop deep structural knowledge of complex domains and engage higher level reasoning processes (Freeman et al., 2011).

In contrast, didactic instruction is an approach in which students are commonly passive listeners to an instructor's lecture. The advantage of the didactic approach is that it allows for the rapid conveyance of knowledge built up over decades of experience to novices. The benefit of time savings for the novice is counterbalanced by a loss of depth; students do not necessarily develop deep understanding of the concepts being taught, nor do they develop proficiency in the processes required to grow expertise. While didactic learning lacks the ability to instill higher levels of understanding, constructivist learning emphasizes student discovery of deep level structures/principles through experience (Duffy & Kirkley, 2007). However, the potential benefit of deeper understanding by the student is offset by the cost to the instructor of structuring learning events so that they guide the student to true domain principles, and by the risk that, in the absence of well-planned events, students may become too frustrated to learn (Freeman et al., 2011).

Schwartz and Bransford (1998) proposed that the two approaches are more effective when used in combination. Specifically, they argued that there is a "time for telling" (or, in other words, a time for lecturing) that occurs after students have had the opportunity to discover principles in the context of a rich problem, and before they engage in deliberate practice that exercises their knowledge.

Exploratory learning has major implications for transfer and may be best illustrated by the use of contrasting cases (e.g., Schwartz, 2008). This strategy is ideal for situations in which what is being trained is not a procedural skill consisting of strict rules, but rather a complex cognitive construct guided by adherence to principles, as is the case with Operational Culture. The general premise is that the two cases (i.e., problems or exercises) contrast within a meaningful dimension, and that through working on the two cases, the student derives the common principle that underlies them both and comes to overlook their surface dissimilarities. In addition, struggling through contrasting cases allows students to identify gaps in their knowledge, and this presumably primes them to receive and understand the subsequent lecture on the subject (Bransford & Schwartz, 2001; Genter et al., 2003; Schwartz, 2008). This facilitates the transfer of knowledge in that it addresses a common explanation for failures of transfer- that surface features are being attended to at the expense of deep structures. Beginning a training session with exploratory learning allows the development of both knowledge of deep structure and the skills of learning itself, thus promoting transfer to novel situations.

Once the students have had the opportunity to wrestle with the problem and identify gaps in their knowledge, they are better prepared to receive and understand a lecture on the subject (Bransford & Schwartz, 2001; Genter et al., 2003; Schwartz, 2008). The lecture/didactic instruction promotes consolidation of knowledge by filling in the gaps in knowledge of domain structure, and by providing answers to students' individual questions.

A substantial body of research demonstrates a direct relationship between expertise and the accumulated amount of deliberate practice with feedback. Extensive deliberate practice in the presence of feedback enables learners to efficiently encode knowledge of key domain concepts and solution procedures and, by extension, facilitates near-automatic retrieval of knowledge and execution of skills (Ericsson, Krampfe, & Tesch-Romer, 1993; Ericsson & Lehman, 1996). Deliberate practice requires that the instructor first identify the elements of expert form; the learner then performs the task while consciously attending to these elements; the instructor notes discrepancies from the expert form and provides remedial feedback to the learner; and the learner repeats cycles of practice and feedback independently or with an instructor or instructional system. Previous applications of deliberate practice-based instruction have produced strong learning outcomes across a wide range of cognitive and psychomotor tasks, including those in military settings (Charness, 1991; Ericsson et al., 1993; Allard & Starkes, 1991; Beaubien et al., 2009; Shadrick et al., 2007).

To be most effective, the deliberate practice should include instructional events that refine the organization and representation of domain knowledge. These events will eventually enable learners to (1) solve domain problems well and (2) assess and grow their own expertise in difficult corners of the domain (Ericsson, 1996). Optimization of instruction must, at a minimum, focus the student on such instructional events (those that refine the organization and representation of domain knowledge). This maximizes deliberate practice (Ericsson & Lehmann, 1996), which reliably predicts expertise.

Based on the literature reviewed above, the learning approach for CAMO combines exploratory learning with didactic instruction. We divided the Operational Culture training into three components: exploratory learning, didactic instruction, and deliberate practice (Figure 1). This combination of approaches allows trainees to develop deep structural knowledge and encourages trainees to develop an expert mix of recognitional and deliberative problem solving methods that should lead to more effective modification of problem solutions, as well as implementation of plans of actions based on these solutions.

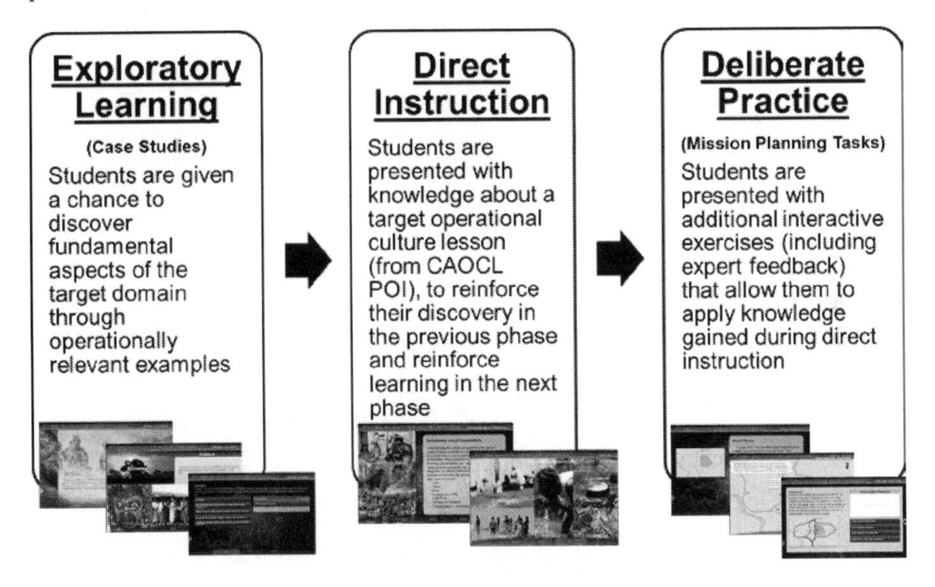

Figure 1. Three phase instructional strategy used in CAMO training.

3. IMPLEMENTATION OF THEORETICAL TRAINING STRATEGY

We have implemented this three-phase approach (exploratory learning, didactic instruction, and deliberate practice) in a computer-based system for training Operational Culture. This program is designed to develop deep, expert-level structural knowledge of complex cultural domains and encourage higher level reasoning processes and problem solving skills.

The structure of the Operational Culture training program is similar to that employed in the previous classroom based program, where the content is broken into six lessons aimed at providing knowledge on topics crucial to Operational Culture - apply Operational Culture, non-verbal communication, communicate with an interpreter, use tactical language, interact with a foreign population, and recognize culture. Within each lesson are sub-sections, each representing an instantiation of the three-phase approach - opening with exploratory learning, followed by didactic instruction, and ending with deliberate practice. This presentation order of the three components is designed to maximize learning goals and is based on previous literature (Schwartz & Bransford, 1998).

Each sub-section begins with exploratory learning experience in the form of contrasting cases. Marines are presented with two problems that appear different at the surface, but that share a common principle. For example, a case study targeted at identifying where to construct a clean water well is contrasted with a case study demonstrating how the location of a Marine exercise had caused hostility. The latter asks what could have been done to prevent the outcome, while the former asks for the identification of the most effective solution. These case studies are designed to encourage the Marine to understand the complex relationship between the environment and other cultural factors. By representing this relationship from different vantage points and including different cultural factors, the hope is that a broader understanding of this principle will be achieved.

The didactic instruction portion of the three-phase approach is instantiated with engagement as a priority. Multiple presentation formats, all of which include relevant and appropriate eye-catching pictures, are used in an effort to keep engagement high. The most basic presentation has material presented in one area of the page and a picture (or pictures) in another. A more elaborate version requires the Marine to mouse over certain areas on the screen to reveal information regarding a particular topic. In this way, the Marine can interact with the training such that portions of the instruction can be revealed or hidden as desired. For example, a graphical representation of the five dimensions of Operational Culture is presented along with instructions to mouse over areas of the screen to learn more about these dimensions. When the Marine mouses over a portion of the graphical representation, a summary of the relevant dimension is displayed in a pop-up window. Videos are also included as another way to present information in a didactic format. These videos are likely to be highly engaging, as

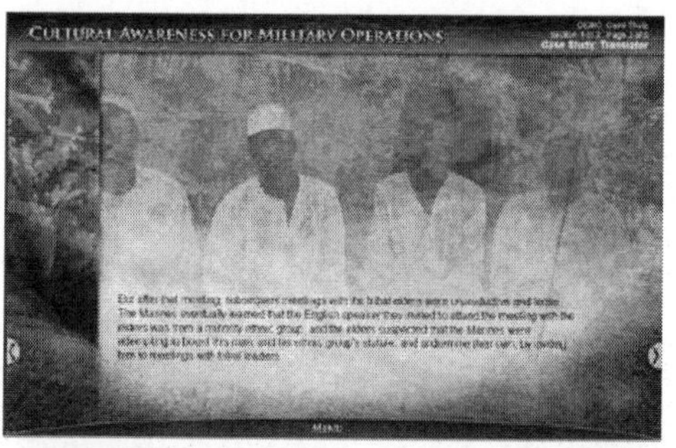

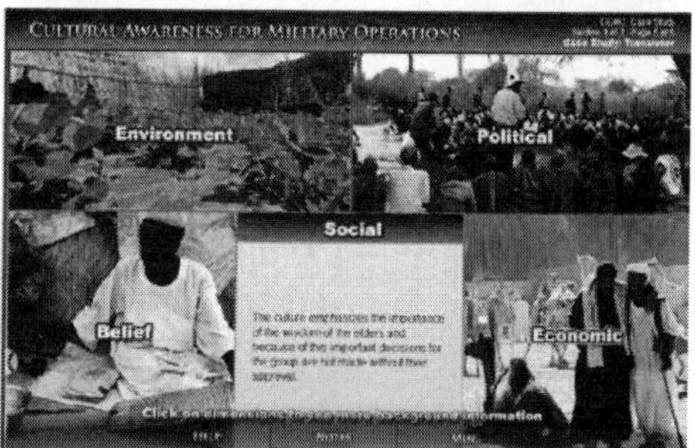

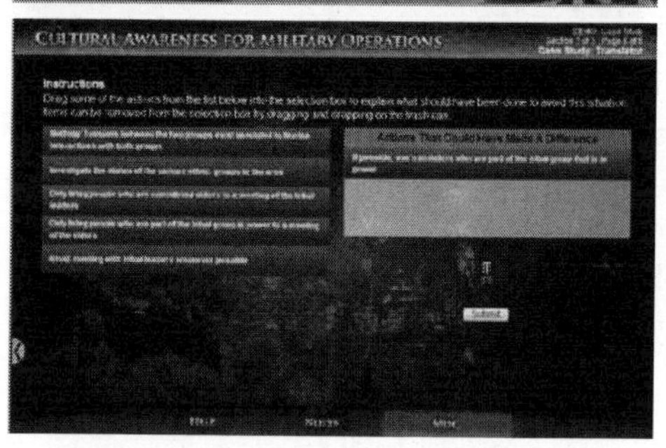

Figure 2. Sample exploratory case study task. Marines first read a scenario, then gather additional information, and finally make a decision based on the information presented.

they are commonly used for the purpose of demonstrating foreign cultural practices. The presentation of the required material in multiple formats is likely to result in a heightened level of engagement, which is crucial given the link between increased engagement and learning (e.g. Carini, Kuh, & Klein, 2006).

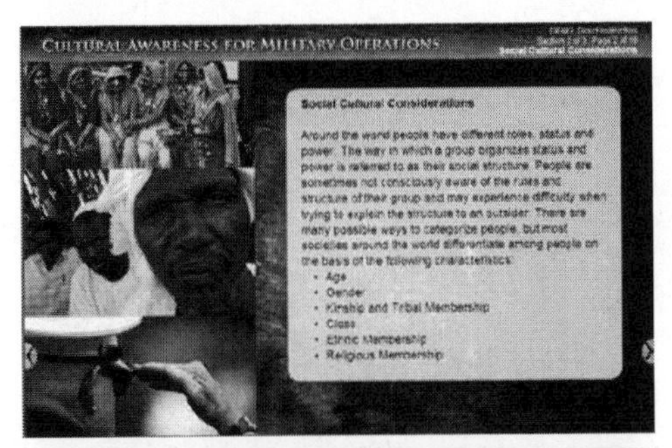

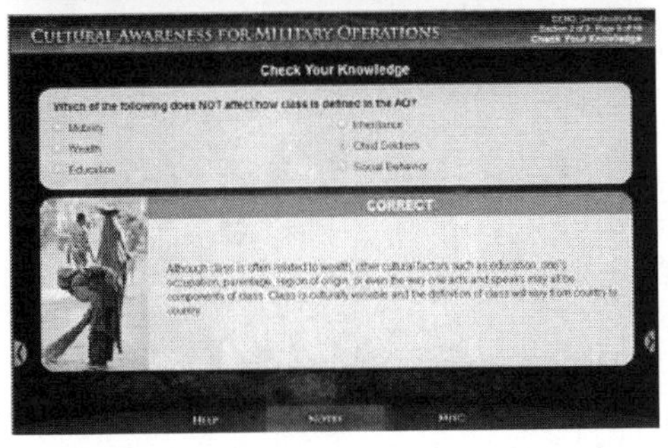

Figure 3. Sample didactic section. This section combines a number of interactive techniques to present direct instruction in an interesting and interactive way.

Throughout the didactic lessons, trainees are presented with periodic “knowledge checks,” designed to encourage critical thinking. Each knowledge check probes the Marines understanding of the previous section and is either derived or taken directly from the course material supplied by the Marines. Knowledge checks can require either the selection of one correct answer or the selection of multiple answers. This variation in testing format is designed to ensure that the Marine has gained an expert level understanding of the material. Feedback is given each time a question is answered, and includes the correct response as well as the rationale of an expert. The ability to compare their own selections and rationale to that of an expert should encourage students to (a) think as an expert would and (b) critically examine their responses.

Following each didactic section, deliberate practice takes the form of mission planning tasks (MPTs), modeled after the Tactical Decision Games (TDGs) that are often used in Marine Corps courses to train decision making and problem solving. Traditionally, during TDGs, a Marine instructor presents a situation to the Marines, often accompanied by a map, and asks the Marines how they would respond at a critical decision point. When the Marines provide a solution, the instructor typically probes further to get them to explain the “why” behind their decisions. In this way, TDGs do not have one right answer; instead it depends on whether the Marines have thought through the problem and provided a plausible and doctrinally sound justification for their solutions.

While the MPTs within CAMO lack the direct interaction of a live instructor, the scenarios are set up to encourage Marines to consider the "why" behind their decisions and apply the cultural knowledge gained during didactic instruction to solve an operational problem. For example, at the conclusion of the Social dimension training, Marines are presented with a fragmentary order describing a disaster relief scenario. They are given their mission (recommend the optimal location to rebuild a demolished school) and commander's intent (maintain stability in the region). To complete their mission, the Marines are presented with an interactive map that allows them to investigate the various locations where a school could possibly be constructed. Through investigation and application of the principles learned during the Social dimension training, the Marines should choose the solution that will maintain stability. After recommending a specific location for the construction of the school, the Marines are presented with expert feedback explaining which choice was optimal, from an Operational Culture perspective. This feedback allows the Marines to understand that while doctrinally there may be several "right" answers, if they take commander's intent into consideration and apply the Operational Culture principles they learned previously, there is one "best" answer.

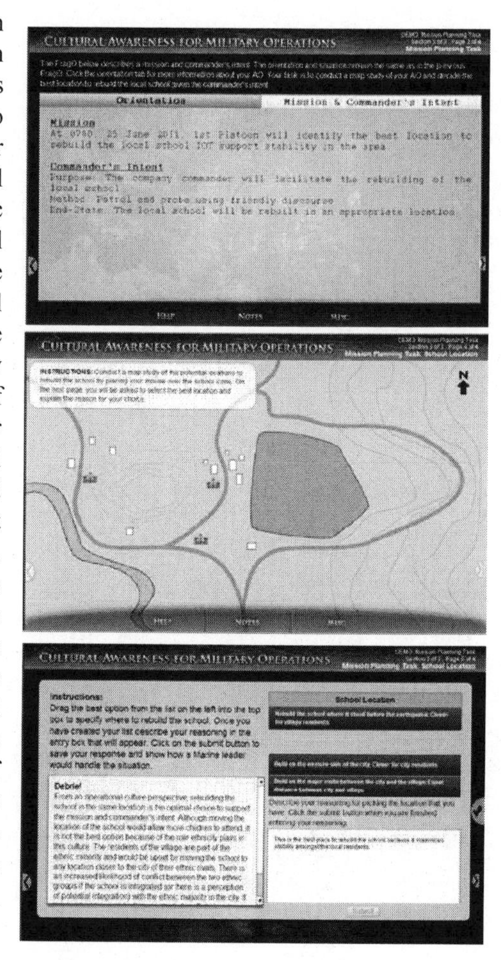

Figure 4. Sample Mission Planning Task.

4. FUTURE DIRECTIONS

The strategy applied within the CAMO training system was developed based on a combination of sound theoretical principles and operational input. While we believe that this is the best approach to maximize the acquisition of the complex cognitive skills required for Operational Culture tasks, it has yet to be validated in its current form. The next step in the development of the CAMO system is to test the training effectiveness of our approach using active duty Marines. This validation will assess how the computer based training in CAMO compares to the current classroom-based instruction provided to Marines before deployment. The current hypothesis is that the CAMO training will perform just as well as the instructor led training in helping Marines to acquire the complex principles of Operational Culture, and will require less time and fewer resources. Upon validation of this hypothesis, the CAMO training will undergo the appropriate certification process to be implemented on the Marine Corps' learning management system, MarineNet.

Additionally, the theoretically based training strategy developed during CAMO will be applied to and tested within other domains requiring the acquisition of complex cognitive skills, such as combat profiling, and team communication and coordination.

REFERENCES

Allard, F., & Starkes, J. L. (1991). Motor-skill experts in sports, dance, and other domains. In K. A. Ericcson, & J. Smith (Eds.), *Toward a General Theory of Expertise: Prospects and Limits* (pp. 126–152). New York: Cambridge University Press.

Beaubien , J.M., Alexander, A.L., Artis, S., Orvis, K.L., and Marc, Y. (2009). Generalizability of the deliberate practice technique for training higher-order skills. *Proceedings of the Interservice/Industry Training, Simulation & Education Conference (I/ITSEC),* Orlando, FL

Bransford, J. D., & Schwartz, D. L. (2001). Rethinking transfer: A simple proposal with multiple implications. *Review of Research in Education,* 24, 61-100.

Carini, R.M., Kuh, G.D., & Klein, S.P. (2006). Student engagement and student learning: Testing the linkages. *Research in Higher Education, 47(1),* 1-32.

Charness, N. (1991). Which problem is being solved? Review of K. J. Gilhooly (Ed.) Human and machine problem solving. *Contemporary Psychology*, 36, 227-229.

Chin, D.B., Dohmen, I.M., Cheng, B.H., Oppezzo, M.A., Chase, C.C., & Schwartz, D.L. (2010). Preparing students for future learning with teachable agents. *2010 Society for Research on Educational Effectiveness (SREE) Conference*, Washington, DC.

Duffy, T.M., & Kirkely, J.R. (2007). Experiential learning. Presentation at Redstone Arsenal, Huntsville, AL.

Ericsson, K. A., Krampe,R.T. & Tesch-Römer, C. (1993). The role of deliberate practice in the acquisition of expert performance. *Psychological Review*, 100: 363-406.

Ericsson, K. A., & A. C. Lehmann, (1996). Expert and exceptional performance: Evidence on maximal adaptations on task constraints. *Annual Review of Psychology*, 47: 273-305.

Freeman, J, Walker, A, Puglisi, M, Geyer, A, Marceau, R, Marc, Y (2011). Cognitive and perceptual skills involved in Combat Hunter expertise: Exploratory evidence for a new training framework. *Interservice/Industry Training, Simulation, and Education Conference (I/ITSEC) 2011, Orlando, FL.*

Gentner, D., Loewenstein, J., & Thompson, L. (2003). Learning and transfer: A general role for analogical encoding. *Journal of Educational Psychology*, 95, 393-408

Slamoin, B.A. & Holmes,-Eber, P. (2008) *Operational Culture for the Warfighter: Principles and Applications.* Quantico, VA: Marine Corps University Press.

Schwartz, D.L., & Bransford, J.D. (1998). A time for telling. *Cognition & Instruction*, 16, 475-522.

Schwartz, D. L., Varma, S., & Martin, L. (2008). Dynamic transfer and innovation. in S. Vosniadou (Ed.), *International Handbook of Research on Conceptual Change* (pp. 479-506). New York: Taylor & Francis.

Shadrick, S. B., Lussier, J. W., & Fultz, C. (2007). Accelerating the development of adaptive performance: Validating the Think Like a Commander training (ARI Research Report 1868). Arlington, VA: U.S. Army Research Institute for the Behavioral and Social Sciences.

CHAPTER 29

Efficient Cross-Cultural Models for Communicative Agents

Alicia Sagae[1], *Emily Ho*[1], *Jerry R. Hobbs*[2]

[1]Alelo Inc.
Los Angeles, CA, USA
{asagae, eho}@alelo.com

[2]Information Sciences Institute
University of Southern California
Los Angeles, CA, USA
hobbs@isi.edu

ABSTRACT

This paper presents a training system that uses compositional models of culture for social simulations involving conversational agents. We compare the compositional framework to a state-of-the-art agent system, in terms of development effort, number of reused and new objects, and flexibility and accuracy of resulting conversational simulations. Resulting trends indicate that the new architecture is more efficient, especially as the number of simulations grows.

Keywords: Hybrid Modeling, Training & Simulation, Conversational Agents, Commonsense Models

1 INTRODUCTION

Cross-cultural competency is a critical need for military personnel. For example, the US Defense Regional and Cultural Capabilities Assessment Working Group has identified the ability to integrate cultural knowledge and skills into mission execution as a critical cross-cultural competency for general purpose forces

(McDonald, et al., 2008). Training of these skills, knowledge, and abilities is resource-intensive for both trainees and organizations. Simulation-based training promises anytime, anywhere access that can allow instructional material designed by a single trainer to be delivered in an effective, interactive way to thousands of trainees at lower cost and higher convenience (Fletcher, 1990). However, when instructors and domain experts encode this material, the current tools offered to them typically produce script-like, monolithic data structures that are culture-specific, non-reusable, and difficult to update or apply to new cultures and missions. As a result, creating training scenarios is costly and inefficient, especially as the number of scenarios grows large.

In the CultureCom project, we address these problems by developing a new system for creating training simulations in cross-cultural competency. Because these simulations encode a variety of linguistic, cultural, and task-level features, we refer to them as *social simulations*. Our system produces flexible, model-driven simulations that use both culture-general and culture-specific rules. As a result, we achieve the novel capability to swap cultural models, in the form of rule sets, in and out of a social simulation to reveal pedagogically relevant differences at the level of behavior (utterances, gestures) and intention (communicative act).

In this paper we evaluate the gains in efficiency that our new architecture provides. We encode multiple social simulations, using the CultureCom architecture and using a state-of-the-art architecture based on finite state automata. We show that the new model-driven architecture requires comparable authoring time for an initial simulation, but allows more objects to be reused, reducing authoring time and total number of objects created for each subsequent simulation.

2 CONVERSATIONAL AGENTS FOR CROSS-CULTURAL COMPETENCY TRAINING AND SIMULATION

Alelo produces language and culture training products on a range of devices, including desktop, web-based, and hand-held platforms. In these products, immersive serious games provide an integrated learning environment in which trainees must make decisions about mission goals and logistics and engage in cross-cultural and interpersonal interactions with socially intelligent virtual agents. Examples include Tactical Iraqi (Johnson & Valente, 2009) and the Virtual Cultural Awareness Trainers (VCATs), both of which were designed using Alelo's Situated Culture Methodology (Johnson & Friedland, 2010). A screen shot is given in Figure 1. In this example, the player controls an avatar (center left) in a simulated meeting with village elders. The player speaks in Dari, and the virtual elders respond to him with speech and gestures.

The architecture for simulating conversations in these systems is based on finite state automata (FSAs) that encode conversational branches at the level of communicative act. An example of such an FSA is shown in Figure 2. The sequence of communicative acts is strictly prescribed by the shape of the graphical FSA, whose objects can be manually re-authored and copied but not reused in new graphs.

Figure 1. Meeting with the malek in Alelo's Operational Dari

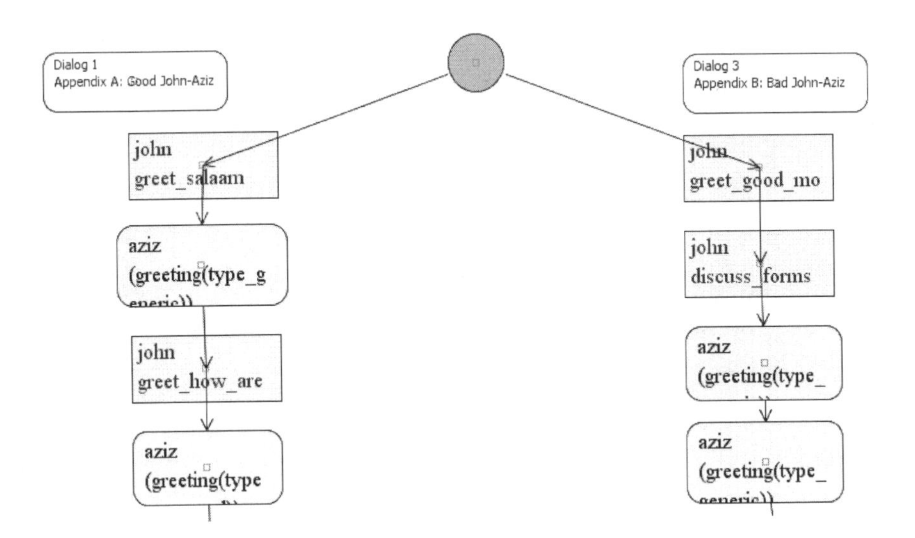

Figure 2. Sample of conversation captured as a finite state automaton

3 IMPROVED MODEL OF CONVERSATION

3.1 Modular Architecture

In contrast to the finite state system described above, CultureCom encodes agent behavior in a group of unsequenced, context-dependent rules, captured in a set of interoperable Protégé-frames ontologies (Gennari, et al. 2002) and executed using the CLIPS expert system (http://clipsrules.sourceforge.net/). This allows culture-general rules, such as "engage counterparts with respect" to be inherited and combined with culture-specific rules such as "in Afghan culture, questions about female family members is disrespectful." Crucially, the culture-general and culture-specific rule sets are stored in separate interoperable files, meaning that the agent's behavior can be adapted to a new culture by loading an American or Colombian culture model in place of the Afghan one. A sample of the inheritance hierarchy that makes this possible is shown in Figure 3.

The modular architecture also allows pieces of language that have already been authored for one simulation to be re-used in subsequent ones, rather than typed in again. This reduces the chance of misspellings and allows global management of the quality of the lexicon. As a result, linguistic behavior is consistent across all simulations that share the same language modules (for example, the "World English Language" file in Figure 3). Systems that require such knowledge to be duplicated or reauthored once per scenario run an increased risk of inconsistency. "Hello" in one simulation may become "Helo" in another. The risk increases as the number of scenarios grows; our architecture mitigates this risk.

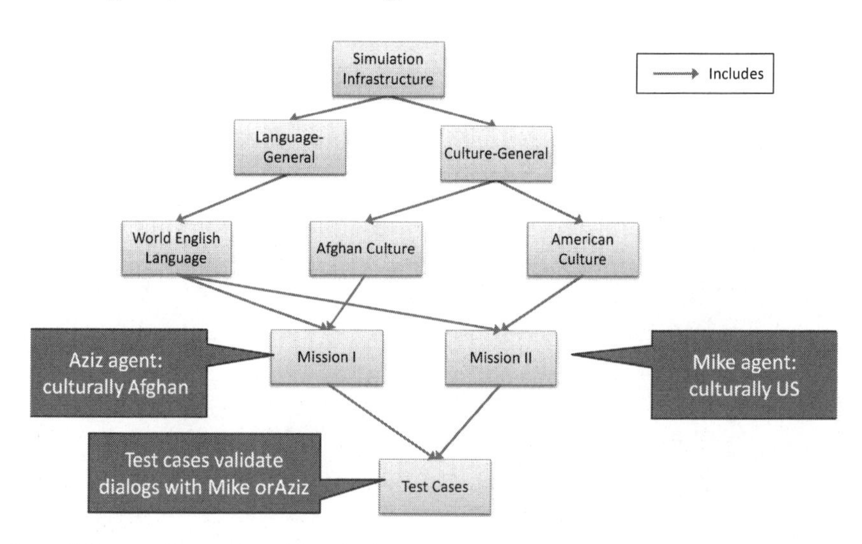

Figure 3. Swap-in Swap-out model hierarchy

3.2 Commonsense Model of Microsociology

Figure 3 shows that files capturing culture-specific knowledge ("Afghan Culture" and "American Culture") rely on data from culture-general files farther up the hierarchy ("Culture-General"). Together they comprise a logical commonsense model of culture that focuses on microsocial interaction (Hobbs & Sagae, 2011). The model is encoded as a set of predicates and axioms that express entities, properties, relations, events, and causal relations among events. It extends the framework described by Hobbs & Gordon (2010). The culture-general model applies to all CultureCom social simulations. An excerpt is given below:

A good reason for demonstrating a real friendship or establishing the pretense of a fictional one in this way is because friends are more likely to help each other out. So politeness is itself a way for people to increase the reliability of the Golden Rule.

```
(forall (p1 p2 e)
  (if (and (polite p1 p2) (goal' e1 e p2) (believe p1 e1) (etc))
    (exist (e2) (and (help' e2 p1 p2 e) (cause e1 e2)))))          (1)
```

This axiom says politeness leads others to help one... many greeting conventions are motivated by exactly this rationalization -- we're friends, we care about each other's desires, and we help each other.

This meta-rule is elaborated in more detail in the Afghan Culture model by an excerpt that explains culture-specific polite conversational openings (greetings):

...Greetings, defeasibly, are required to initiate an interaction:

```
(forall (e p1 p2)
  (if (and (interact' e p1 p2) (etc))
    (exist (e1)
      (and (greet' e1 p1 p2) (intBegins e1 e)))))          (2)
```

The specifically Afghan form of the greeting has three exchanges. First there is the generic exchange "Salaam Alikum". Then each asks the other how they personally are. Then each asks the other about the well-being of their families. We can call the first turn the generic greeting, the second the personal greeting, and the third the family greeting. The generic greeting is defined simply as the specific utterance.

```
(forall (e i u)
  (iff (genericGreet' e i u)
    (utter' e i u "Salaam Alikum")))
```

That is, a speaker i utters to a hearer u the string "Salaam Alikum".

3.3 Data for Training Cross-Cultural Communication

The content of the logical model was developed in coordination with a data development and validation process (Wertheim & Agar, in press). This process was conducted by a team of Cultural and Linguistic Anthropologists, who interviewed subject matter experts from two cultures of focus: Dari-speaking urban Afghanistan, and Spanish-speaking Colombia. The interview material is annotated with ethnographic and sociolinguistic observations.

Based on this material, example *dialogs* are composed representing the performance target at which the final training system aims. A dialog is a script for the verbal communication that occurs in a social simulation. In keeping with the task-based nature of the training system as a whole, we developed dialogs with better (more successful) and worse (less successful) outcomes. A description of the developed dialogs is given in Table 1. Excerpts from dialogs 1a and 3a are shown in Table 2 and Table 3, respectively.

Table 1. Dialogs developed for evaluation. Language context indicates native language of interviewees. All dialogs are encoded in World English (W.E.).

ID	Player Name	Non-Player Name	Outcome	Culture Context	Language Context	Encoding	Length in Turns
1a	John	Aziz	Better	Afghanistan	Dari	W.E.	29
1b	John	Aziz	Worse	Afghanistan	Dari	W.E.	14
1c	John	Mike	Better	America	English	W.E.	10
2a	John	Aziz	Better	Afghanistan	Dari	W.E.	12
2b	John	Aziz	Worse	Afghanistan	Dari	W.E.	14
3a	John	Diego	Better	Colombia	Spanish	W.E.	31
3b	John	Diego	Worse	Colombia	Spanish	W.E.	31
4a	John	Diego	Better	Colombia	Spanish	W.E.	29
4b	John	Diego	Worse	Colombia	Spanish	W.E.	29

4 EXPERIMENTS

We conducted a series of experiments to evaluate the authoring efficiency gained by using the CultureCom system to instantiate these dialogs. In the CultureCom condition, objects are created using the Protégé ontology editor and saved into a file structure parallel to the one shown in Figure 3. The resulting files are ready to be used in the social simulation framework described by Sagae, et al. (2011). However in this evaluation we load the files into a text-based interaction loop where the author types conversational turns ("John" turns from dialogs 1-4) and views the system response, printed to the screen. These responses are produced in real time by the dialog engine, given currently-loaded models. To validate whether the models accurately capture one of the dialogs, the system can run each input turn sequentially against the current models and compare the predicted output

("Aziz" or "Diego" turns) to actual output (real-time system response).

We compare the CultureCom condition to a baseline condition where the same dialogs are authored using the FSA formalism. In this condition, dialog accuracy is tested using a tool similar to the CultureCom batch-load function. The FSA tester provides a pass/fail result, depending on whether actual output matched the predicted output exactly, or not.

4.1 Efficiency in the Number of Files and Build Process

Our first hypothesis was that instantiating a given scenario in a new culture is simpler, in terms of file changes and build process, for the CultureCom condition.

Table 2. Excerpt from dialog 1a: Better outcome in Afghanistan

Turn	Speaker	Line	Cultural Observations
1	John:	Salaam Alikum, Aziz.	Good: customary local greeting in local language.
2	Aziz:	Salaam Alikum, John.	Customary response.
3	John:	How are you today?	
4	Aziz:	I am well. And how are you?	
5	John:	I'm fine, things are going well. And how are things with your family?	Asking about family (in general, not women) before getting down to business.
		…	
9	John:	We have some forms that need to be filled out…	
		…	
24	Aziz:	I promise you that I will have the forms for you…	Fixed "promise" phrase is required to imply commitment; "ok" would not.

Table 3. Excerpt from dialog 3a: Better outcome in Colombia

Turn	Speaker	Line	Cultural Observations
1	John:	Buenas tardes, Diego.	Good: customary local greeting in local language.
2	Diego:	Buenas tardes, John. How are you today?	Customary response.
		…	
5	John:	I stopped by so we can set up a meeting …	
		…	
11	John:	Do you think it's too early for people… I suppose we could meet from 9 to 10.	Accommodates change in timing.
16	Diego:	I think it is best to give us the time.	Culturally appropriate indirectness. Doesn't come out and talk about lateness.

To test this hypothesis, we created the file hierarchy shown in Figure 3, and tested against dialogs 1b and 1c. In Dialog 1b, the Aziz character models Afghan cultural norms but John fails to observe them. John greets Aziz only once, failing to build rapport with a three-stage greeting. In dialog 1c, John has an American interlocutor and his directness results in a better, not worse, outcome. To accomplish dialog 1c given a working 1b, we create a new character, Mike, in the Mission II file. This character inherits existing data from the World English language model and new data from the American culture model. The test cases remain unchanged and the build process is unaffected. To accomplish the same behavior in the FSA case would require a new FSA, duplicates of the language objects, and duplicate communicative act objects. Since these objects are stored in a single file, total files changed is 1, but the change affects a large percentage of the file. In addition, the FSA formalism requires a build step to compile these objects into a runnable object, unlike in the CultureCom condition. As a result, such changes cannot be made on the fly to the FSA.

4.2 Efficiency in the Time Required to Author

Our second hypothesis was that, as the number of scenarios being authored grows, the CultureCom condition exhibits more efficiency than the baseline condition in terms of time required to author each scenario. To test, we encoded the same dialogs (3a, 3b, 4a, 4b) using both methods and compared authoring time for a number of steps, as well as overall. The results show that total time to author with CultureCom was greater for these four dialogs, however the trend in terms of scalability was favorable to the CultureCom condition. Time per dialog fell consistently in the CultureCom case from dialog 3a to dialog 4b, while in the baseline condition, time fell only when adapting a given dialog for a new outcome, as in 3a-3b or 4a-4b. There was no scalability in the baseline case when adapting from dialog 3 to dialog 4. An example of this trend is shown in Table 4, which shows authoring time for creating Communicative Act objects.

4.3 Efficiency in the Number of Authored Objects

Our third hypothesis was that, as the number of scenarios being authored grows, the CultureCom condition exhibits more efficiency than the baseline condition in terms of the number of objects that must be instantiated. To test, we used the same authoring task as for hypothesis 2, but evaluated on object counts rather than authoring time. The result shows that object reuse is greater in the CultureCom condition. In particular, language data, communicative act data, and higher-level behavior rules (the equivalent of transitions in the FSA) are all reused to greater advantage in the CultureCom condition. Object reuse in the baseline condition occurs, but as with authoring time, reuse is limited to dialogs that share the same language, culture, and topic (as in 3a-3b), but when adapting to a new topic (as in 3a-4a) there is much greater reuse in the CultureCom case. Table 4 bears out this trend in the case of Communicative Act objects. In the CultureCom condition, the total number of Acts authored for dialog 3a is high, since we break each dialog turn into a greater number of Acts in this condition. Acts can cover a portion of a turn, and Acts which recur often (modeling “Okay”, “Bye”, or “No”) do not have to be reauthored. In the baseline case, monolithic Acts represent entire turns (“Okay, I’ll get the papers to you by Monday”). This yields fewer acts, but each of them occurs in a limited context and can rarely, if ever, be reused.

As a result, in the baseline case we see the number of Acts required for dialog 3a (31) is nearly the same as the number required for 4a (29). In the CultureCom case, we see a significant drop from 3a (77) to 4a (63)

Table 4. Time and object counts for Communicative Act authoring. Lowest time for each condition is shown in bold.

	CultureCom			Baseline		
Dialog	**Time**	**# Objects**	**Time/Obj**	**Time**	**# Objects**	**Time/Obj**
3a	00:38:04	77	00:00:30	00:12:07	31	00:00:23
3b	00:14:33	31	00:00:28	**00:08:06**	16	00:00:30
4a	00:28:53	63	00:00:28	00:14:21	29	00:00:30
4b	**00:13:52**	34	00:00:24	00:11:23	17	00:0031

5 CONCLUSIONS AND FUTURE WORK

The results presented here show that a compositional, model-based approach to social simulation development can result in greater efficiency, in terms of authoring time and reuse of linguistic and cultural resources that are expensive to develop. As the number of simulations increases, the advantage of authoring with reusable objects becomes more and more evident.

In addition to efficiency, another advantage is increased consistency. In the case of FSAs, there is no centralized data structure where cultural cues, norms, expectations, or rules can be saved. Two different authors working on FSAs for the same system must agree informally on these features, and there is no formal method for validating that a given FSA upholds the agreement. The CultureCom system identifies precisely which culture-general and culture-specific rules are in force for a given simulation, supporting consistency and formal validation.

In future work, we would like to investigate the accuracy of the CultureCom framework and the tradeoffs that exist between efficiency and word-level accuracy with respect to a given dialog. In the experiments described here, accuracy for the FSA condition was greater than for the CultureCom condition at the surface level, meaning that the FSA did a better job of replicating the dialog turns word-for-word. This effect is partly caused by the fact that CultureCom communicative acts (e.g., *greeting-response*) can be linked to multiple surface-level utterances (“I’m fine, thanks” “I’m doing well”). At evaluation time, the intent planning module of the dialog engine (Sagae et al., 2011) may select any of these utterances. The same features of the system that lead to greater object reuse and efficiency contribute to this perceived drop in accuracy, when we would like to optimize for both.

In addition, our current work focuses on dialogs, which encode verbal behavior. However the social simulation engine supports rules that capture non-verbal behavior as well. A natural extension of this work could apply the same data development, logical modeling, and social simulation architecture to model non-verbal behavior. Like verbal behavior, gestures are made and interpreted in culture-general and culture-dependent ways that make them well suited to an approach like ours.

ACKNOWLEDGMENTS

The authors would like to acknowledge the Office of Naval Research for their support of this work under contract NOOO14-09-C-0613.

REFERENCES

Fletcher, D. J. (1990). Computer-Based Instruction: Costs and Effectiveness. In A. Sage (Ed.), Concise Encyclopedia of Information Processing in Systems and Organizations. Elmsford, NY: Pergamon Press.

Gennari, John H., et al. (2002). "The Evolution of Protégé: An Environment for Knowledge-Based Systems Development." in International Journal of Human-Computer Studies. 58:89-123.

Hobbs, Jerry R., and Gordon, A. (2010). "Goals in a Formal Theory of Commonsense Psychology", in A. Galton and R. Mizoguchi (eds.), Formal Ontology in Information Systems: Proceedings of the Sixth International Conference (FOIS 2010), IOS Press, Amsterdam, pp. 59-72.

Hobbs, J. R., & Sagae, A. (2011). "Toward a Commonsense Theory of Microsociology: Interpersonal Relationships" in *Proceedings of the 10th Symposium on Logical Formalizations of Commonsense Reasoning, AAAI Spring Symposium Series*. March 21-23, 2011. Stanford, California.

Johnson, W. L., & Friedland, L. (2010). Integrating Cross-Cultural Decision Making Skills into Military Training. In D. Schmorrow & D. Nicholson (Eds.), Advances in Cross-Cultural Decision Making. London: Taylor & Francis.

Johnson, W.L. & Valente, A. (2009), "Tactical Language and Culture Training Systems: Using AI to teach foreign languages and culture." AI Magazine, 30 (2), 72-83.

McDonald, D. P., McGuire, G., Johnston, J., Selmeski, B., & Abbe (2008), "Developing and Managing Cross-Cultural Competence Within the Department Of Defense: Recommendations For Learning and Assessment." US Defense Regional and Cultural Capabilities Assessment Working Group.

Sagae, A., Johnson, W. L., & Valente, A. (2011). Conversational Agents in Language and Culture Training. In D. Perez-Marin & I. Pascual-Nieto (Eds.), Conversational Agents and Natural Language Interaction: Techniques and Effective Practices (pp. 358-377). Madrid: IGI Global.

Wertheim, S. and Agar, M. (in press) "Culture that Works" in *Proceedings of the Second Conference on Cross-Cultural Decision Making*. July 23-25, 2012. San Francisco, CA.

CHAPTER 30

Operational User Requirements and Example Use Cases for Human Social Culture Behavior (HSCB) Technologies

Ronald A. Moore, Chiesha M. Stevens, Heather M. Oonk, Marlin G. Averett

Pacific Science & Engineering Group
San Diego, California, United States
RonMoore@Pacific-Science.com

ABSTRACT

The Human Social Culture Behavior (HSCB) modeling program is working to develop reliable, validated forecasting/predictive models of human behavior based on social and cultural factors; to integrate HSCB processes and technologies into strategic, operational, and tactical level operations; and to provide socio-cultural understanding and skills needed for individuals and small teams engaged in irregular warfare operations. Understanding the requirements of potential HSCB technology users can help inform and guide research and development efforts.

Presented here are user requirements and example problems/use cases gathered from several operational sites that consider and use human, social, and cultural information.

Results from the site visits revealed many similarities (and a few differences) across the potential HSCB user community. HSCB-related information is often used to help guide decision makers as they determine how best to interact with specific groups. This information is most often used to develop and share information, and to present recommendations and alternative courses of action. A wide range of products are developed at these sites, but typical products include summaries, recommendations, technical reports, presentations, and statistical information and projections.

Keywords: HSCB, user requirements, site visit, visualization, analysis

1 BACKGROUND

The Office of the Deputy Undersecretary of Defense (OSD) directs and oversees the Human Social Culture Behavior (HSCB) modeling program in direct support of the nation's irregular warfare efforts. The HSCB modeling program is working to:

- Develop reliable, validated forecasting/predictive models of human behavior based on social and cultural factors,
- Demonstrate the feasibility of integrating HSCB processes and technologies into strategic, operational, and tactical level operations – specifically with regard to conflict resolution and regional stability operations, and
- Integrate and demonstrate training technologies to provide socio-cultural understanding and skills needed for individuals and small teams engaged in IW operations.

These goals are being achieved through a wide variety of basic and applied research, and technology development and demonstration managed by the Office of Naval Research (ONR).

Pacific Science & Engineering (PSE) is a research and development company specializing in human factors and human-systems integration. PSE currently supports ONR on the HSCB modeling program to assist with determining user requirements, and transitioning tools, systems, and technologies to potential users at a variety of operational commands. This effort has involved meeting with potential HSCB users and technology developers to align goals and capabilities between the two groups. Due to normal variations between the sites visited, methods of interaction and results varied slightly but have typically included:

1. Conducting initial and follow-up site visits to meet with, interview, and/or observe the operations of a wide range of staff members to gather information about their current and future/desired HSCB-related operations, information requirements, work processes, and products,
2. Establishing and maintaining a longer-term relationship between the user site and the HSCB modeling program to facilitate the evaluation and adoption of prototype HSCB datasets, models, and technologies,
3. Assisting in integrating HSCB-related data, models, and products into their processes, workflows, and tools (where possible), and
4. Providing site- and user-related information to the HSCB research and development community to support their research and development.

2 SITES VISITED

Five different military installations were visited – some on multiple occasions – from June 2009 to September 2011. Each of the sites visited engage in planning for, supporting, monitoring, or commanding/coordinating military operations involving interactions with other groups and cultures. Many of these sites also share information with US and international non-governmental partners. The personnel staffing the visited sites had diverse backgrounds and filled different roles within their organizations. Some collected data, conducted analyses, and developed reports and recommendations. Others reviewed information and recommendations and then made high-level operational plans and decisions. Many were active duty military personnel or had served in or with the military previously. Others were government employees or contractors. Some had formal or informal training and experience in the social sciences (e.g., anthropology, psychology, sociology, economics, linguistics, international relations), and most had specific subject matter expertise with particular cultures or groups. All of the personnel participating in the site visits actively engaged in using human, social, culture, and behavior information to solve operationally-relevant problems. Personnel at these sites served as the subject matter experts (SMEs) for the information obtained.

One of the sites has the additional mandate to innovate analytic methods, processes, and technologies.

2.1 U.S. Africa Command (AFRICOM)

The United States Africa Command (AFRICOM) is a regional headquarters and fully unified combatant command located in Stuttgart, Germany and focusing on Africa. Its stated mission is to:

- Protect and defend US security interests by strengthening the defense capabilities of African states and regional organizations,
- When directed, conduct military operations, in order to deter and defeat transnational threats, and
- Provide a security environment conducive to good governance and development. (United States Africa Command, 2011)

AFRICOM has administrative responsibility for and oversight of US military support associated with US government policy in Africa, to include fostering and maintaining military relationships with 54 African nations. AFRICOM's area of responsibility (AOR) is large, diverse, and incredibly complex. AFRICOM commanders must continuously consider a wide range of social, political, economic, and environmental challenges facing African nations that are of strategic interest to the United States. Ongoing internal conflicts and wars, economic instability, poverty, ethnically disparate populations, and the lack of reliable and trustworthy governance are just a few of the many problems faced by African nations.

On several occasions during 2011 PSE visited AFRICOM and met with representatives of two related groups supporting AFRICOM's mission.

The first group, AFRICOM's Intelligence and Knowledge Development Plans and Architectures (IKD-P&A) group gathers and analyzes large quantities of socio-culture, economic, and geographic data and knowledge, prepares reports and recommendations related to the EUCOM commander's evolving operational requirements, and develops and manages technical solutions associated with gathering, analyzing, visualizing, storing, and sharing data with its US and African partners. The IKD-P&A group is staffed primarily by active duty, prior-service, and civilian analysts and technologists.

The second group, AFRICOM's Social Science Research Center (SSRC), performs in-depth socio-culture analyses with a focus on understanding social substructures, influences, power, and resources throughout the continent. The SSRC also deploys small advisory teams in-theatre with the explicit approval of the host nations being visited. The SSRC is staffed primarily by social scientists and is supported by active duty, prior-service, and civilian analysts and technologists.

2.2 U.S. European Command (EUCOM)

Like AFRICOM, the United States European Command (EUCOM) is a regional headquarters and fully unified combatant command located in Stuttgart, Germany. EUCOM however, focuses specifically on the European region, as well as some parts of Asia and the Middle East, and the Atlantic and Arctic oceans.

EUCOM's mission is to conduct military operations, international military partnering, and interagency partnering to enhance transatlantic security and defend the interests of the United States and its allies in the European AOR (United States European Command, 2012).

Example EUCOM activities include providing pre-deployment training to US and NATO forces, providing logistical support to equipment and personnel moving between the continental US and various destinations overseas, participating in numerous international exercises and security assistance projects, and interacting with NATO, academia, and numerous international and non-governmental organizations (NGOs) on a wide range of issues (Stavridis, 2010). Additionally, EUCOM plays a major role in the fight against transnational terrorism by engaging with its many partners and sharing information and intelligence in the region.

There are currently 51 nations within the EUCOM AOR, and with them come a wide range of near- and long-term opportunities and challenges associated with the many diverse languages, cultures, and religions of the nations involved, and the ever-changing political, economic, and military environment in the region. To help deal with some of these opportunities and challenges, EUCOM's Deep Futures group was established to provide insight into the long-range risks and opportunities in EUCOM's strategic environment through the analysis of socio-cultural dynamic futures to inform command decision-making and planning (Busch, 2010).

During 2011 PSE visited EUCOM on several occasions and met with Deep Futures representatives to discuss available current and desired data sets, models, and supporting technologies that would help them in their efforts to address and provide insights into EUCOM's long-range operational mission.

2.3 Air Force Targeting Center (AFTC)

The mission of the AFTC is to enable the integration of capabilities across air, space, and cyberspace to deliver precise coercive effects in support of national defense and the global interests of the United States. The AFTC produces geospatial intelligence, targeting, combat identification products, and unit support to better prepare combat air forces for operations (Brown, 2011). As part of the broad support that the AFTC provides, AFTC personnel are currently considering a number of human-, social-, and culture-related data sets and models to supplement their more traditional products. For example, the AFTC is currently exploring the concept of behavioral influence analysis (BIA) as a means of extending their analytic capabilities and products.

PSE staff interviewed numerous AFTC staff members in 2011 to gather information on their current and desired future operations, and to determine the extent to which HSCB data and tools could support the AFTC mission. The goals of these visits were to assist the AFTC in developing a capacity for integrating human, social, and culture information and products into the targeting process, and to provide AFTC-related information to the HSCB modeling program performers to help inform and guide their research and development efforts.

2.4 Marine Corps Information Operations Center (MCIOC)

The MCIOC mission is to provide the Marine Air Ground Task Force (MAGTF) commanders and the Marine Corps a responsive and effective full-spectrum information operations (IO) planning and psychological operations (PSYOPS) delivery capability by means of deployable support teams and a comprehensive general support reach-back capability in order to support the integration of IO into Marine Corps operations (United States Marine Corps, 2010).

Supporting this mission often involves training and supporting mobile training teams, supporting forward-deployed units during exercises and real-world operations, providing an information and analysis reach-back resource for forward-deployed teams, and deploying IO planning specialists to be integrated with joint and coalition forces. MCIOC staff include a combination of active duty military, and government and civilian contractors.

PSE staff spent two weeks onsite at the MCIOC and worked closely with MCIOC personnel to understand how information and technology resources available at the MCIOC are currently being used when responding to requests for information (RFI) from deployed units, and to discuss available current and desired data sets, models, and supporting technologies to help support MCIOC's mission.

2.5 Skope: An Analytic Center for Innovation

Skope is an intelligence analysis group that performs sophisticated analyses of various information sources in direct support of forward deployed teams. Using small, multi-disciplinary teams of experts – known as Skope cells – and a range of high-tech analytic tools and methods, it produces a variety of synthesized, aggregated, easy-to-understand analytic products custom-tailored to each forward deployed customer. (Kenyon, 2011).

In addition to supporting forward-deployed forces, the Skope center has the additional mandate to actively look for ways to innovate their own technologies and work processes. Therefore, in addition to various sorts of intelligence analysts, Skope employs "in-house" software developers and technologists who focus on finding, adapting, or creating time-saving technologies that support analysts in tasks they preform repetitively over time, in particular those that involve the filtering and analyzing of massive amounts of data. Skope developers work closely with analysts to ensure that the technologies meet specific analysis needs and they frequently make changes to the technologies "on-demand," as analysts request them. A number of these tools have been made available to, and have been adopted by other analysis organizations and forward-deployed units.

PSE spent several weeks onsite at Skope in 2009 and 2010 to understand Skope operations and assist Skope developers with technology development and evaluation. PSE conducted informal interviews and observations of users to understand their tasks and operational environment, and used the results to expand and tailor tool help files so that they were more complete, usable, and operationally-relevant. Additionally, PSE provided human factors feedback recommendations for two Skope tools, and designed and conducted empirical assessments that measured the tools' effectiveness and suitability for use by the larger intelligence community.

3 USER REQUIREMENTS

The site visits revealed many similarities across the potential HSCB user community. At every site visited, HSCB-related information is collected, analyzed, and used to help guide decision makers and tactical personnel as they determine how best to interact with specific groups. To best support their "customers" (i.e., the various military personnel and partner-organizations and -nations that use their products), and to develop the kinds of products expected of them (e.g., summaries, recommendations, technical reports, presentations, and statistical information and projections), users at the visited sites described the need for more and better quality data, for more powerful data analysis, management, and visualization tools, and for better sharing and collaboration technologies.

Presented here is a summary of some of the most consistent findings from the site visits. Unless otherwise noted, the findings were consistent across all user sites. Numerous other important findings cannot be included here due to space limitations and therefore will be reported separately elsewhere.

3.1 Data and Access

Potential HSCB data and technology users reported numerous issues with the data that they have available to them to include human, social, cultural, tribal, economic, religious, and other types of human-related data. Notably, the problem is not simply a matter of quantity – in fact, analysts often report having *too much* data to properly analyze and consider – instead, the problem is getting enough "good," reliable, *relevant* data. Analysts report that much of the data available to them is incomplete, obsolete, inaccurate, or difficult to validate. Additionally, they report that data sources were simply not available to answer many of the questions posed by their customers.

With the partial exception of the Skope center, many of the users reported a heavy reliance on open-source data. This was perceived as both a blessing and a curse. On the one hand, open source data was sometimes more readily available on the open internet, and open source data (and products derived from it) can easily be shared with all customers. On the other hand, open source data tended to be incomplete, difficult to validate, and was sometimes considered untrustworthy by customers more familiar with and confident in classified intelligence sources.

Access to data was another challenge common to most analysts. There are two problems associated with data access. First, there is the problem of simply knowing whether the data exists at all, and knowing where it is and how to access it. The second problem is that there is no single, comprehensive clearing house of HSCB-type data. As a result, numerous sources must be accessed to assemble a complete data set. Many of these sources – including open-source data sources – are limited access sites, or they require individual log-ins or specialized systems to access the data. As a result, even if the analyst knows that data exists somewhere, he or she may still not be able to access it easily.

In summary, potential HSCB data and technology users stated a significant requirement for more and better *relevant, reliable* data, and for improved methods of accessing and managing data.

3.2 Analytic and Visualization tools

Analysts and technologists at the visited sites typically use *Microsoft Excel* to organize and analyze their data – particularly when analyzing very large data sets. They often augment *Excel*-based analyses with other analytic tools such as Esri's *ArcGIS* or GeoEye Analytic's *Signature Analyst* for geographic analyses, i2's *Analyst Notebook* or Carnegie Mellon University's *ORA* and *AutoMap* tools for social network analysis, and other specialized modeling, simulation, and analytic tools for other kinds of specialized analyses. Analysts report that these tools can prove extremely useful and powerful when used by experts who fully understand both the data and the analytic tools, but each of these tools requires a great deal of training and expertise to truly master. At the same time, users report that these tools have significant drawbacks with regard to analyzing many kinds of data, and that a *significant* amount of pre-processing/pre-analysis work must be performed to make some data usable. This makes performing rapid-turnaround analyses difficult. Users report the requirement for more powerful, easier-to-use tools that can perform much of the pre-processing/pre-analysis work automatically. Additionally, most users report a strong desire to better understand what's going on "under the hood" of these tools – currently they must accept at face value the tools' outputs/products.

Many of these same tools have a visualization capability in addition to their analytic capabilities. For example, *ArcGIS* and *Signature Analyst* can be used to show important geographic and geo-temporal relationships; *Analyst Notebook* and

ORA can be used to visualize complex social networks; and various models and simulations allow for the visualization of likely outcomes or scenarios; however, similar to their use as analytic tools, to be valuable these tools must be used by expert analysts with significant training and experience. Users report some difficulty understanding the visualizations and using these them as a basis for developing recommended courses of action. Analysts frequently report difficulty explaining some of the more complex visualizations to their customers who often do not have a background in sophisticated analytic techniques or modeling and simulation. Therefore, users unanimously state the requirement for improved visualization tools to perform analyses and explain their results to others.

3.3 COA Development and Analysis

Oftentimes the customers of potential HSCB data and technology users demand multiple recommended courses of actions (COA) – in other words, customers want several realistic options to choose from, and they want to understand the risks and tradeoffs associated with each. This means that analysts at the visited sites are required to develop multiple potential COA and analyze the tradeoffs between each, and then explain these to the customer. Analysts reported that, depending on the situation and available data, it could prove extremely time consuming to manually develop and analyze several different COA; and in some cases the analytic tools simply did not support anything other than an optimized solution, so analysts had to figure out ways to "trick" the system into developing alternate COA. Analysts voiced a desire for tools that would automatically or semi-automatically develop multiple COA and highlight the differences between them.

3.4 Information Sharing and Collaboration Tools

Scientists and analysts at the visited sites discussed the need for improved methods and technologies to support information sharing and collaboration. Currently, there are a number of challenges to efficient and effective information sharing and collaboration, primary among them being that analytic work is conducted on several different networks (e.g., local intranets, the internet, NIPRNET, SIPRNET, JWICS, and various closed systems) while information sharing and product distribution often takes place on other networks causing a cross-domain challenge.

Information sharing *policy* can also be an issue. Oftentimes multiple sources are used to develop aggregated end-products; yet government and/or organizational policy limits the sharing of some information. Analysts sometimes aren't aware of these policies and have difficulty determining what can be shared.

3.5 Model and Product Validation

Many customers ask scientists, analysts, and technologists to explain or validate their analytic results and recommendations. This can be extremely difficult or impossible without insight into the reliability of the data sources, the analytic methodologies used by some tools, and the validity of various models and simulations employed. Therefore, potential HSCB data and technology users require the ability to fully understand – and explain to others in easy-to-understand terms – how various models and technologies work, and to discuss their reliability, and their extensibility to other data sets and problem domains.

4.0 EXAMPLE USE CASES

As a common context for ongoing discussions, and research and development activities at potential user sites, several real-world problems/use cases were considered and/or selected and are briefly summarized below. These problems/use cases were suggested and/or vetted by scientists, analysts, and technologists.

4.1 The Lord's Resistance Army

The Lord's Resistance Army (LRA) use case focuses on understanding and developing recommendations associated with the LRA, a rebel group that is responsible for Africa's longest running armed conflict and the displacement of hundreds of thousands of people in central Africa.

4.2 Migration into Europe

The migration use case explored current and future migration patterns into and within Europe. For one particular discussion and analysis, African migration into France was studied. Topics of interest include migration patterns and trends specific to different groups and sub-cultures, and a projection of likely destinations.

4.3 Key Leader Engagements

The key leader engagement use case is somewhat more "tactical" in nature and attempts to understand and develop specific recommendations associated with US and coalition forces identifying and engaging (i.e., meeting with) key individuals such as government, tribal, organizational, or group leaders. Topics of interest include leader identification, customs and cultural norms and taboos, and specific recommendations regarding conducting the meetings.

4.4 The Arab Spring

The Arab Spring use case explores events and issues associated with Egypt's Tahir Square, the Libyan Uprising/civil war, and other cultural and political unrest in the Middle East, and attempts to develop specific recommendations regarding understanding the confluence of factors leading to unrest, predicting unrest, influencing the peace process or other group actions or decisions, and using social media as a source of HSCB data.

5 CONCLUSION

Five potential HSCB user sites were visited by PSE representing approximately 60 scientists, analysts, and technologists. Personnel at the various sites were actively engaged in using human, social, and cultural information to solve real-world, operationally-relevant problems.

The user requirements gathered from these visits were highly consistent across all five sites. While the future HSCB user community may eventually number in the thousands, and user requirements will no doubt evolve over time, the findings from these site visits should prove useful in informing and guiding HSCB research and development in the future.

ACKNOWLEDGMENTS

The authors would like to acknowledge the time, dedication, and insights provided by each of the participating groups at AFRICOM, EUCOM, AFTC, MCIOC, and the Skope center. Their input, expertise, and unique perspectives made this report possible.

REFERENCES

Brown, J. (April 2011). AFTC experiences change of command. Accessed May 2011 from http://www.jble.af.mil/news/story.asp?id=123251695

Busch, B. (October, 2010). Deep Futures Social Science Research & Analysis Mission Model: A KIBS-based approach. Brief presented to EUCOM.

Kenyon, H. (May, 2011) Skope cells help dispel fog of war. Defense Systems. Retrieved December 21, 2011 from http://defensesystems.com/articles/2011/05/03/cover-story-sidebar-agile-analysis-teams.aspx

Stavridis, J. G. (March, 2010). Testimony of Admiral James G. Stavridis, United States Navy Commander, United States European Command before the 111th Congress 2010. Retrieved December 20, 2011 from http://armed-services.senate.gov/statemnt/2010/03%20March/Stavridis%2003-09-10.pdf

United States Africa Command (August, 2011). About United States Africa Command. Accessed August, 2011 from http://www.africom.mil/AfricomFAQs.asp

United States Africa Command (August, 2011). Fact Sheet: United States Africa Command. Accessed August, 2011 from http://www.africom.mil/getArticle.asp?art=1644

United States European Command (unknown date). United States European Command Mission. Retrived January 16th, 2012 from http://www.eucom.mil/

United States Marine Corps (July 2010). Marine Corps Information Operations Center Command Brief.

CHAPTER 31

VRP 2.0: Cross-Cultural Training with a Hybrid Modeling Architecture

W. Lewis Johnson, Alicia Sagae, and LeeEllen Friedland

Alelo Inc.
Los Angeles, CA, USA
{ljohnson, asagae, lfriedland}@alelo.com

ABSTRACT

This chapter describes a novel hybrid modeling approach for training and simulation of full-spectrum missions. This approach applies knowledge of an area of operation at the macrosocial, mesosocial, and microsocial levels, to provide an accurate, up-to-date, and compelling simulation that includes kinetic and non-kinetic action. These simulations run at a variety of time scales and enable practice for individual or coordinated small group activities, as well as support the unique capability to practice real-time conversations with Alelo's VRP® (Virtual Role-Players) framework and to see the consequences of those interactions affect larger mission objectives.

Keywords: Hybrid Modeling, Training & Simulation, Conversational Agents, Serious Games

1 INTRODUCTION

Operators of simulation-based training systems identify a critical capability gap in conducting training and mission rehearsal in full-spectrum engagements with virtual counterparts. The solution we present is a framework for social simulations where trainees make decisions about mission goals and logistics, and engage in cross-cultural and interpersonal interactions with socially intelligent virtual agents developed using the VRP® (Virtual Role-Player) framework.

The VRP 2.0 framework uses a novel hybrid modeling approach that incorporates macrosocial (e.g., large group phenomena), mesosocial (e.g., social networks), and microsocial (e.g., interpersonal) factors from diverse data sources, in the form of composable models. These models can be updated with new data reflecting facts on the ground, ensuring up-to-date training. They are exposed to the operator as libraries of reusable characters and scenarios that can simply be dropped into a new mission rehearsal session and run. The full lifecycle of training simulations is supported: importing data, configuring and running new simulations, logging, and learning from completed simulation runs.

This work contributes to ongoing HSCB research on hybrid modeling and patterns of life. Matthieu (2011) describes how hybrid models capture spreading influence from factors at the macrosocial level to the microsocial level. In VRP 2.0, this means that features of the population as a whole, such as the amount of trust the community has in its leaders, influence the behavior of individual members of that population. In addition, mesosocial connections among agents allow information to spread from one to another, so that any interpersonal encounter between a trainee and a communicative agent affects not only that agent, but the disposition of his friends, relatives, and community.

Finally, in state-of-the-art systems for training and simulation at a conversational level, behavior of non-player characters (NPCs) is often based on a strictly prescribed sequence of behaviors. There is also no centralized data structure for cultural knowledge. In contrast, the VRP 2.0 hybrid modeling architecture includes composable ontologies that encode macro-, meso- and micro-social behavior rules. These rules can be mixed and matched, and may appear in a wide variety of combinations and orderings. The result is a new hybrid modeling framework for full-spectrum mission simulations that has more flexible, believable behavior and that draws on a variety of sociocultural models to drive that behavior.

2 CONVERSATIONAL AGENTS FOR CROSS-CULTURAL COMPETENCY TRAINING AND SIMULATION

VRP 2.0 builds on and extends Alelo's culture and language training products and technologies, such as Tactical Language courses (Johnson & Valente, 2009) and the Virtual Cultural Awareness Trainers (VCATs) (Johnson et al., 2011), both of which were designed using Alelo's Situated Culture Methodology (Johnson & Friedland, 2010). Immersive serious games provide an integrated learning environment in which trainees must make decisions and engage in cross-cultural and interpersonal interactions with socially intelligent virtual agents. A screen shot is given in Figure 1.

The architecture for simulating conversations in these systems is based on finite-state automata (FSA) that encode conversational branches at the level of communicative function (Sagae et al., 2011). An example of such an FSA is shown in Figure 2. Although adequate for set-piece training scenarios, the FSA approach has shortcomings for broader full-spectrum training. The sequence of actions is strictly prescribed by the FSA, without respecting changes in the surrounding sociocultural context. Cultural knowledge is embedded in FSA states and transitions, making it difficult to adapt behavior to new cultural contexts.

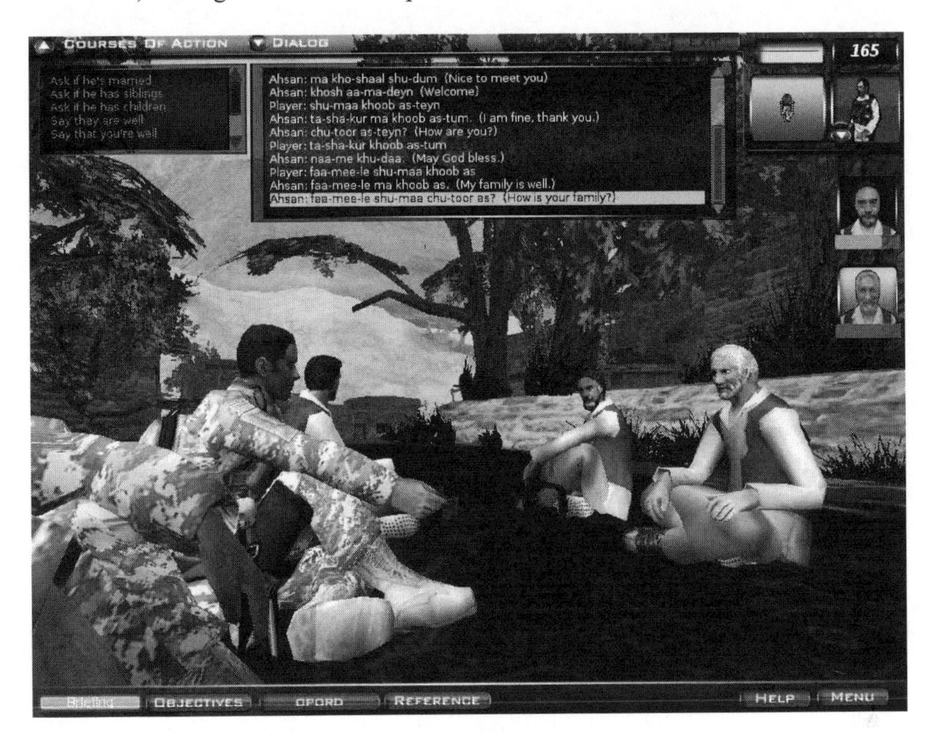

Figure 1. Meeting with the malek in Alelo's Operational Dari

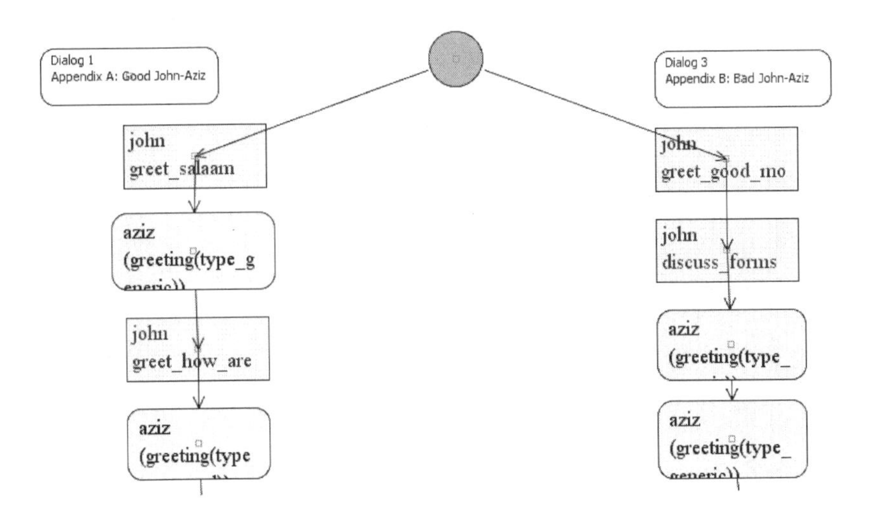

Figure 2. Sample of conversation captured as a finite state automaton

3 A NEW FRAMEWORK FOR SOCIAL SIMULATION

A high-level view of the software and data components of VRP 2.0 is given in Figure 3. A central feature of the architecture is a data or knowledge hub, where integrated models capture the sociocultural, linguistic, operational, and immersive contexts that apply in a given social simulation, or *scenario*. The hub is surrounded by tools that enable cultural or operational subject matter experts (SMEs) to develop and validate sociocultural models, and enable training experts to select, configure, and integrate models into playing scenarios. Multiple immersive simulations are supported, through a common application-programming interface (API). Components shown in green comprise the sociocultural modeling framework. It includes knowledge in the form of macro-, meso-, and micro-social models, tools for humans to create and modify sociocultural models, and APIs that allow other programs to query our models and add knowledge to them. It thus can serve as a common framework for integrating best-of-breed sociocultural models in an immersive training framework.

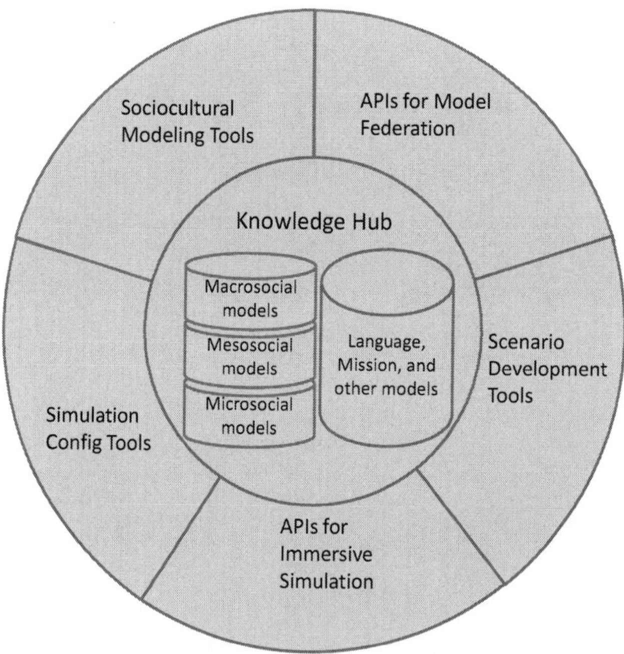

Figure 3. Knowledge and software components of the VRP 2.0 architecture

3.1 KNOWLEDGE HUB

The VRP 2.0 knowledge hub is a flexible graph-based data warehouse. The types of objects stored there are defined in a meta-model called the domain model or domain schema. This schema identifies the major subdomains of knowledge that are required for building an immersive training system, including:

- Purpose: work context, mission/task details, policies, doctrines, and directives.
- Language: forms (phrases, sentence patterns, vocabulary, etc.) and functions (communicative functions, discourse structures, etc.).
- Culture: social frames (e.g., greetings), macro/meso/microsocial levels, and sociolinguistic practices in language use (e.g., politeness strategies).
- Instructional design elements: hierarchical, typed competencies and scaffolding strategies for adjusting levels of difficulty.
- Scenarios: encounters between trainees and agents, and the process by which agents interact with their environment, modeled as an observe-orient-decide-act (OODA) loop (Osinga, 2007).
- Learner history: training context and trainee interaction history.
- Learner model: performance assessments based on the learner history.

Collectively these subdomain models are called social simulation models. They capture the factors influencing interpersonal relationships and social networks, and the mission, region, language, and culture of interest for a given scenario. For example, a populated scenario model might include reusable roles ("Village Elder") and characters ("Aziz, the local Malek") that appear in a mission-specific scenario ("Engage regional leaders in support of an educational campaign on clean water"). A detailed, file-level look at these models is shown in Figure 4. Without the model-based architecture, instantiating this scenario would require a scenario description that is single-use, with few hooks for making connections between elements of that scenario and population models that may already be in use by scenario authors.

A feature of the knowledge hub is that it supports annotation, so that concepts defined in VRP models may be mapped to elements from other lightweight social simulation models and to HSCB population models. For example, the Culture subdomain may include structures like families, regional communities, and other social groups. To associate new information from an external mesosocial model (e.g., "Corruption is the main grievance of heads-of-households in this region") with these structures, we must ensure that the external concepts are captured in the VRP model and that naming differences are resolved. We may want to allow values from the external concept "household" to be stored in the VRP concept "family." We accomplish this through annotating the family concept with a pointer to the household concept. This process makes the equivalence clear to computer systems that may use both knowledge sources as input, without making unintended changes to either original knowledge source. In addition to concept-to-concept mappings, the annotation framework will support structure-to-structure mappings, so that concepts that are modeled in different ways by two different frameworks can still interoperate. This concept mapping process is a critical step toward automatically updating scenarios when external models are updated.

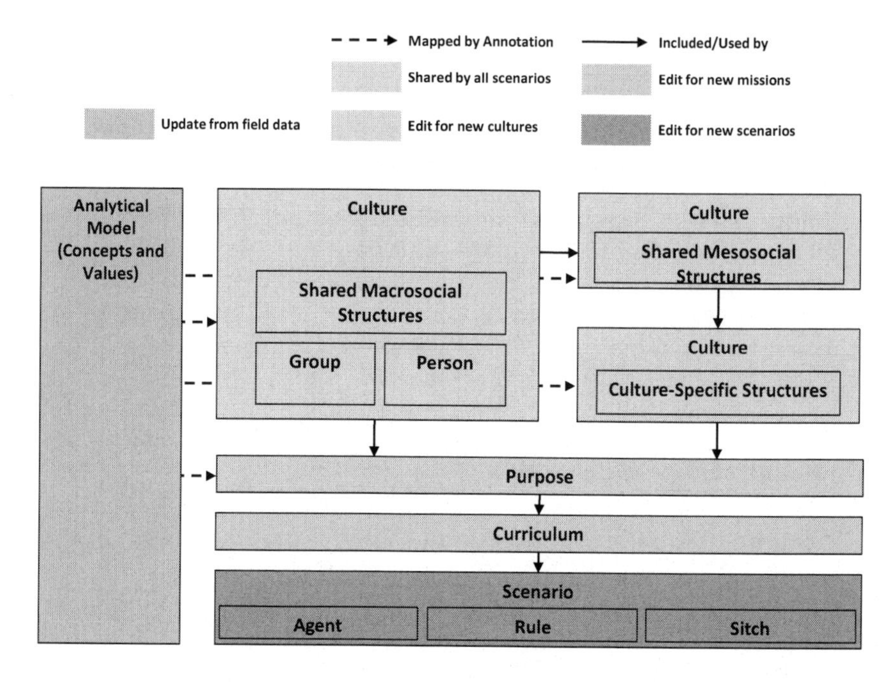

Figure 4. File organization of the social simulation models stored in the VRP knowledge hub

4 SOCIOCULTURAL MODEL DEVELOPMENT

One critical component of the VRP 2.0 capability is the architecture described above. A second component of equal importance is the methodology that uses this architecture to develop sociocultural models and to apply them in training scenarios.

4.1 Situated Culture Methodology and Hybrid Modeling

Alelo has a well-developed methodology for developing sociocultural knowledge called the Situated Culture Methodology (SCM). This methodology results in validated culture-specific knowledge related to the macrosocial, mesosocial, and microsocial features of the culture of interest. An example of how these layers of knowledge fit together is shown in Figure 5.

The VRP 2.0 architecture makes it easy to integrate best-of-breed HSCB population models and data sets. To date, these models have focused primarily on the macrosocial and mesosocial levels. The VRP 2.0 architecture integrates macro- and meso-social factors with microsocial behavior, so that trainee interactions with virtual role-players can result in population effects (microsocial influence spreading to macrosocial level) and population dynamics can affect virtual role-player behavior (macro/mesosocial influence spreading to microsocial level).

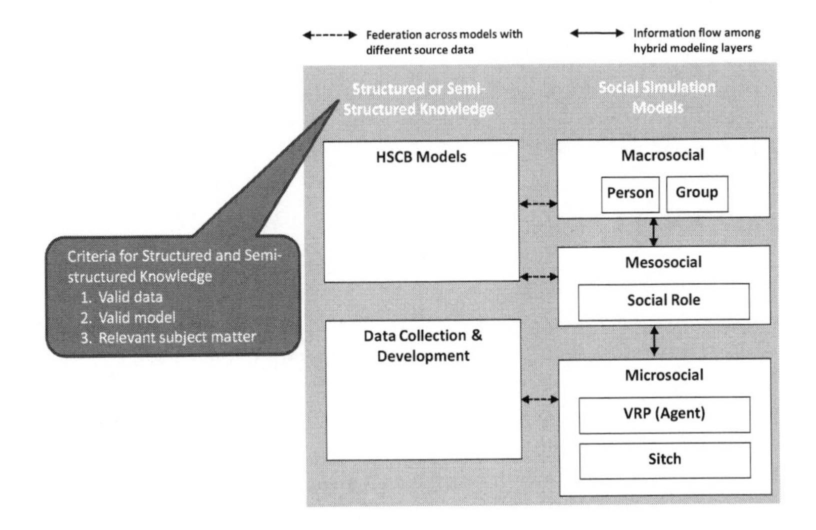

Figure 5. High-level interaction among Alelo hybrid sociocultural models

4.2 Microsocial Modeling

Although we use the situated approach to develop validated models at all levels of this hybrid structure, Alelo has particular expertise in microsocial models that describe verbal and nonverbal behavior of conversational agents during real-time interactions with trainees. The VRP 2.0 framework expands the notion of situation into a hybrid object called a Sitch^{TM}. The Sitch^{TM} combines local and non-local features that affect a particular scenario, in addition to the software framework required to instantiate the connections between macro-, meso- and micro-social models described here.

A typical microsocial model includes Conversational Agents and character-level Dialog Rules for each agent, e.g., "when trainee suggests a location, agree IF trust at the regional level is high AND rapport with this player is also high". In addition to these default rules that are equipped for each agent, we require additional objects that capture more detailed, situated paths that the dialog may take. Alternative contexts may become active when the agents become frustrated by trainee mistakes, when a virtual-world event like an IED explosion pushes all participants in the interaction into a new stress level, or in general when the "pattern of life" in the simulation dynamically changes. Subsets of dialog rules that are active for an agent in a particular situated context are stored in the Sitch^{TM} object. The Sitch^{TM} contains dialog rules that apply there, including rules that exit the Sitch^{TM} and return control to the background Scenario-level rules, or to a subsequent Sitch^{TM}.

This sociocultural modeling work also allows us to capture and control the features of a scenario that make it more or less difficult for a trainee to complete. If challenges are too easy, trainees may get bored and engage with minimal effort; if challenges are too difficult, trainees may experience frustration. The features we designed for trainers to control levels of difficulty include: cooperativeness of the virtual community where the mission takes place, tolerance level of characters in the community for mistakes made by the trainee, the number of "good" and "bad" paths that the trainee is allowed to pursue, time allowed for a trainee to make decisions during the simulation, and storing and resetting learner history.

5 EXAMPLE SCENARIO

To bring the architecture and modeling work together, we now present an end-to-end description of the lifecycle of a social simulation, or scenario: creating social simulation models (scenario modeling), integrating and updating these models with new data sources (model alignment), selecting elements from the models to configure a new scenario (scenario configuration), deploying these models in a training-ready executable (scenario generation), and tracking trainee performance in order to adjust subsequent simulations (scenario execution). The example scenario deals with a Humanitarian Assistance and Commander's Emergency Response Program (HA/CERP) mission, taking place in Colombia, South America. The trainee must negotiate with a local counterpart in the Colombian military, in order to secure collaboration on a joint project.

In the scenario modeling phase, the sociocultural research team performs basic research on the Colombian region of interest, including primary sources such as interviews with SMEs and investigation of existing macro- and meso-social models for Colombian culture. This research is encoded in a variety of social simulation models, including Spanish words and phrases in the Language subdomain; HA/CERP procedures in the Purpose subdomain; learning objectives like "Hold a planning meeting" in the instructional design subdomain; and a conversational agent, Col Suarez, is created and enabled with verbal and nonverbal behavior rules.

If existing macro- or meso-social models are available, then a model alignment phase allows concepts from such models to be added, along with an annotation that connects the new simulation-ready interpretation to the original concept. These macro- and meso-social model elements are encoded as meta-behavior rules, e.g., "In social networks in this region, information flows quickly across familial and business relations." If this model is active, then the trainee may find later that information he shared with Col Suarez is also known to the Colonel's brother and business partner. All of this scenario modeling work is done using the sociocultural modeling tools shown in Figure 3. The result is new instances of objects in the knowledge hub that are then accessible to other tools and build processes.

At scenario configuration time we create a new empty Scenario object and include, by reference, the agent we created for our Colombian military counterpart. We also create at least two Sitches: one for the context where the agent is cooperative, and another for the context where he is not. Although the conversation may enter the uncooperative context for a variety of reasons (low trust in the community, previous bad actions by the trainee or his teammates or general high difficulty setting) the Sitch™ object captures the subset of behaviors that the agent should display once this context is in force.

After selecting the appropriate scenario components, a build process collects the artwork, recordings, maps, and other assets that are required to execute these components in a simulation. The build process is a fully automatic step that creates an executable plug-in for a game engine. Once the plugin has been created, player interaction with the scenario can begin. The trainee starts the game engine and selects our Colombia HA/CERP scenario.

Finally, at scenario execution time the system allows player actions to generate second- and third-order effects that surface in later simulation runs. The system captures learner history and saves it to an object that can be loaded when the learner re-enters the scenario. The learner history includes the system's long-term memory of the state of the virtual world after a trainee has affected it. Col Suarez may start the meeting in a cooperative mood, but cultural or procedural errors by the trainee trigger a transition into the uncooperative Sitch™. At this point, if the trainee asks Col Suarez to confirm his participation in the shared project, the Col will defer, claiming "There are many local factors to consider; the plan will take time." After feedback from the system that gives him clear feedback on what he did well and what he did poorly, the trainee exits the simulation and allows it to fast-forward, so that he can return a few days later in the simulated timeline. He starts the conversation again, applying what he learned. This time Col Suarez confirms that his unit will support the trainee in the joint project.

Although this example describes a Colombian HA/CERP scenario, it also illustrates how the VRP 2.0 model-driven approach leverages reusable components. Humanitarian assistance missions carried out in different countries share many operational details and learning objectives, though the relevant sociocultural data often varies. Therefore, HA missions in Colombia are often similar to those conducted in Afghanistan in their mission structure and instructional design, although the micro-, meso-, and macro-social data represented in the scenario is culture-specific. Our model-driven approach to building social simulations allows us to reuse and replace data as needed in order to create an authentic scenario that accurately represents the culture and supports the instructional goals.

6 CONCLUSIONS AND FUTURE WORK

The architecture described here presents a model-driven approach to integrating the wide range of sociocultural background knowledge that is required to achieve realistic social simulations. It improves on the state-of-the-art frameworks for simulated interpersonal interaction by using a flexible, modular rule engine that can encode behavior at the macrosocial, mesosocial, or microsocial level. Using this architecture, it will be possible to more effectively model second- and third-order effects, so that mistakes that trainees make in a simulated conversation with one character will have realistic consequences for the rest of his or her mission. This architecture is currently under development, with parts of the system including sociocultural modeling tools and the rules engine currently in testing. This development makes possible new collaborations that are already underway among HSCB teams that specialize in different layers of sociocultural granularity (macro-, meso-, micro-social modeling), yielding a hybrid-modeling training and simulation solution.

ACKNOWLEDGMENTS

The authors wish to thank Sara Zaker and others on the VRP project for their contributions to the work presented here. The authors would like to acknowledge the Office of Naval Research for their support of this work under contract NOOO14-09-C-0613.

REFERENCES

Fletcher, D. J. 1990. Computer-Based Instruction: Costs and Effectiveness. In A. Sage (Ed.), Concise Encyclopedia of Information Processing in Systems and Organizations. Elmsford, NY: Pergamon Press.

Gennari, John H., et al. 2002. "The Evolution of Protégé: An Environment for Knowledge-Based Systems Development." In International Journal of Human-Computer Studies. 58:89-123.

Hobbs, Jerry R. and A. Gordon. 2010. "Goals in a Formal Theory of Commonsense Psychology", in A. Galton and R. Mizoguchi (Eds.), Formal Ontology in Information Systems: Proceedings of the Sixth International Conference (FOIS 2010), IOS Press, Amsterdam, pp. 59-72.

Johnson, W. L. and L. Friedland. 2010. "Integrating Cross-Cultural Decision Making Skills into Military Training." In D. Schmorrow & D. Nicholson (Eds.), Advances in Cross-Cultural Decision Making. London: Taylor & Francis.

Johnson, W.L., L. Friedland, P. Schrider, A. Valente, and S. Sheridan. 2011. The Virtual Cultural Awareness Trainer (VCAT): Joint Knowledge Online's (JKO's) Solution to the Individual Operational Culture and Language Training Gap. In Proceedings of ITEC 2011. London: Clarion Events.

Johnson, W.L. and A. Valente. 2009. "Tactical Language and Culture Training Systems: Using AI to teach foreign languages and culture." *AI Magazine*, 30 (2), 72-83.

Mathieu, J. 2011. Hybrid Models. *HSCB Newsletter, Spring 2011*(9), 19-20.

McDonald, D. P., G. McGuire, J. Johnston, B. Selmeski, and A. Abbe. 2008. "Developing and Managing Cross-Cultural Competence Within the Department Of Defense: Recommendations For Learning and Assessment." US Defense Regional and Cultural Capabilities Assessment Working Group.

Osinga, F.P.B. 2007. *Science, strategy, and war: The strategic theory of John Boyd.* Abingdon: Routlege.

Sagae, A., W. L. Johnson, and A. Valente. 2011. Conversational Agents in Language and Culture Training. In D. Perez-Marin & I. Pascual-Nieto (Eds.), Conversational Agents and Natural Language Interaction: Techniques and Effective Practices (pp. 358-377). Madrid: IGI Global.

CHAPTER 32

Theoretical and Practical Advances in the Assessment of Cross-Cultural Competence

William K. Gabrenya Jr., Richard L. Griffith, Rana G. Moukarzel, Marne H. Pomerance

Institute for Cross-Cultural Management
School of Psychology
Florida Institute of Technology
Melbourne, FL USA
gabrenya@fit.edu

Patrice Reid

Defense Equal Opportunity Management Institute
Patrick Air Force Base, FL USA

ABSTRACT

Organizations engaged in international operations must navigate complex intercultural dynamics for successful performance, necessitating identification of individuals who are likely to succeed in these environments and training personnel in cross-cultural competence (3C). To do so, adequate competency models of 3C need to be developed and valid 3C assessment instruments must be identified or generated. The present chapter reviews issues and challenges in 3C model development and illustrates these problems in an analysis of the Defense Language Office's Framework for Cross-Cultural Competence. The comparative advantages of competency versus causal models of 3C are discussed and an integration of competency and causal models is suggested. An examination of 34 instruments that have been recommended in the 3C civilian and military literatures for assessing cross-cultural competencies and their antecedent factors showed that existing 3C measures suffer from poor construct validity and have not been empirically linked to important outcome variables. A measurement strategy that eschews self-report methods and broadly assesses KSAOs and behavioral competencies is advocated.

Keywords: cross-cultural competency, assessment, competency models, military

1 COMPETENCIES FOR CROSS-CULTURAL READINESS

Cross-cultural competence (3C), often defined as the ability to quickly understand and effectively act in a culture different from one's own (e.g., Abbe, Gulick, & Herman, 2007), has garnered increasing attention within the U.S. Military. Military missions have become more complex, blending traditional military operations with nation building goals that require a broader set of competencies over a greater range of specialties and ranks, than the Military was previously required to sustain (Abbe, 2008). The U.S. Department of Defense, through entities such as the Defense Equal Opportunity Management Institute (DEOMI), the Defense Language Office (DLO), and the Army Research Institute (ARI), has devoted substantial resources to understanding and enhancing 3C. Doing so requires resolving three issues: (1) What is 3C, or more specifically, which competencies are needed by which personnel, and at what level of performance? (2) How can these competencies be assessed, for both selection or training purposes? and (3) How can these competencies be trained? The present chapter addresses the first two issues.

Cross-cultural competence has been studied actively in the civilian sector since the 1950s. More recently, the Department of Defense has attempted to develop a conceptualization of 3C for the Military through several official programs, including the National Security Language Initiative (U.S. Department of State, 2008), the Department of Defense Strategic Plan for Language, Regional, and Cultural Capabilities (2010), and the Department of Defense Language Transformation Roadmap (2005). The Roadmap states that language, culture, and regional expertise are not only important "defense core competencies" but also "critical weapons systems."

Extant conceptualizations of 3C generally include four components: (1) relatively stable characteristics of the individual, such as personality traits, cognitive capabilities, social competency, and cognitive styles; (2) culture-general and region-specific knowledge; (3) attitudinal and motivational dispositions such as ethnocentrism, interest in culture, and motivation to learn; and (4) skills such as communication, language, culturally appropriate behaviors, and executive functions such as emotional regulation and metacognition.

Spitzberg and Changnon (2009) reviewed the many models of 3C developed for business, education, the Peace Corps and migration. The most carefully constructed conceptualization of 3C for the U.S. Military at this time is the "Framework for Cross-Cultural Competence." The Framework was developed in several stages beginning in 2008 within the Defense Regional and Cultural Capabilities Assessment Working Group (RACCA WG) (McDonald, McGuire, Johnston, Selmeski, & Abbe, 2008). RACCA subject matter experts (SMEs) identified a set of 40 general cross-cultural learning statements consisting of knowledge, skills, and personal characteristics (also called "core competencies"). A second, overlapping group of culture experts reduced the RACCA competencies to a smaller number and drew a distinction between antecedent variables, which were termed "enablers," and competencies (Johnston, Paris, McCoy, Severe, & Hughes, 2010). Johnston et al. (2010) was revised in 2011 and the Framework was subsequently revised again in a series of communications within a group of DLO culture experts in March, 2011.

1.1 Competency Models of 3C

The Framework is one of the most recent 3C models and reflects two mutually supportive trends: (a) an increased emphasis on addressing 3C as an applied psychology problem, the solution to which has important, material implications for the success of businesses and the Military, rather than a theoretical area of pursuit; and (b) the inclusion of behavioral capabilities or performance domains using the language, if not the full development cycle, of competency modeling (Shippmann, Ash, Battista, Carr, Eyde & Hesketh, et al. 2000). In a competency model of 3C, *core competencies* are identified in a hierarchical categorization system. In this system, general competencies such as "cultural perspective taking" are used to form categories encompassing more specific competencies that are defined behaviorally, for example, "understands how one's own group is viewed by members of another group." *Competency potential* dimensions (Bartram, 2005) are also identified in such models, incorporating variables that have been called antecedents or precursors in the expatriate and overseas study literatures. These variables are termed *enablers* in the Framework, and include a wide range of characteristics such as personality traits, cognitive abilities, cognitive style, and attitudes. However, unlike antecedents in earlier models, enablers are conceptualized in a competency modeling (i.e., behavioral) style.

Competency models of 3C such as the DLO Framework may be subject to some of the unresolved problems in competency modeling in general. Although competency modeling is popular in human resource management, it suffers from a great deal of ambiguity concerning its core construct—competency—as well as how it differs from traditional job analysis (Shippman et al., 2000). Morgeson, Delaney-Klinger, Mayfield, Ferrara, and Campion (2004) state, "perhaps one of the most vexing issues involves actually defining a competency" (p. 676). For example, are competencies composed of KSAOs (knowledge, skill, ability, other), or are KSAOs antecedent to competencies? If the latter, what is antecedent to KSAOs? As Van de Vijver and Leung note, "It could be argued that intercultural competence is no exception to the rule that there are no widely shared definitions of crucial concepts in psychology" (2009, p. 406). Specifying the correct number of competencies and their organization poses a problem for competency modeling that is also present in the Framework. For any given military occupational specialty (MOS), mission, or action, which competencies are important and how many can be practically assessed?

Competency models of 3C share several limitations that persist within the intercultural adjustment and performance literature: (1) imprecision in defining constructs, often in the absence of operationalization; (2) conceptual overlap and unsatisfactory distinctions among key model components such as antecedents, KSAOs, and performance outcomes; (3) imprecision in specifying the causal order among constructs; and (4) imprecision or poor articulation of competencies with respect to the U.S. Military's practical selection and training needs due to insufficient attention to MOS, rank, and service variables. These shortcomings limit the predictive and explanatory ability of existing 3C models, and consequently, limit the predictive ability of existing 3C assessments, making them less than ideal for military use. It is critical that future work on 3C focuses explicitly on addressing as many of these concerns as possible.

Gabrenya, Moukarzel, Pomerance, Griffith and Deaton's (2011a) in-depth analysis of the DLO Framework competency model can be used to illustrate some of these issues. The current version of the DLO Framework includes five competencies and seven enablers. Each competency is defined by or explained by sets of more specific behaviors and skills. These definitional items are themselves considered competencies, introducing a hierarchical structure. Some definitional items are in turn comprised of more than one relatively distinctive component competency. For example, the competency Culture General Concepts and Knowledge includes three parts:

- Acquires…culture-general concepts and knowledge
- Applies culture general concepts and knowledge
- Comprehends…
    - …and navigates…
- …intercultural dynamics

Gabrenya et al. (2011a) refer to components at the most granular level as *elements*. The imprecision of these elements stems from the Framework's use of colloquial terms, a long-recognized problem throughout social and behavioral science. Most readers would believe they know what "navigates" means in this context, but they would probably disagree about how to operationalize or assess the competency. Similarly, "intercultural dynamics" has strong referents in common life experiences but is not useful from a conceptual or measurement perspective.

The overlapping competency issue is illustrated by considering the competency Culture Perspective Taking, which includes four distinguishable components and some subcomponents:

- Demonstrates an awareness of one's own world view (i.e. cultural perceptions, assumptions, values, and biases) …
    - …and how it influences our behavior and that of others
- Understands how one's own group is viewed by members of another group

- Understands …
    - …and applies…
    - …perspective-taking skills to detect, analyze, and consider the point of view of others and recognizes how the other will interpret his/her actions
- Takes the cultural context into consideration when interpreting situational cues

An observer, trainer, or evaluator would be hard-pressed to distinguish "understanding and applying culture general concepts for navigating intercultural dynamics" in the Culture General Concepts and Knowledge competency from "applying perspective-taking skills to recognize how others will interpret one's actions" in this competency.

Competency models do not usually attempt to specify causal relationships among competencies or between competencies and antecedent variables. Nonetheless, such causal relationships are assumed in a model that includes sets of competencies and enablers, and they are implicit in competency hierarchies that specify both the acquisition and subsequent application of knowledge or skills. We argue in a later section that causal models should be explicitly included in 3C conceptualizations.

Most serious for the practical application of cultural expertise in the Military is the fourth criticism, insufficient attention to MOS, rank, and service variables. This problem essentially involves establishing the appropriate content of the model, analogous to establishing content validity in test development. Whether performed from a classic job analysis perspective (Brannick, Levine & Morgenson, 2007) or from a competency modeling approach, information must be obtained to attend to the competencies, selection factors, and appropriate training of 3C at a greater level of detail than found in existing models. Several empirical studies have been utilized to inform 3C requirements in the Military. A study by the RAND Corporation (Hardison, Sims, Ali, Villamizar, Mundell, & Howe, 2009) was conducted to help conceptualize training program content to improve cross-cultural performance within the Air Force. Unlike most such studies, the RAND study incorporated MOS (AFSC-Air Force Specialty Code) and rank, finding considerable variability in the overall importance rating of 3C across specialties. Spencer (2010) reported a qualitative analysis of special operations forces (SOF) personnel that combined features of competency modeling and job analysis. The most recent investigation of 3C in the Military at the time of this writing was carried out by McCloskey, Grandjean, Behymer, and Ross (2010) using respondents who had returned from various overseas postings. Theoretical approaches to establishing the content of 3C have been carried out by Caligiuri, Noe, Nolan, Ryan, and Drasgow (2011) for the Military and by Deardorff (2006) and Hunter, White and Godbey (2006) using a Delphi approach based on the combined perspectives of multiple subject matter experts (SMEs) concerning civilian 3C.

1.2 An Analysis of the Adequacy of the DLO Framework

Gabrenya et al. (2011a) used these and other sources to analyze the content of the DLO Framework. They concluded that the Framework competencies adequately reflect military and civilian conceptions despite problems involving overlapping and imprecisely articulated competencies. The Framework was found to have 19 enabler elements, 12 of which were well represented in other conceptions while four were poorly or not at all represented (e.g., "avoid stress-induced perspectives that oversimplify culture," and "acts as a calming influence"). A few competencies were not found in the Framework despite their importance in the civilian competency literature, such as ability to manage family obligations. While not relevant for some deployments, family and spouse adjustment in the foreign context is the strongest predictor of success/failure in overseas assignments (Caligiuri, Hyland, Joshi, & Bross, 1998). The issue of language ability has been discussed in various documents. Caligiuri et al. (2011) note that teaching and maintaining language skills is not cost effective for the Military, but training 3C may yield a better payoff. The RAND Air Force study (Hardison et al., 2009) found both low valuation of language skills and low language capabilities: 4% claimed a working knowledge of the language of the place to which they had been deployed, and 10% claimed a working knowledge of any foreign language. The authors suggest that low proficiency may have led to low valuation, suggesting that self-reported valuation of competencies may not provide a good measure of their actual importance. Language acquisition as a culture competency will undoubtedly remain an ongoing locus of debate in the Military.

The McCloskey et al. (2010) study, although an admirable effort, suffers from some of the limitations of this kind of research and illustrates some of the dilemmas in using active duty service personnel as SMEs. The study included an overly small and insufficiently broad sample, which from the start limited its ability to analyze its SME data in terms of MOS and rank. Like all qualitative research, it had to navigate the problem of developing conceptual categories (in this case, competencies or antecedents) inductively from its respondents' information, while using the large existing 3C literature to interpret the data. Hence, they used a KSAO set generated from previous research to organize the respondents' ideas about what caused or constituted effective 3C when they were deployed. In any given time frame, most deployed personnel will experience only a small set of culture regions. For example, at the present time most deployed personnel are in the Middle East or at bases in Korea, Japan, and Europe. While it should be feasible to generate a competency model that is not overly biased toward a particular culture region or kind of deployment, it is not assured, so over time new data should be continually obtained and used to revisit 3C models. A fundamental problem in such a bottom-up generation of KSAOs and competencies is the veridicality of the relationship between the KSAOs reported by respondents and actual performance. Respondent data such as that generated in the McCloskey study privileges the respondent's point of view, based as it is on the respondent's implicit theories about the relationships between antecedents and outcomes. Research in social cognition and decision making has shown that people are imperfectly aware of the causes of their own and others' behavior and success/failure outcomes (Fiske & Taylor, 2008). Therefore, research is needed to establish the causal links in integrated models once valid competency and performance measures are developed.

The problem of identifying the correct source information on which to build 3C models cannot be separated from the bottom-line question of which antecedents do in fact "enable" which competencies, and which competencies in turn actually affect performance. The situationally labile nature of performance, especially in complex missions involving many units and on-the-ground factors, poses considerable difficulties to researchers who prefer to settle the antecedent-competency-performance relationships empirically. Gabrenya et al. (2011a) called on the large civilian literature to examine the relationships between DLO Framework enablers and competencies on the one hand, and performance on the other. Their strategy was to interpret the competency and enabler elements in terms of constructs or variables that have been the subject of research on intercultural effectiveness (i.e., map elements to constructs), find good measures of these constructs, and then review the research on the criterion validity of the measures that used performance or adjustment criterion variables. The strategy looks like this:

Competency or Enabler Element → Representative construct → Valid measure → Performance/Adjustment Criterion

For example, the Framework enabler element *receptive to new ways of doing things* maps to constructs such as flexibility and openness which in turn can be assessed by several instruments. Due to the manner in which elements were often expressed behaviorally, vaguely, or in mission-specific terms, the mapping phase was imprecise.

We performed a comprehensive search of the sojourner adjustment/performance literature to identify measures that could be used in this evaluation. Our search capitalized on other attempts to create comprehensive lists of instruments, for example Fantini (2009), Thornson and Ross (2008), the website of the Institute for Intercultural Training (www.intercultural.org), and the websites of several consulting companies. In addition to instruments that were developed specifically for cultural research purposes, we also looked at studies that used familiar personality instruments in the large sojourner adjustment literature, such as the NEO, coping style scales, and measures of social interaction individual differences (e.g., the Self-Monitoring Scale). We return to our evaluation of the state of measurement in this field in another section.

The constructs represented by five of the 12 competency elements and five of the 19 enabler elements could not be measured because either (1) no suitable instrument has been developed, or (2) the available instruments have never been used in concurrent or predictive validity studies involving intercultural adjustment of performance variables. Of the seven measurable competency elements, six were supported and one received mixed evidence. Of the 12 measurable enabler elements, four were well supported, six received mixed support, and two received no support. It must be noted that this is a conservative analysis in that (1) better instrumentation would afford evaluation of more elements; (2) additional research might be able to produce more conclusive findings when evidence was mixed by sorting out which instruments best assessed the construct; and (3) most importantly, elements that did not map well against constructs (in one-to-many and many-to-one relationships) need to be additionally decomposed if model adequacy is to be established empirically.

The findings of this evaluation of an important military 3C model are not encouraging in several respects. First, we could not find valid measures of a large number of elements despite a wide-ranging search, and many of the instruments that we did find did not demonstrate adequate psychometric properties. Second, the competency model nature of the Framework rendered it poorly amendable to empirical validation. Third, our mapping project illustrated an intrinsic weakness of competency models, an issue to which we now turn.

2 COMPETENCY MODELS, CAUSAL MODELS, AND INTEGRATED MODELS

The DLO Framework may be termed a *compositional model* in the sense used by Spitzberg and Changnon (2009) in their attempt to categorize the disparate models of sojourner adjustment and expatriate performance. Such models are primarily lists of KSAOs that comprise 3C, usually organized in logical sets in a way that implies a causal sequence. The Framework uses descriptions of job-related behaviors, in the manner of a typical competency model, rather than constructs or variable names, to describe the set of desired qualities. In contrast to list-like compositional models, *causal path models* represent a linear causal system that may or may not involve feedback paths; such models can usually be tested using multivariate methods (Spitzberg & Changnon, 2009). Causal path models are familiar to social scientists who primarily create models to represent individual and social processes and to generate testable hypotheses; hence, they are probably the most common type of model found in this field. Good models approximate miniature theories; they are tentative and falsifiable (Graziano & Raulin, 2004).

Competency models have advantages and disadvantages compared to causal models. A competency model provides generalizable guidance for training, selection, and assessment, and is therefore directed to solving an applied psychology problem, such as enhancing 3C capabilities in the Military. Causal models, however, provide conceptual, theoretical, and research advantages that can guide selection and training by showing where in the antecedent-to-competency relationship they should be used to greatest effect. A competency model without causal validation may be difficult to generalize to different contexts because it lacks a nomological network of meditational and moderator constructs that can take into account contextual or situational variation.

2.1 An Integrated Model Approach

These two approaches may be reconciled, however, if competencies and antecedents were integrated in models that show causality, mediation, and/or moderation. Each higher level or general competency would be embedded in a model, and the competencies would be related to each other in larger models. Valid measures of the competencies would need to be identified or developed that assess a wide range of competencies and antecedents beyond cognitive measures. Figure 1 illustrates how such an integrated model could be developed using competencies and enablers in the DLO Framework.

One final critical advantage of the integrated model approach is that it provides multiple assessment points for evaluating the 3C competency of individuals. An integrated model reveals antecedents that, because they are often more easily measured than complex competencies, can be assessed, along with the competency itself, to provide a *profile* of the knowledge, skills, abilities, and other characteristics of the individual that "triangulates" on the person's overall level of cross-cultural proficiency. Although the ideal tool for assessment of behavioral competencies is the classic assessment center, this method is costly and time consuming. By including antecedent measurement in the mix of 3C assessments, the need to use full assessment center methods to obtain behavioral observations of competencies is reduced. Instead, simpler and more efficient "mini assessment centers" can be developed to add behavioral assessment to the available information. These simplified behavioral measures would not require assessors and would be designed for administration via the Internet in batteries that include self-report measures of antecedent variables.

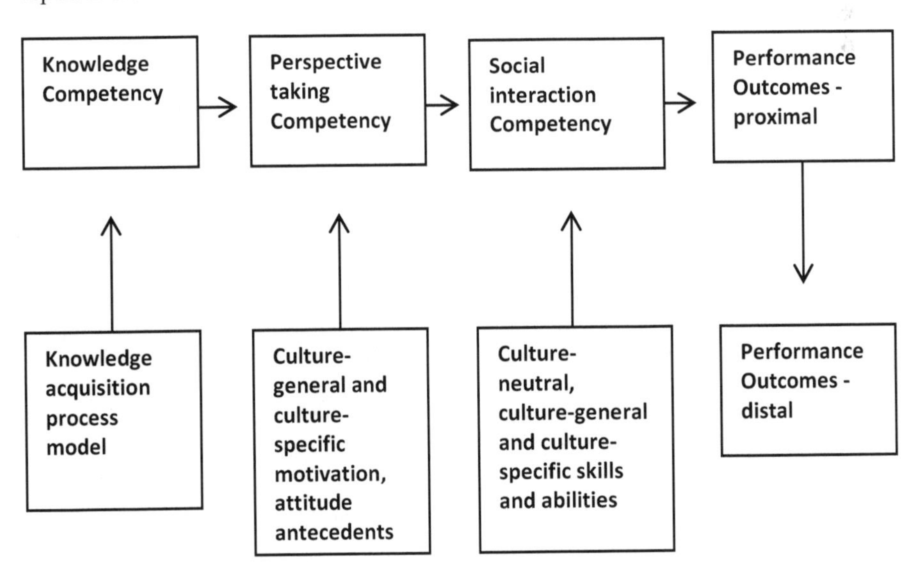

Figure 1. Example of integrated model surrounding perspective taking as a competency

Therefore, we suggest that an integrated causal-competency approach should be utilized in developing a new 3C model. The integrated 3C model should be built as a set of causal models drawn from existing theory and research, each designed to understand a single 3C competency. Each competency will be conceptualized in a nomological network of antecedents, competencies, and performance outcomes drawn from prior research that includes the most important moderators, articulated as needed by MOS/rank/service. Cross-sectional and longitudinal correlational studies, as well as experimental training studies, can be conducted to test such an integrated model. In this way, the applied usefulness of a competency model can be retained, while a research capability is added that can test the adequacy of the proposed competencies. Through a process of refinement the result should be a robust competency framework that can guide selection and training efforts.

The integrated model we advocate is measurement intensive: we want to measure everything in the model, if possible, in addition to moderators that prove important. A serious problem we encountered in our attempt to establish the criterion validity of the elements of the DLO Framework was the paucity of good measures for elements that one would expect should be measureable, and the complete absence of measures of many behavioral competencies. In the next section, we discuss the state of measurement and suggest some new avenues for assessment.

3 ASSESSING CROSS-CULTURAL COMPETENCE

3.1 Validity of Existing Instruments

Our evaluation of the criterion validity of the DLO Framework necessitated identifying instruments to assess the constructs to which the Framework elements were mapped and then finding studies that used these instruments alongside acceptable criterion variables, i.e., performance and adjustment measures. In this second step, we were essentially evaluating the quality of much of the existing 3C instrument domain. Because of the manner in which we tracked down instruments (from others' lists, primarily), we were secondarily evaluating the quality of these ubiquitous lists. We discovered that a variety of performance measures are used in this literature, including job performance (manager ratings, peer ratings, self-ratings) and several informal ratings of overseas "success" or "effectiveness" in non-job situations. Expatriate "performance" is also assessed indirectly through self-reports of intent to remain on the job, job attitudes, and occupational citizenship behaviors (Mol, Born, Willemsen, & van der Molen, 2005; Thomas & Lazarova, 2006). Proprietary instruments were difficult to evaluate: validation studies have been published in peer reviewed journals for only a few such instruments and satisfactory validation reports are rarely published on consulting companies' websites. Altogether, we identified 34 instruments.

Gabrenya et al. (2011a) present the results of this analysis in detail. The criterion validity with respect to intercultural performance or adjustment of five of the instruments was good, for two it was moderate (subscales differed in quality), for 25 it was poor or no validation information could be found, and two turned out not to be quantitative instruments "as advertised" in instrument lists. In several cases, the instruments appeared to have good construct validity but unproven or untried criterion validity, holding out hope that future research would lead to discovering some additional usable instruments. The best overall instrument in terms of face, construct, and criterion validation appears to be the Multicultural Personality Questionnaire (MPQ; Van Oudenhoven & Van der Zee, 2002).

3.2 Need for New Instrumentation

The list of instruments evaluated in the previous section attests to the tremendous effort that researchers have devoted to developing instruments assessing 3C, but our analysis demonstrates that the results of their efforts are discouraging, so far. The following overlapping problems can be identified in the existing instrument armamentarium: (1) nearly all use self-report methods that appear unsuitable for assessing most competencies; (2) only declarative, cognitively accessible, and self-referent information is usually obtained; (3) the potential for faking ranges from subtle to severe; (4) affective states or processes are poorly assessed; (5) behavior is rarely measured; (6) the instruments map poorly to DLO Framework competencies; and (7) few were found to be adequately validated using performance criteria.

These issues significantly limit the validity and construct coverage of existing measures. For instance, relying on assessments that target explicit, declarative knowledge neglects important characteristics that are more implicit or procedural in nature (e.g., affective and skill-based variables). The self-report approach used in existing measures may not provide sufficiently valid assessments of key constructs. Faking, for example, is one major issue with self-report measures in many situations where respondents are motivated to make a good impression. Numerous studies have indicated that individuals are able to substantially alter their scores these measures (e.g., Birkeland, Manson, Kisamore, Brannick, & Smith, 2006; Viswesvaran & Ones, 1999) and that a sizable proportion do fake in motivating situations (e.g., Griffith & Converse, 2011). Further evidence has indicated this behavior has negative implications for construct and criterion-related validity (e.g., Converse, Peterson, & Griffith, 2009). Another concern with these types of self-reports is that respondents may not be able to provide accurate self-assessments of knowledge-, skill-, ability-, or competency-related constructs relevant to cross-cultural competence. Thus, even if there is little motivation to intentionally distort responses, self-report measures of these variables may not provide valid assessments. Indeed, a number of studies have indicated self-reports of knowledge, skills, abilities, or competencies have limited validity (e.g., Carter & Dunning, 2008).

Consistent with these ideas, self-reports of cross-cultural skills and abilities have been criticized on methodological grounds and may have questionable validity (see Brackett, Rivers, Shiffman, Lerner, & Salovey, 2006; Gabrenya, van Driel,

Culhane, Turner, Pathak, & Peterson, 2011b; Thomas et al., 2008). For example, Gabrenya, et al. (2011b) examined the validity of the Cultural Intelligence Scale (CQS; Ang & Van Dyne, 2008). The CQS uses self-reports of competencies, attitudes, and behaviors to infer cultural intelligence, the ability to generate appropriate behavior in new cultural settings (Earley & Ang, 2003). They found that the CQS fails to mediate between antecedent variables assessed using objective and behavioral measures and criterion variables such as situational judgment tests and adjustment. The CQS appears to unintentionally assess constructs such as self-efficacy rather than cultural intelligence or competence.

The considerations listed above indicate there is a critical need to develop alternative measurement approaches to assess 3C. Comprehensive measurement of 3C involves three sets of variables: (1) the antecedent variables that comprise each competency's causal model, such as the DLO Framework enablers; (2) the individuals' competencies; and (3) the performance outcomes through which criterion-related validation of the competency models is accomplished.

We propose that the development of new measures should be based on two general principles for improving overall measurement: (1) broadening the measurement spectrum and (2) moving toward more dynamic measurement. As noted earlier, a major limitation of existing 3C measures is that they tend to have a strong cognitive focus, with most involving self-reports of declarative, cognitively accessible, self-referent information. Cultural competence involves not only cognitive factors, but also affective (e.g., emotion regulation) and behavioral (e.g., interpersonal skills) components. Thus, future measure development should expand the focus of assessment to include cognitive, affective, and behavioral approaches in order to more closely and accurately match the full set of competencies and antecedents in a well developed competency model.

Another key limitation of existing 3C measures is that they are also largely static in nature, where individuals respond to sets of items under typical testing conditions. However, cross-cultural experiences and interactions are generally much more complex and dynamic, requiring more active behavior (e.g., decision making and self-regulation) and involving elements such as distraction, stress, ambiguity, and emotion. Thus, instrument development must incorporate more dynamic assessment approaches. In this context, we use the term *dynamic* in a general sense to refer to alternatives to traditional self-reports that involve richer stimuli, changing task characteristics, and/or more active involvement from the individual, i.e., behavioral responding.

4 CONCLUSION

In summary, many models of 3C have been proposed across decades of military and civilian research and theorizing. Most recently, competency frameworks of 3C that are aimed at informing selection and training of personnel have taken the forefront. These models have the potential to solidify an amorphous research area and aid multi-national organizations as they encounter new cultures. However, the existing frameworks suffer from conceptual ambiguity, a lack of causal linkages between enablers and competencies, and inadequate assessment measures. To advance the study of 3C and provide organizations with the tools to make important personnel decisions, integrated models of 3C that incorporate competencies with traditional causal modeling are necessary. In addition, psychometrically sound measures that rely less on self-report methods must be developed if these models are to be tested. While the construct of 3C has intuitive appeal to address many of the challenges facing 21st century organizations, operational measures and models of 3C have far to go before they meet that promise.

ACKNOWLEDGMENTS

This research was carried out with the support of a grant from the Defense Equal Opportunity Management Institute. We thank Dr. Daniel McDonald for his support.

REFERENCES

Brannick, M. T., Levine, E. L., & Morgeson, F.P. (2007). *Job and work analysis: Methods, research, and applications for human resource management* (2nd ed.). Thousand Oaks, CA: Sage Publications.

Abbe, A. (2008). *Building cultural capability for full-spectrum operations.* (ARI Study Report 2008-04). Arlington, VA: U.S. Army Research Institute for the Behavioral and Social Sciences.

Abbe, A., Gulick, L. M. V., & Herman, J. L. (2007). *Cross-cultural competence in army leaders: A conceptual and empirical foundation.* Technical Report, U.S. Army Research Institute.

Ang, S., & Van Dyne, L. (2008). Conceptualization of cultural intelligence: Definition, distinctiveness, and nomological network. In S. Ang & L. van Dyne (Eds.), *Handbook of cultural intelligence: Theory, measurement, and applications.* Armonk, NY: M. E. Sharpe.

Ascalon, M. E. (2005). Improving expatriate selection: Development of a situational judgment test to measure cross-cultural social intelligence. (Doctoral dissertation, University of Tulsa, 2005). *Dissertation Abstracts International, 65,* 9-B.)

Bartram, D. (2005). The great eight competencies: A criterion-centric approach to validation. *Journal of Applied Psychology, 90*(6), 1185-1203.

Birkeland, S., Manson, T., Kisamore, J., Brannick, M., & Smith, M. A. (2006). A meta-analytic investigation of job applicant faking on personality measures. *International Journal of Selection and Assessment. 14,* 317-335.

Brackett, M. A., Rivers, S. E., Shiffman, S., Lerner, N., & Salovey, P. (2006). Relating emotional abilities to social functioning: A comparison of self-report and performance measures of emotional intelligence. *Journal of Personality and Social Psychology, 91*(4), 780-795.

Brannick, M. T., Levine, E. L., & Morgeson, F. P. (2007). *Job analysis: Methods, research, and applications for human resource management* (2nd ed.). Thousand Oaks, CA: Sage Publications.

Caligiuri, P. M., Hyland, M. M., Joshi, A., & Bross, A. S. (1998). Testing a theoretical model for examining the relationship between family adjustment and expatriates' work adjustment. *Journal of Applied.Psychology. 83*, 598–614.

Caligiuri, P., Noe, R., Nolan, R., Ryan, A. M., & Drasgow, F. (2011). *Training, developing, and assessing cross-cultural competence in military personnel.* Technical report. Technical Report, U.S. Army Research Institute.

Carter, T. J., & Dunning, D. (2008). Faulty self-assessment: Why evaluating one's own competence is an intrinsically difficult task. *Social and Personality Psychology Compass, 2*, 346-360.

Converse, P., Peterson, M., & Griffith, R. (2009). Faking on personality measures: Implications for selection involving multiple predictors. *International Journal of Selection and Assessment, 17*, 47–60.

Deardorff, D. K. (2006). Identification and assessment of intercultural competence as a student outcome of internationalization. *Journal of Studies in Intercultural Education, 10*, 241-266.

Department of Defense (2005). *Department of Defense Language Transformation Roadmap.*

Department of Defense (2010). *Department of Defense Strategic Plan for Language, Regional, and Cultural Capabilities.*

Earley, P. C., & Ang, S. (2003). *Cultural Intelligence: Individual interactions across cultures.* Stanford University Press: Stanford, CA.

Fantini, A. E. (2009). Assessing intercultural competence: Issues and tools. In J. K. Deardorff (Ed.), *Sage handbook of intercultural competence* (pp. 456-476). Thousand Oaks, CA: Sage.

Fiske, S. T., & Taylor, S. E. (2008). *Social cognition: From brains to culture.* New York: McGraw-Hill.

Gabrenya, W. K., Jr., Moukarzel, R. G., Pomerance, M. H., Griffith, H., & Deaton, J. (2011a). *A validation study of the Defense Language Office Framework for Cross-cultural Competence.* Technical report, Defense Equal Opportunity Management Institute.

Gabrenya, W., van Driel, M., Culhane, E., Turner, S., Pathak, J., Peterson, S. (2011b). *Validating the Cultural Intelligence Scale: What does it really measure?* Under review.

Graziano, A. M., & Raulin, M. L. (2004). *Research methods: A process of inquiry* (5th Ed.). New York: Pearson.

Griffith, R. L., & Converse, P. D. (2011). The rules of evidence and the prevalence of applicant faking. In M. Ziegler, C. McCann & R. Roberts (Eds.), *New perspectives on faking in personality assessments. Oxford, UK: Oxford University Press.*

Hardison, C. M., Sims, C. S. Ali, F., Villamizar, A., Mundell, B., & Howe, P. (2009). *Cross-cultural skills for deployed Air Force personnel.* Santa Monica, CA: RAND Corporation.

Hunter, B., White, G. P., & Godbey, G. C. (2006). What does it mean to be globally competent? *Journal of Studies in International Education, 10*(3), 267-285.

Johnston, J. H., Paris, C., McCoy, C. E. E., Severe, G., & Hughes, S. C. (2010). *A framework for cross-cultural competence and learning recommendations.* Technical Report, Naval Air Warfare Center, Training and Systems Division.

McCloskey, M. J., Grandjean, A., Behymer, K. J., & Ross, K. (2010). *Assessing the development of cross-cultural competence in soldiers.* Technical Report, U.S. Army Research Institute.

McDonald, D. P., McGuire, G. M., Johnston, J., Selmeski, B., & Abbe, A. (2008). *Developing and managing cross-cultural competence within the Department of*

Defense: Recommendations for learning and assessment. Technical Report, Defense Language Office.

Mol, S. T., Born, M. Ph., Willemsen, M. E., & van der Molen, H. T. (2005). Predicting expatriate job performance for selection purposes: A quantitative review. *Journal of Cross-Cultural Psychology, 36*(5), 590-620.

Morgeson, F. P., Delaney-Klinger, K., Mayfield, M. S., Ferrara, P., & Campion, M. A. (2004). Self-presentation processes in job analysis: A field experiment investigating inflation in abilities, tasks, and competencies. *Journal of Applied Psychology, 89*(4), 674-686.

Shippmann, J., Ash, R., Battista, M., Carr, L., Eyde, L., & Hesketh, B., et al. (2000). The practice of competency modeling. *Personnel Psychology, 53*(3), 703–740.

Spencer, E. (2010). Solving the people puzzle: Cultural intelligence and special operations forces. Toronto: Dundurn Press.

Spitzberg, B. H., & Changnon, G. (2009). Conceptualizing intercultural competence. In D. K. Deardoff (Eds.), *The SAGE Handbook of intercultural competence,* (pp. 2-52). Newbury Park, CA: Sage.

Thomas, D. C., & Lazarova, M. B. (2006). Expatriate adjustment and performance: A critical review. In G. K. Stahl & I Björkman (Eds.) *Handbook of research in international human resource management* (pp. 247-264). Cheltenham, UK: Edward Elgar.

Thomas, D. C., Elron, E., Stahl, G., Ekelund, B. Z., Ravlin, E. C., Cerdin, J.-L., Poelmans, S., Brislin, R., Pekerti, A., Aycan, Z., Maznevski, M., Au, K., & Lazarova, M. B. (2008). Cultural intelligence: Domain and assessment. *International Journal of Cross-Cultural Management, 8*(2), 123-143.

Thornson, C. A., & Ross, K. G. (2008). *Identification of measures related to cross cultural competence.* Technical report: DEOMI.

U.S. Department of State (2008). *Enhancing foreign language proficiency in the United States: Preliminary results of the National Security Language Initiative.*

Van de Vijver, F. J. R., & Leung, K. (2009). Methodological issues in researching intercultural competence. In J. K. Deardorff (Ed.), *Sage handbook of intercultural competence* (pp. 404-418). Thousand Oaks, CA: Sage.

Van Oudenhoven, J. P., & Van der Zee, K. I. (2002). Predicting multicultural effectiveness of international students: The Multicultural Personality Questionnaire. *International Journal of Intercultural Relations, 26*, 679–694.

Viswesvaran, C., & Ones, D. S. (1999). Meta-analyses of fakability estimates: Implications for personality measurement. *Educational and Psychological Measurement, 59*, 197-210.

CHAPTER 33

Metacognitive Underpinnings of 3C Development

Louise J. Rasmussen
Winston R. Sieck
Global Cognition

ABSTRACT

This chapter discusses a perspective on the development of Cross-Cultural Competence (3C) that regards it as an activity that students do for themselves in a proactive way rather than as an event that happens in reaction to teaching. It argues that metacognitive, self-regulatory learning strategies provide the basis for the efficient and effective development of 3C over time. In the chapter, the authors define the essential qualities of self-regulation, describe the structure and function of self-regulatory processes in the context of cultural learning, and, finally, give an overview of approaches for guiding students to learn on their own.

INTRODUCTION

For an increasing number of people, the world is becoming their workplace. Whether employed by a multinational organization, the government, or the military, people in professions that require them to work in different parts of the world across their careers, share a critical job characteristic. Mainly, they inevitably encounter situations in other cultures where their current knowledge does not suffice. That is, they are confronted with situations that do not meet their expectations. In order to perform effectively in situations where the answers are not immediately known, people must acquire new knowledge on the fly rather than merely reproduce knowledge they have already learned.

In this chapter we discuss the results of a number of expertise studies aimed at characterizing the competencies that enable highly experienced cross-culturalists to learn about and become effective in new cultural environments quickly. These studies indicate that cross-cultural experts develop certain metacognitive strategies that help them self-regulate their own learning about new cultures—and that enable them to learn more effectively than novices. In other words, cross-culturally competent individuals are experts at becoming experts in new cultures. In this chapter, we define the essential qualities of self-regulatory learning, describe the structure and function of metacognitive self-regulation processes in the context of cultural learning, and, finally, give a brief discussion of approaches for empowering students to learn about cultures on their own.

Cross-Cultural Competence

3C refers to the knowledge, skills, and affect/motivation that enable individuals to adapt effectively in cross-cultural environments. A central aspect of 3C involves comprehending individuals from distinct cultural backgrounds, as well as the ability to convert this knowledge into action (Selmeski, 2007). Selmeski defines 3C as:

> *The ability to quickly and accurately comprehend, then appropriately and effectively engage individuals from distinct cultural backgrounds to achieve the desired effect; despite not having an in-depth knowledge of the other culture, and even though fundamental aspects of the other culture may contradict one's own taken-for-granted assumptions/deeply-held beliefs. (p. 12)*

Therefore, in addition to acting and engaging effectively within a variety of different cultures, 3C subsumes the abilities to quickly acquire new cultural knowledge and skills and to extend or transfer knowledge and skills acquired in one cultural environment to another. The inclusion of requirements for effective learning and transfer is what fundamentally distinguishes 3C from regional competence or, culture-specific competence. However, effective characterization of the cognitive and metacognitive learning processes that these cross-cultural skills entail has yet to be fully developed. Such characterization is fundamental to developing and validating education that promotes development of these skills, as well as strategies for assessing the effectiveness of such efforts.

Based on our past work, in this chapter we will outline a process-focused metacognitive framework for characterizing knowledge and skills that enable effective learning and transfer.

Self-Regulated Learning Processes

The immense scope of the learning space combined with the requirement to acquire knowledge and adapt on the fly entails that the majority of cultural learning takes place outside of formal learning environments, i.e. classrooms. To be effective, inside and outside the classroom, learners must be self-regulated. That is, they must be self-directed and self-motivated to develop and improve their skills and knowledge. Self-regulation refers to self-generated thoughts, feelings, and behaviors that are aimed at attaining goals (Zimmerman, 2000). Self-regulated learners are proactive in their efforts to learn because they are guided by personally set goals and task strategies, and because they are aware of their strengths and limitations.

Self-regulation requires metacognitive awareness of or knowledge about ones' own thinking and learning (Flavell, 1979). These include awareness of how one learns and learning preferences; setting specific learning goals for oneself; selecting effective strategies for attaining these goals, which includes knowledge of how to use available information to achieve a goal; monitoring one's performance for signs of progress, which requires awareness of when one does and does not understand; restructuring one's physical and social context to make it compatible with one's goals; managing one's time use efficiently; and self-evaluating one's strategies.

From a theoretical perspective, self-regulation is not a mental ability or a skill; rather it is the self-directive process by which learners transform their mental abilities into skills. Self-regulated learning therefore is an activity that students do for themselves in a proactive way rather than as a covert event that happens to them in reaction to teaching. In recent years, there have been exciting discoveries regarding the nature, origins, and development of how students regulate their own learning processes (Zimmerman & Schunk, 2001). Although these studies have identified ways in which self-regulatory processes lead to academic success, few existing instructional programs prepare students to learn on their own. Further, only few studies address metacognitive learning and reasoning strategies specifically related to cultural competence (Sieck, Smith, and Rasmussen, 2008).

CROSS-CULTURAL LEARNING

In the following we will provide an overview of our past applied research efforts aimed at understanding the cognitive processes involved in practicing and developing 3C within military contexts. Next we will discuss the general results of these studies as they speak to the metacognitive elements of 3C that enable self-regulated, self-motivated cultural learning.

Cross-Cultural Expertise Studies

The majority of our research involves field studies focusing on ground operators who have repeated and extensive interaction with foreign populations, however at least one study examined elements of 3C in the context of intelligence analysis as well. Although these specific scientific objectives and methodologies vary slightly across these studies, in one way or another all studies addressed the cognitive aspects of cross-cultural expertise. Altogether 140 members of the military participated in these studies, representing the full spectrum of rank (from junior enlisted to 4-star general) and cultural exposure (ranging from none at all to spending decades overseas). Aside from the experienced intelligence analysts all participants were non-specialists.[1]

All studies involved in-depth, incident-based interviews. About half of the interviews were semi-structured, following a Critical Decision, incident-based method that relies on recollection of tough cases and challenging events (Crandall, et al., 2006). The starting point for such interviews were critical incidents in which the interviewee personally experienced (inter)cultural challenges—focusing on their most recent experience abroad. In the interviews, the interviewee's own examples of recent challenging interactions were used as a point of departure for eliciting detailed information about ways in which competencies were used within specific intercultural situations. The other half of the interviews used a think-aloud procedure with pre-specified scenarios that involved intercultural interactions. All such scenarios were developed based on past CTA interviews and as such represented authentic intercultural situations. For a large portion of these studies qualitative analyses were conducted in which teams of analysts noted emerging themes, distinct categories, and commonalities across the data set. For other studies, quantitative approaches were used in which analysts annotated excerpts with replicable codes that described knowledge and skill-based strategies employed to understand, decide, and engage within intercultural situations. Frequency of code use was employed as a standard, quantitative measure of strategy employment.

Across these studies our objectives have been to characterize the types of situations and interactions that make demands on a person's interpersonal and intercultural competencies; characterize in detail the strategies experts use to manage and learn from these kinds of situations; and, finally, characterize differences between novice and expert strategies.

Metacognitive Cultural Learning Processes

In general, our results indicate that cross-cultural experts have developed certain metacognitive strategies that support their continual acquisition of declarative, conceptual and procedural knowledge and skills, as well as provide the affective and motivational foundation needed to attain high levels of expertise.

We will organize the overview around four high level domains of metacognitive processing. (For a discussion of the ways in which these strategies relate to military practice, see Rasmussen & Sieck, 2012). Mainly, we have found that self-regulated learning in the culture domain occurs most effectively on the backdrop of 1) self-awareness which includes an acknowledged cultural conceptualization of the self; 2) self-motivation which involves conceptualization of the self in relation to the learning domain; 3) a meta-understanding of general kinds of cultural knowledge that are most useful; and finally 4) generative behavioral

---

[1] Non-specialists are individuals in career paths that for which specialized language and culture training is not provided. Examples of specialist careers include civil affairs-, foreign area-, liaison-officers and some types of intelligence analysis/collections.

skills which support the acquisition and integration of new knowledge into the individual's existing conceptual system.

Self-awareness

The cross-cultural experts in our studies were aware that they see the world in a particular way because of their own background, personal history, and culture. They appeared to intuitively anticipate that in an interaction with someone who has a different background, each person brings a unique perspective to the situation. Interestingly, we have found that both novices and experts are able to consider other people's perspectives on events and behaviors. However, novices are less likely to integrate alternative viewpoints into their decision making and strategies for managing interactions; and they are less likely to compare and contrast their own and others' perspectives (Rasmussen & Sieck, forthcoming).

Recognizing potential mismatches in perspectives appear to drive cross-cultural experts to continually explore commonalities and differences between themselves and the people around them. Further, extensive experience living in multiple locations led experts to develop their own theories about ways in which Americans differ from other people in the world. Cultural researchers often attempt to frame cultural differences objectively (Boas, 1948). However, the cross-cultural experts interviewed in our studies generally appreciated that they were likely to encounter differences in most overseas assignments, and had learned to frame these differences in terms of the uniqueness of their own perspective. This self-awareness and way of conceptualizing cultural difference, in turn, appears to support an innate motivation for learning.

Self-motivation

We have found consistent support for three main metacognitive processes that support self-motivated learning: 1) continually framing intercultural experiences as opportunities to learn, 2) developing justification for the value of cultural understanding, 3) setting personal and manageable expectations about what and how much to learn about a culture. In the following we will discuss each in detail.

In order to effectively use experiences as opportunities for practice; people must explicitly frame the experience as an opportunity to learn (Ericsson et al., 1993). The cross-cultural experts we interviewed deliberately sought out experiences and relationships that they could learn from. Conceptually, these experts all had the expectation that they would continue to learn new things about a culture the whole time they were in it. Further, they would explicitly think about the knowledge and skills that they acquired in training simply as a springboard for continuing learning.

The cross-cultural experts interviewed for our studies used their own personal interests as the starting point for learning about new cultures. When learners formulate their own questions—they are defining their own learning objectives. These self-defined learning objectives reflect areas that they are personally interested in—and which they are intrinsically motivated to learn about. In our studies, such self-defined learning objectives were often developed either from long-term interests or from immediate needs to improve or adapt action. Common across the experts in the sample was the practice of defining manageable expectations with regard to how knowledgeable or proficient they wanted and needed to become in a culture and language.

Some of the experts had life-long curiosity about human social, cultural and psychological dynamics which motivated their learning. However, many instead had deep, intrinsic interests in history; some were interested in religion, others in sports, yet others again in weapons. All used these personal interest areas as a basis for formulating questions about new regions or cultures. They would seek thee answers through research prior to deployment, or through conversation with locals once on the ground. For example, one cross-cultural expert had a personal interest in knives, and would take every opportunity to discuss knife-making practices with Afghans. In this way he used his personal interest to both learn and to establish a personal connection to a new culture.

Meta-knowledge

We found that relative to novices, experts appear to possess a meta-conceptualization that is critical to stimulating self-motivation; namely, justification for the value of cultural knowledge. In other words, the experts tended to have internalized explicit reasons that allowed them to justify the importance of cultural learning to themselves (and others). These justifications served to motivate learning each time they entered a new culture.

For example, almost without exception, every cross-cultural expert we interviewed described using cultural knowledge as a foundation for building relationships with natives by using it to demonstrate interest. Some experts also illustrated ways in which cultural knowledge was invaluable for assessing risk in the operational environment. For example, one Colonel noted that to him learning some things about a culture helps to increase confidence and therefore motivation to engage members of the culture. Importantly, the experts themselves identified the information they wanted and needed to have, including words and phrases they wanted to learn, in order to achieve self-identified goals—indicating that they have a meta-level understanding of what constitutes useful cultural knowledge. Plainly put, cross-cultural experts know what they need to know and why.

Generative behavioral skills

In this section we will describe a class of behavioral skills which we have found that explicitly serve to enhance cultural knowledge and skills. These include strategies for information seeking such as identifying, interacting with, making sense of and evaluating information and information sources.

Cross-cultural experts know who to ask and where to look for information about other cultures. We noted several variations on the practice of deliberately establishing relationships with “cultural insiders” (i.e. natives) or other cultural experts to support learning. For example, several interviewees described to us how they use their interpreters as cultural mentors. They would engage in ongoing discussions with trusted interpreters to assess and improve their knowledge of a region's history, culture, and language. At times, they even sought feedback on how they performed in specific interactions, after the fact. Many were very creative in both the sources they identified and the strategies they used for obtaining information they felt would be useful for them.

By using their own questions as the starting point for learning about culture, and by developing and using their own strategies for getting information, the interviewees were making the information and their learning relevant to themselves. Although enacted in different ways, the primary objective, though, was shared among many interviewees: Making culture learning meaningful to oneself.

The literature on metacognition has identified "inquiry-based learning" as an especially effective learning strategy, especially in open-ended learning environments and situations that afford experiential learning. Few studies have examined inquiry learning specifically in in the cultural domain. However, in our studies of expertise in cultural sensemaking we have found that, in the context of surprising, or unexpected intercultural behaviors, expert cultural sensemakers use inquiry strategies that are akin to those used by effective scientists who encounter experimental evidence inconsistent with their original hypothesis (Sieck et al., 2008; Osland & Bird, 2000). That is, they changed their goal to one of determining the cause of the unexpected behavior. This pattern of results closely mirrors studies of scientific reasoning strategies (Dunbar, 1993). Dunbar found that individuals who maintained a goal of finding support for existing (incorrect) hypotheses failed to discover the actual mechanisms underlying a complex biological process (gene regulation). Those who instead set a new goal of attempting to explain the cause of the inconsistent findings tended to generate the correct hypothesis. This study provides evidence for generalized inquiry strategies as a key metacognitive component of 3C.

Culture is in many ways subjective and any one individual or source's account is therefore likely to be biased. Many of the experts we have interviewed were aware of this, and would critically evaluate information provided to them either by native mentors or by other sources, such as the web. They might look for a second opinion, or at times go online after a discussion to check facts they had been provided. This served both as a check on the validity of the information itself but allowed them to assess the general reliability of their source.

DEVELOPING SELF-REGULATED CULTURE LEARNERS

In this chapter, we have described a number of metacognitive processes and strategies that support effective, self-regulated culture learning. As such, these provide a template for how to think about culture in a manner that promotes longer term learning. Although there is significant research and development underway across DoD that is focused on enhancing the effectiveness of language and culture training, preponderance of the pre-deployment culture training that is currently available and accessible to warfighters focuses on teaching what-to-think within specific cultures, by providing facts and do's & don'ts rule-sets (Salmoni, Hart, McPherson & Winn, 2010). While this form of training may meet requirements for efficiency in the short term—it fails to do so over the long term because the learning content does not transfer and it neglects to develop a foundation for further knowledge and skill acquisition. We propose that providing formal educational and organizational support for the development of metacognitive learning strategies is a way to efficiently and effectively cultivate 3C.

Teaching Cultural Learning Skills

Research has demonstrated that metacognitive skills *can* be improved through training in both children and adults (see Palincsar, 1986; Salas & Cannon-Bowers, 2001; Cohen, Freeman & Wolf, 1996). So, what are ways that metacognitive skills can be improved in warfighters? In the following we will provide suggestions for instructional as well as organizational approaches for doing so.

Culture-focused thinking and learning skills training offers the promise of meeting requirements for both efficiency and transferability. Skills training that provides students with metacognitive, cross-cultural learning strategies early on offers the promise of not only providing strategies for how-to-think, but at the same time accelerating expertise development. Further, this kind of training could potentially allow warfighters to take maximum advantage of the richest learning opportunities presented to them outside formal learning institutions and environments: namely, their experiences. Instructional objectives that aim to enhance metacognitive learning skills can be integrated into existing culture curriculums—even those with a culture-specific focus, existing curriculums focusing on strategic thinking, or even combat competencies (as learning skills are important in all domains) or they can be achieved as stand-alone modules. A combination of domain-specific (i.e. culture) and domain-independent instruction likely provides the most fertile foundation for the development of these skills.

Cultivating Cultural Learning Communities

Several lines of cognitive theory and research point toward the idea that "people develop habits and skills of interpretation and meaning construction though a process more usefully conceived of as socialization than instruction" (Lave, 1993). In the present context, the suggestion is that it may not be useful to only conceive of cross-cultural education in the traditional sense of formally teaching specific, well-defined skills or items of knowledge. Instead, within a socialization conception, people develop skills and long-term patterns of interaction from their participation in a social environment that supports and encourages the development of these skills and patterns. The question then is: from an organizational perspective, how can military leadership provide support and encouragement for the development and practice of cross-cultural competence?

There are a number of kinds of tactics that leaders can employ to create safe, productive learning environments in the field (Schein, 1996). By listening for and responding to elements of naturally occurring discussions or interactions that relate to culture, leaders can demonstrate that they value consideration of cultural factors, and learning about same. Leaders at all levels are influencers. A leader's ideas, beliefs, and values set the standard for subordinates. If a leader deliberately engages subordinates in dialogue around culture and cultural issues, they are demonstrating that they value cultural skills and knowledge, and in doing so they are laying the foundation for establishing a community of practice (Lave, 1993). Enabling ongoing dialogue within a unit about intercultural experiences, perhaps in the context of After Action Reviews (AARs) or through systematic interaction with interpreters, can allow sensemaking to occur as a social activity. Social sensemaking is a particularly valuable learning activity seeing that an increased number of alternative perspectives are introduced and considered that may challenge the individual's a-priori understanding.

Providing safe practice fields entails that leaders, in the context of such ongoing dialogue, treat mistakes and misunderstandings as learning opportunities instead of merely as occasions to evaluate performance. Further, by setting a positive vision; i.e. through descriptions of what the outcomes can be if subordinates engage thoughtfully, leaders can provide encouragement towards continuing the dialogue, and continuing to improve skills, knowledge, and ultimately performance.

CONCLUSION

In the 3C literature, metacognitive skills are often talked about as the outcome of a long learning process—i.e. they require additional training and education (MacDonald, et al., 2008). Training and educational programs do not, and cannot, we argue, produce cross-cultural experts. Training and education can support the development of expertise by providing the foundational skills required to maximize learning on the job, or learning from experience. To do that, they must support the development of generative, metacognitive learning processes.

We suggest that a generative approach to supporting the development of cross-cultural expertise. Namely, we suggest that providing students with metacognitive, cross-cultural learning strategies early on in their careers can, not only allow them to develop strategies for how-to-think within challenging intercultural situations, but at the same time accelerate their expertise development.

ACKNOWLEDGEMENTS

We appreciate support received at various stages of this research program from ARA, Rababy & Associates, DLO, ARI, ARL, OSD-HSCB, as well as the military personnel who participated in our research.

REFERENCES

Boas, F. (1948). Race, Language and Culture. New York: Macmillan.

Cohen, M.S., Freeman, J.T. & Wolf S. (1996) Meta-recognition in time stressed decision making: Recognizing, critiquing, and correcting. Human Factors, 38(2):206-219.

Crandall, B., Klein, G., & Hoffman, R. R. (2006). Working minds: A practitioner's guide to Cognitive Task Analysis. Cambridge, MA: The MIT Press.

Dunbar, K. (1993). Scientific reasoning strategies for concept discovery in a complex domain. *Cognitive Science, 17*, 397-434.

Ericsson, K. A., Krampe, R., & Tesch-Romer, C. (1993). The role of deliberate practice in the acquisition of expert performance. Psychological Review 100, 363-406.

Gourgey, A. F. (2001). Developing students' metacognitive knowledge and skills. In H. J. Hartman (Ed.), *Metacognition in learning and instruction: theory, research and practice* (pp. 17-32). Dordrecht, The Netherlands: Kluwer Academic Publishers.

Lave, J. (1993). Situating learning in communities of practice. In L. B. Resnick, J. M. Levine, & S. D. Teasley (Eds.) Perspectives on socially shared cognition (pp. 17-36). Washington, DC: American Psychological Association.

McDonald, D. P., McGuire, G., Johnston, J., Selmeski, B. R., & Abbe, A. (2008). Developing and managing cross-cultural competence within the Department of Defense: Recommendations for learning and assessment. Washington, DC: DLO.

Osland, J.S., & Bird, A. (2000). Beyond sophisticated stereotyping: Cultural sensemaking in context. Academy of Management Executive, 14, 65-79.

Palincsar, A. (1986). Metacognitive strategy instruction. Exceptional Children, 53(2), 118-24.

Rasmussen, L. J., Sieck, W. R., & Osland, J. (2010). Using cultural models of decision making to develop and assess cultural sensemaking competence. In D. S. a. D. Nicholson (Ed.), Advances in Cross-Cultural Decision Making. Boca Raton, FL: CRC Press, Taylor & Francis Group.

Rasmussen, L. J., & Sieck, W. R. (manuscript in preparation). Expert novice differences in intercultural perspective-taking.

Salas, E., & Cannon-Bowers, J.A. (2001). The science of training: A decade of progress. Annual Review of Psychology, 52, 471-499.

Schein, E. H. (1996). Kurt Lewin's change theory in the field and in the classroom: Notes toward a model of managed learning. Systems Practice, 9(1), 27-47.

Selmeski, B. R. (2007). Military cross-cultural competence: Core concepts and individual development (Final Report prepared under U.S. Air Force Culture and Language Center Contractor, Report #2007-01). Kingston, Ontario, Canada: Centre for Security, Armed Forces & Society, Royal Military College of Canada.

Sieck, W. R., Smith, J. L., & Rasmussen, L. R. (2008). Expertise in making sense of cultural surprises. *Proceedings of the Interservice/Industry Training, Simulation, and Education Conference (I/ITSEC)*, December 2008, Orlando, FL.

CHAPTER 34

Outreach: Building Cross-Cultural Competence in the Total Force

Karen J. Gregory

HRSS Consulting Group
Merritt Island, FL
kgregory@hrssconsultinggroup.com

Marinus van Driel, PhD

Van Driel Consulting, Inc.
Satellite Beach, FL
marinus@vandrielconsulting.com

Allison Greene-Sands, PhD

Defense Language and National Security Education Office
Arlington, VA
allison.greene@wso.whs.mil

Daniel P. McDonald, PhD

Defense Equal Opportunity Management Institute
Patrick Air Force Base, FL
daniel.mcdonald@patrick.af.mil

ABSTRACT

Cross-cultural competence (3C) has become an imperative for successful execution of security and stabilization missions around the world. On 10 August 2011, the SECDEF released a memo stating, "As a minimum, both military and civilian personnel should have cross-cultural training to successfully work in DoD's richly diverse organization and to better understand the global environment in which we operate." However, only a limited number of avenues exist to build and sustain 3C throughout the Department of Defense (DoD). To address this need, the Defense Equal Opportunity Management Institute (DEOMI) and the Defense

Language and National Security Education Office (DLNSEO) developed a knowledge portal that provides contemporary 3C resources tailored to the operational needs of DoD personnel. To this end, the portal facilitates 3C through education and training, assessments, and leadership resources, as well as relevant 3C news and operationally relevant research. The portal content is based on 3C-related theory as well as empirical inquiries conducted by DEOMI and DLNSEO in regard to the needs of DoD personnel, primarily military.

Work began in February 2008, when the Office of the Secretary of Defense (OSD) directed the Defense Regional and Cultural Capabilities Assessment Working Group (RACCA WG) to establish a common terminology and typology for identifying, developing, measuring, and managing regional and cultural capabilities. Chaired by DEOMI, a RACCA subgroup consisting of trainers, operators, practitioners, and scientists from DoD and all Service components produced an initial cross-cultural development and assessment model. Subsequently, this model was refined based on a comprehensive analysis of the literature, previous task analyses, and feedback received from Combatant Command (COCOM) leadership. This information was then organized by Naval Air Warfare Center Training Systems Division (NAWCTSD) into a draft 3C Framework that grouped the various competencies and enablers into broader categories for simplification and ease of use. This framework was the basis for the 3C knowledge portal during Phase 1 design and development.

The functionality of the portal is derived from a variety of tools and resources. The 3C portal is designed to help organizations develop cross-culturally competent leaders by assessing their personal knowledge, skills, attitudes, and behavioral preferences. The 3C assessment tools that are available assist with increasing self-awareness and measuring perceptions of team performance as well as diversity and inclusion. Based on these assessments, feedback reports with actionable recommendations are provided to personnel to improve 3C individually and within their units. To complement the assessments, the education and training area of the portal offers updated theory-based training materials including culture clips, e-learning, and simulation training. To further promote basic and applied cultural research within and for the DoD, contemporary research papers are available for download, along with links to relevant Service programs and current 3C-related news. The portal also offers the ability to communicate and share with the operational community via a protected collaboration forum.

Keywords: cross-cultural competence, 3C, culture clips, e-learning, simulation training, assessments, collaboration forum, leadership, operations, resources

1 INTRODUCTION TO THE 3C KNOWLEDGE PORTAL

Cross-cultural competence (3C) has become an imperative as the Total Force continues to face complex security and stability operations around the world. It is essential for both leaders and operators to understand their own and others' cultures. Those with the necessary 3C knowledge and skills will be better equipped to navigate the cultural landscape and mitigate risk to achieve operational success.

So, what is cross-cultural competence? Draft DoD Policy defines 3C as "a set of culture-general knowledge, skills, abilities, and attributes (KSAAs) developed through education, training, and experience that provide the ability to operate effectively within a culturally complex environment. 3C is further augmented through the acquisition of cultural, linguistic and regional proficiency, and by their application in cross-cultural contexts."

The 3C knowledge portal facilitates development of cross-cultural competence across the Total Force. The portal is located at www.defenseculture.org (Figure 1). The purpose of the 3C portal is to provide tools and resources that promote discovery and learning to produce more effective leaders and operators. It is imperative that the DoD builds a Total Force that is not only globally aware, but also adept at interacting with people from a variety of cultures. 3C is about acquiring this knowledge and honing the skills necessary to relate, negotiate, influence, motivate, manage, adapt, plan, and execute effectively across cultural lines both domestic and abroad.

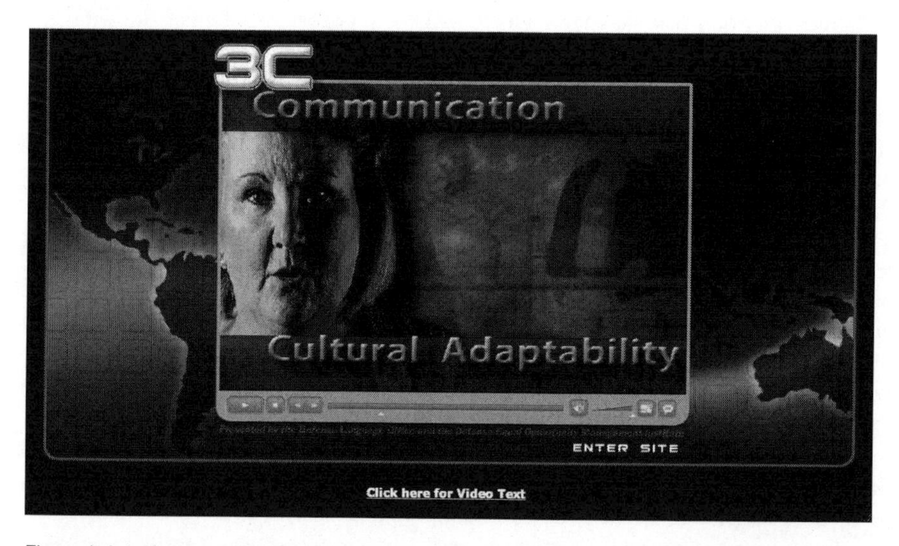

Figure 1 Introductory webpage (www.defenseculture.org) to the 3C, Cross-Cultural Competence, knowledge portal, hosted by DLNSEO and DEOMI.

Current initiatives are being employed to further develop and refine the core competencies and enablers for 3C. However, work began in February 2008, when the Office of the Secretary of Defense (OSD) directed the Defense Regional and Cultural Capabilities Assessment Working Group (RACCA WG) to establish a common terminology and typology for identifying, developing, measuring, and managing regional and cultural capabilities. Chaired by DEOMI, a RACCA subgroup consisting of trainers, operators, practitioners, and scientists from DoD and all Service components produced an initial cross-cultural development and assessment model (McDonald et al., 2008). Subsequently, this model was refined based on a comprehensive analysis of the literature, previous task analyses, and feedback received from Combatant Command (COCOM) leadership. This information was then organized by Naval Air Warfare Center Training Systems Division (NAWCTSD) into a draft 3C Framework that grouped the various competencies and enablers into broader categories for simplification and ease of use (Johnston, 2011). This framework was the basis for the 3C knowledge portal during Phase 1 design and development (Figure 2).

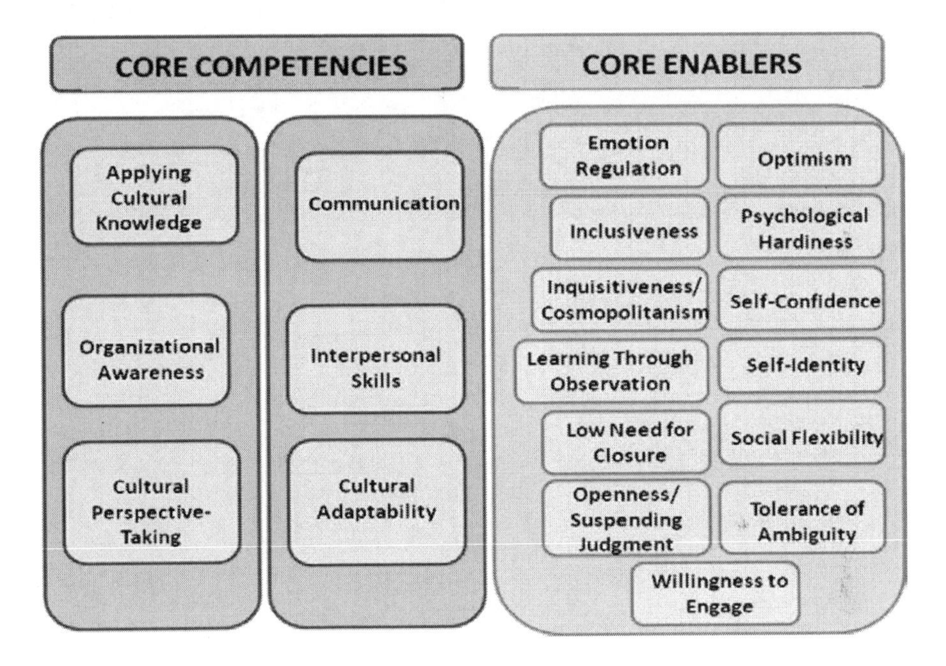

Figure 2 3C Framework presented by NAWCTSD in 2011.

Furthermore, each of these cross-cultural competencies and enablers permeate the model below (Figure 3), beginning with understanding one's own culture, and continuing with understanding a team's, organization's, partner's, and/or adversary's culture.

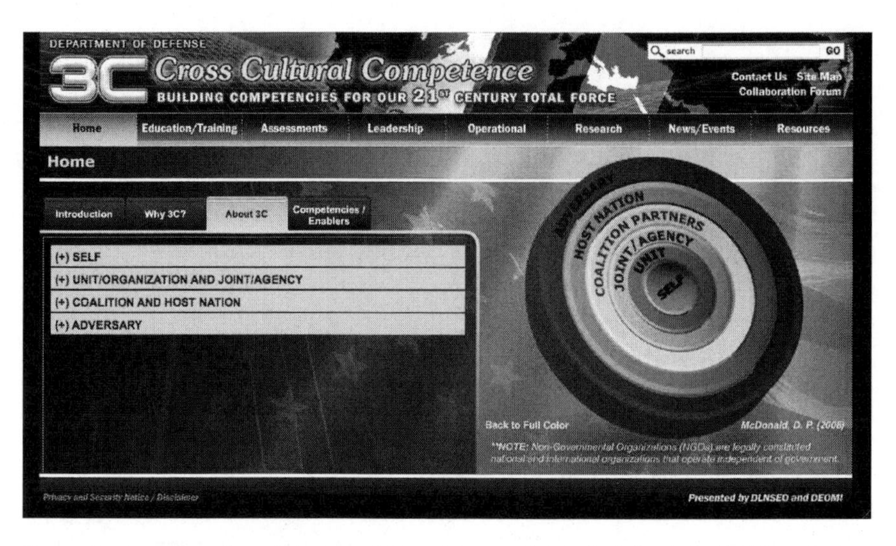

Figure 3 3C permeates all layers of the above model to include Self, Unit/Organization, Joint/Agency, Coalition and Host Nation Partners, and the Adversary (developed by Dr. Daniel P. McDonald, DEOMI).

In 2011, a subset of competencies from the 3C Framework were identified and considered the minimum requirement for all military and select DoD civilian personnel (Table 1). This subset, or 3C Baseline, includes four basic competencies that are life-long processes that continue to develop or mature over time. Policy for implementing this 3C Baseline is currently in coordination and, once finalized, will be included in DoD Instruction 5160.70.

Table 1 3C Baseline includes four foundational competencies and is a common starting point for the Total Force.

Competencies	Definition
What is Culture?	Acquiring and Applying Culture-General Knowledge: The ability to learn and employ culture-general concepts and knowledge, to include the notion of worldview and universal behavioral domains such as religion, kinship, and cultural heritage; use an understanding of the various components of culture to make sense of situations or environments.
Who am I?	Demonstrating Cultural Self-Awareness: The ability to be aware of one's own worldview and understand the beliefs, values, cultural perceptions, assumptions, and biases that shape one's worldview; recognize how worldview influences one's own behavior as well as one's perception of others' behavior;

	understand the second and third order effects of one's behavior.
What makes them who they are?	Cultural Perspective Taking: The ability to understand and apply perspective-taking skills to detect, analyze, and consider the point of view of others; recognize how others will interpret and react to one's actions; demonstrate an understanding of others' needs and expectations.
What's going on around me?	Cultural Learning through Observation: The ability to gather and interpret information about people, surroundings, and important social/cultural cues; recognize and retain cultural information and behavior that is perceived/observed; continually learn and update one's own knowledge base as new situations are encountered as part of cultural learning.

The 3C knowledge portal cultivates the cross-cultural competencies through various mechanisms, to include: e-learning, video vignettes, avatar and virtual training, science and literature, podcasts, assessments, knowledge sharing, research development and publications, and collaboration across the spectrum of 3C.

1.1 Education and Training

The current education and training opportunities include the use of video vignettes, e-learning (Figure 4), simulation training, and non-verbal behavior, as well as a plethora of additional education and training resources from organizations and universities outside of DEOMI and DLNSEO's official purview. The education and training content addresses several important cross-cultural leadership challenges to include awareness of one's own culture, attitudes toward cultural differences, knowledge of different cultural practices and worldviews, and a variety of cross-cultural interaction skills.

Figure 4 3C Education and Training includes Culture Clips, E-learning, Simulation Training, and additional resources provided from external organizations and universities.

1.2 Assessments

The 3C website is designed to help organizations develop cross-culturally competent leaders by assessing their personal knowledge, skills, attitudes, and behavioral preferences. The 3C assessment tools will increase self-awareness, measure perceptions of team performance, as well as diversity and inclusion, and empower organizations in building culturally competent teams (Figure 5).

Figure 5 3C Assessments include the 3C Self Assessment (3CSA) and the Unit Cross-Cultural Competence Survey (U3CS).

1.3 Research and Resources

The 3C Research area of the knowledge portal enables researchers and academicians to share data, information, and discussions about ongoing research activities and interests (Figure 6). Through promotion of both basic and applied cultural research, 3C research offers the opportunity to inform DoD policymakers and collaborate with researchers and professionals around the world.

Figure 6 Contemporary research includes emerging 3C-related research and recent research publications.

Additionally, a multitude of information relevant to 3C is provided throughout the 3C knowledge portal, to include: congressional reports, strategic guidance, leadership & 3C papers, lessons learned, world religion & 3C, 3C-related news, upcoming events, monthly book summaries, and a plethora of links to additional relevant resources.

1.4 Collaboration Platforms

Collaboration is highly encouraged utilizing the current Collaboration Forum or the up-and-coming R-Space platform. The Collaboration Forum is a tool designed to provide interested parties with a way to share information on 3C in general, as well as the overlap between 3C and leadership and operations. Anyone can access the Collaboration Forum to review public discussion threads. Once they are members, users are allowed to create their own group forums and read, upload, and comment on 3C-related topics, documents, and threads.

R-Space is currently under development. However, once complete, it will be a social networking and collaboration tool specifically designed for researchers. Members will be able to read, upload, and comment on various research documents;

create and participate in discussion threads; create, post to, and manage groups; post upcoming calendar events; and manage personal bios and professional contacts.

2 CONCLUSIONS

In an ever more complex operational environment, military members at all levels must have well-developed cross-cultural competencies in order to exercise their military missions. Although these are ambitious aspirations, their implementation within the development of the Total Force is likely the most critical aspect of present and future military successes. Asymmetric warfare against non-conventional adversaries - as well as the dire need for military forces to assume the roles of nation builders, peacekeepers, and humanitarian aid providers - makes cross-culturally sound leadership imperative.

Long gone are the days in which military success was measured by the ability to overpower adversaries with firepower and material. Success in today's military can only be measured by the extent to which military and civilian personnel develop and utilize their cross-cultural competence. These competencies are essential to building consensus, exerting influence, and negotiating mutually beneficial outcomes while simultaneously allowing our forces to maintain the ability to exert military force, when necessary. The 3C knowledge portal facilitates this learning, enabling our Total Force to develop the critical cross-cultural competence required to accomplish mission success.

REFERENCES

Cross-Cultural Competence (3C) Portal. 2011, www.defenseculture.org, Presented by Defense Language and National Security Education Office (DLNSEO) and Defense Equal Opportunity Management Institute (DEOMI).

Johnston, J. 2011. A Framework for Cross-Cultural Competence and Learning Recommendations. Presentation at Human Social Culture Behavior (HSCB) Focus 2011 Conference. Chantilly, Virginia.

McDonald, D. P., McGuire, G., Johnston, J., Selmeski, B., & Abbe, A. 2008. *Developing and managing cross-cultural competency within the Department of Defense: Recommendations for learning and assessment.* Defense Equal Opportunity Management Institute (DEOMI/J-9), 366 Tuskegee Airmen Drive, Patrick Air Force Base, FL 32925-3399.

CHAPTER 35

Intercultural Competence in Global Collaboration Courses in Computer Engineering

Helena Bernáld[1], Åsa Cajander[2], Mats Daniels[2], Can Kultur[3], Anette Löfström[2], Roger McDermott[4], and Lori Russell-Dag[3]

[1]Helena Bernáld Communications
Wiltshire, UK
Helena_bernald@yahoo.co.uk
[2]Uppsala University
Uppsala, Sweden
asa.cajander@it.uu.se, mats.daniels@it.uu.se, anette.lofgren@it.uu.se
[3]Bilkent Universiy
Ankara, Turkey
kultur@bilkent.edu.tr, russell@bilkent.edu.tr
[4]Robert Gordon University
Aberdeen, UK
roger.mcdermott@rgu.ac.uk

ABSTRACT

With the rapid and ever expanding globalization of the workforce, international collaborations are becoming part of everyday life for people of many professions. Intercultural competence has therefore become one of the central professional competencies needed for students in higher education, in order to equip them for their individual careers as well as contribute to a prospering society as a whole. We will start by discussing some definitions of intercultural competence and then describe two annual international collaboration projects between students from the US, Sweden and Turkey. We will present how intercultural competence is addressed through the introduction of an external lecturer and the use of reflections and discuss the outcomes of this approach in relation to the definition of intercultural competence.

Keywords: intercultural competence, open-ended group projects, global collaboration

1 INTRODUCTION

With the rapidly expanding globalization of the workforce, international collaborations are becoming part of daily life for people of many professions. Intercultural competence has therefore become one of the central professional competencies needed by students generally in higher education, in order to equip them for their individual careers as well as contribute to a prospering society as a whole. For computer engineering students such a competence is perhaps particularly crucial, as companies in the computing and IT areas are at the forefront in terms of globalization. Outsourcing, off-shoring, and similar processes, are endeavors that often have to be taken into account and where intercultural competence is central. However, even though intercultural competence is crucial for students, few courses at the university have the development of intercultural competence as a learning goal. As a counterweight to this deficiency we present our experiences from working with intercultural competence in two collaborative courses.

We start out by discussing models of competence in general and especially that of intercultural competence. This is followed by a presentation of two student collaborations. Our experiences from addressing intercultural competencies in the courses through seminars, project work, and assignments are presented and related to our definition of intercultural competence. We substantiate our experiences with citations of comments from students and conclude with a discussion of how we plan to proceed with our efforts to create a learning environment where students develop intercultural competence.

2 INTERCULTURAL COMPETENCE

2.1 Professional competencies

There is general agreement that students of higher education need to possess a variety of professional competencies when they graduate. However, what constitutes these competencies is often unclear, as is how to develop or assess them. The matter is further complicated by the fact that very few course units specify professional competencies explicitly. Guidance can be found in definitions made by international organizations of professional competencies, e.g. by the Organization for Economic Co-operation and Development (OECD). The Definition and Selection of Competencies (DeSeCo) project (OECD 2005) is intended to provide a framework for an understanding of competencies in general and is based on definitions and assessment methods. It is designed to set overarching goals for education systems and lifelong learning.

The view in the DeSeCo project is that a competence is more than knowledge, it is also the ability to deal with complex situations in particular contexts. The idea is to capture what is needed to deal with such situations in general through the definition of a few key competencies. These key competencies must:

- Contribute to valued outcomes for societies and individuals;
- Help individuals meet important demands in a wide variety of contexts;
- Be important not just for specialists but for all individuals.

The key competencies are classified in three broad categories: being able to use tools for interacting with the environment, being able to engage with others in heterogeneous groups, and being able to take responsibility for ones own life in a broad social context and act autonomously. Central to all categories is the ability to think and act reflectively. The interacting in heterogeneous groups key category contains three competencies that address the need to keep up to date with technologies, to adapt tools to ones own purposes, and to conduct active dialogue with the world. The first competence is the ability to use language, symbols, and text interactively, which concerns using spoken and written language skills, computation and other mathematical skills effectively in multiple situations. This is associated with communication competence and literacy. The second competence is the ability to use knowledge and information interactively, which requires critical reflection on the nature of information itself. This competence is needed in order to understand and form opinions, make decisions, and carry out informed and responsible actions. The third competence is the ability to use technology interactively, which is based on an awareness of new ways technology can be used in everyday life. Harnessing the potential of information and communication technology (ICT) is part of this competence.

2.2 Defining Intercultural Competence

Intercultural competence can be described as the ability to communicate effectively and appropriately in intercultural situations based on one's intercultural knowledge, attitudes and skills. In the past, the focus may have largely been limited to knowledge about the other culture. Today, and in line with the DeSeCo definition of competence described above, the concept stresses both knowledge and comprehension of one's own culture and other cultures, an attitude of openness, curiosity, respect and inclusion, as well as the skills that one may acquire based on this knowledge and an open, inclusive attitude. These skills would allow an individual to select and use appropriate communication styles and behaviour in different intercultural situations, adapt to new cultural environments and successfully interact with people from different cultures.

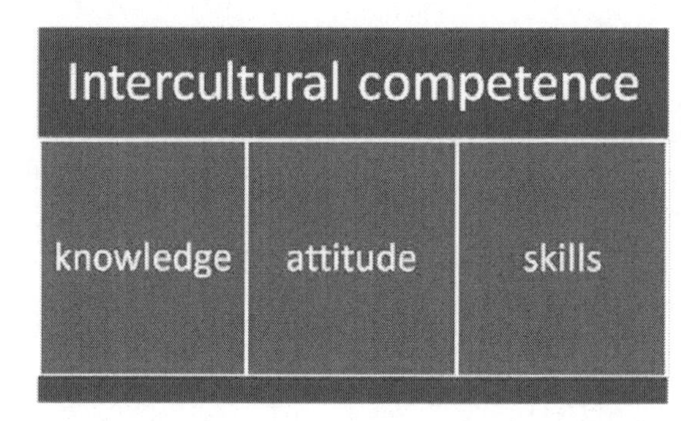

Figure 1 Intercultural component model based on Byram's Model of Intercultural Communicative Competence (1997)

This model of a competence as a combination of knowledge, attitude and skills reflects for example that of Byram (1997), and the Intercultural Competence Components Models by Howard-Hamilton et al. (1998).

We will in this paper adopt a view of what constitutes intercultural competence based on Byram's model depicted in Figure 1. We also stress the importance of capturing the invisible aspects ("codes") of what a culture is. These aspects are discussed by Hofstede and Hofstede (2005) and Trompenaars and Wooliams (2003) described them in terms of layers of an onion. A similar analogy regarding the invisible aspects is that of an iceberg, where only a very small part is visible above the surface, e.g. language and traditions, and most is invisible below the surface, e.g. values, emotions and attitudes. Thus, we have a model of intercultural competence where the components are:

- *Knowledge* about different cultures and culturally inherent values in general and the target cultures in particular, about social processes and "codes", e.g. how people from other cultures may perceive you and interpret your behaviour, as well as knowledge about your own cultural identity.
- *Attitude*: to emphasize and encourage openmindedness and a desire to continue to learn, to be curious about other cultures, continually reflecting on your own behaviour and be open to adapt it (without giving up your own identity).
- *Skills* : these build on knowledge and attitude and concerns the ability to choose and develop appropriate behaviour and communication style when interacting with people from other cultures, to create trust and build intercultural relationships.

3 THE STUDENT COLLABORATIONS

This paper describes experiences from addressing intercultural competence in two student collaborations. These collaborations are shortly presented below.

3.1 The Collaboration Between Uppsala University - Rose-Hulman Institute of Technology

The local setting for this study is a global collaboration between 4th year Computer Science (CS) and IT engineering students at Uppsala University, Sweden, and 2nd and 3rd year CS and Software Engineering students at Rose-Hulman Institute of Technology, Indiana, USA. This setting and different aspects of it have been described elsewhere, as for example in (Cajander et al. 2009a, Cajander et al. 2009b, Cajander et al. 2011, Daniels 2010, Daniels et al. 2010, Daniels 2011, Laxer et al. 2009), but a short summary is given here to provide the reader with a quick overview.

The educational setting is aimed at developing professional competencies that are essential in relation to working in a global collaboration setting. This setting is based on the Open-Ended Group Project (OEGP) concept (Daniels 2011), where complexity and many options for how to approach a problem are central issues. OEGP is hence a suitable concept for preparing our students for working on a global arena.

An important aspect of the educational setting is that the project is placed in a real environment with a real client. This adds to the complexity but is also shown to increase motivation for the students (Marra and Wheeler 2000). An issue with real clients is that they also have other obligations and it can be hard to get reasonable access. This has been addressed by only using one reliable client and putting all students, typically around 25 with a majority in Uppsala, into one project. Another issue with a real client might be that some students feel ethically constrained to help certain clients, e.g. for political, religious, or competition reasons. We have for that reason chosen to work with a client in the public health sector, i.e. the Uppsala County Council and the associated academic hospital.

3.2 The Collaboration Between Bilkent University and Rose-Hulman Institute of Technology

The collaboration between the Rose-Hulman Institute of Technology and Bilkent University, Ankara, Turkey, is the other experience this paper is based on. The setting and student feedback from the initial collaboration has been described in detail elsewhere (Chidanandan et al. 2010). The Rose-Hulman students are drawn from the same cohort as in the Uppsala - Rose-Hulman collaboration and the students from Bilkent involved in the project are 3rd or 4th year students studying at the Department of Computer and Instructional Technology Teacher Education at the Faculty of Education.

In this course, using their Computer Science (CS) background, students were expected to work with a team to analyze, design and develop an IT solution in cooperation with a real client (International Children's Center). Although the Bilkent students are not CS majors, they do follow a CS track in their program and the IT project is well matched with the educational goals of their program. The client was selected based on the criteria that the organization should ideally be a non-profit organization with an international focus in order to minimize language and client communication issues. The class sizes in the two experiences of the Bilkent - Rose-Hulman collaboration were 11 (8 from Bilkent, 3 from Rose-Hulman) and 17 (10 from Bilkent, 7 from Rose-Hulman).

The collaboration experiences in 2009 and 2011 have similarities and differences. Both included external professionals; during the first collaboration the invited speaker gave a presentation on Turkish and American cultural differences, and in the second collaboration the invited professional gave a workshop on intercultural communication at the beginning of the course. In addition, while the supervisor for each project course was different, it was for both their first involvement in a global collaborative project course. One of the course supervisors was a Canadian who had lived in Turkey for 10 years. In the first experience, there was one project and one client, in the second one there were two clients, three user groups and one common project theme.

There is a difference in the actual or perceived focus of the project at the respective institutions. At Bilkent University the course has a strong product focus and Bilkent students usually perceive this course as their senior project course in which they utilize their IT skills, before starting to take the pedagogical courses. At Rose-Hulman the course is offered as an elective course with a focus on the process and global/intercultural awareness.

4 EXPERIENCES FROM ADDRESSING INTERCULTURAL COMPETENCE

A common objective in the courses is to cover the different components of intercultural competence as defined above. The components are addressed in several ways, for example by the students acting in a globally distributed project with team members with different cultural background, by the introduction of an expect on cultural awareness, and by having assignments in which the students are asked to reflect on cultural aspects of their collaboration. In this section we present our experiences in these three areas and relate them to our definition of intercultural competence. We substantiate our experiences with citations of comments from students.

4.1 Introduction of an External Lecturer in Intercultural Collaboration

The external lecturer was introduced in the Uppsala - Rose-Hulman collaboration in 2007 (and has been returning annually since). The same lecturer worked with the students in the Bilkent - Rose-Human collaboration in 2011. In the collaboration in 2009, a presentation on cultural differences was given by a different speaker (as described above). Even though all of the teachers involved in these global collaboration courses have extensive intercultural experience and could bring a great deal of empirical competence to the table themselves, we have found that introducing an external expert in the field has not only provided additional credibility and some extra "gravitas" to the subject, but also sent the signal to the students just how important we, the teachers and faculty, view this competence. The following quote represent a common view among the students and illustrates this point.

"After the seminar with Helena Bernáld, I realized that intercultural skills are more important than I thought they were. Two months on a part-time basis is too short to learn another culture, but I think it will be long enough for me to at least get a general idea, and experience how another culture does business."

The quote also address the issue of learning intercultural competence as a life-long process. Starting the collaboration with a boost of the knowledge component is essential in our experience, since it is noticeable that some of the students have little intercultural knowledge. It is not unusual that their expectation before entering the project course was that the only difficulty in a global distributed project would be handling the time difference, as in this quote:

"I have never worked on a team of people from a different culture as my own. Before this class started I didn't expect to get much out of it. I don't know what I expected, but it wasn't what this class turned out to be. I had no idea what it was like to work on a international team, and I didn't expect it to be different or more complex than just dealing with a time zone difference. From what I have seen in just four weeks has changed my opinion of this course. There are so many issues and items to consider when dealing with international teams that I never thought about. I hope the amazing opportunity presented in this class will help me in both my professional and personal development.".

4.2 Introduction of Reflection Assignments

Intercultural interaction provides excellent opportunities for reflection and learning. Schwartz, Xiaodong, and Holmes (2003) note that intercultural experiences, "in addition to alternative models of practice, [---] provide contrasts that help people notice aspects of their own practice". Reflections can be used to address specific areas to enhance learning and in the courses it has been used as a way to make the students reflect on their intercultural experiences as well as on other aspects of the collaboration. Moreover, reflecting on practice or in practice can be seen as one of the key components of life-long learning in professions, as defined by for example Schön (1983). In the Uppsala - Rose-Hulman collaboration reflections have been used since 2007, and a written reflection on their experience of cultural differences has been a part since then.

Regularly reflecting on cultural differences, on one's own development and intercultural behaviour, is imperative in the process. Some of the students express how they really discussed and reflected on the cultural differences during the weeks when the students met, as in this example:

"During the RH week we experienced some cultural clashes despite the seminar by Helena. The clashes I'm referring to are the way we Swedes communicate (we don't always say what we think), some bad manners at the restaurants and attitudes towards women. I think all of them were harmless though, partially because of everyone's will to be open minded."

The experience from the teachers is that reflection assignments definitely have a positive impact on the students' acquiring intercultural competence. It has not least drastically reduced the tendency to blame "the other side" when things didn't work out well, since most of them realized that they most likely had a part in why things worked as it did themselves (and possibly also realized that they were likely to get a question about what they had tried to do to make things better).

4.3 The Actual Collaboration Project

Creating opportunities for students to have intercultural experiences is seen as essential by most researchers in the field, but it is also pointed out that spending time in a different culture is not enough, as it takes more than mere contact with, or presence in, a different culture to develop intercultural competence. As Deardorff (2009) points out, building authentic relationships through dialogue, listening, observing, asking those from a different background to teach and share their cultural values etc, is key. It is the attitude and skill components of intercultural competence that really matters in achieving this in the real collaboration project.

To get the students to truly interact is however not something that comes easy and is one important reason for arranging an introductory week when most of the students meet face-to-face. The work during this week, and the social activities that are organized by the students in the evenings are generally mentioned as an invaluable aspect for the outcome of the project and the acquisition of intercultural competence in the reflections on intercultural experiences. This positive effect has impacted the organization of the introductory week, and since two years all the American students come to Sweden in the beginning of the semester. Another change made to improve the possibility for intercultural learning was to offer housing of American students in the Swedish students' homes. This made it possible for the students not only to meet during working hours, but to actually spend a whole week together. This has been much appreciated by the students, as described in the following quotations:

"The first week together with the Americans was one of the, if not the best week yet at the university. It wasn't just all fun though, it was also an eye opening experience regarding cross cultural communication and cooperation. This is something I most probably will have use for later on in my life, since I want to work with projects across the borders in the future."

and

"I feel like everyone in my team went into this project with a very good attitude and willingness to do their very best for it to be a success. As I have mentioned earlier I think this has a lot to do with the introduction week in the beginning of the project where we get to know each other. You feel a lot closer afterwards and when you care about the people you work with you are even more prone to do you best work since you do not want to let your friends down."

4.4 Discussion

Our objective for the intercultural aspect of the courses, is to bring the necessity of intercultural competence into focus and to raise the students' awareness of culturally inherent values in general and their implications for successful communication and collaboration. Our aim is for the students to acquire a more clearly defined picture of their own cultural identity, including inherent Swedish, Turkish and American values and how these manifest themselves in communication patterns, decision making and choices made in professional and social situations. Discussing the concept per se with the students seems to have a positive effect on the competence development process, as well as focusing part of the seminar on how they can learn from each other, how to keep an open and inclusive mind if they encounter challenging or frustrating situations in their interaction with the students from the other culture, and how to adapt their own communication style and behavior to overcome these challenges. As some of the quotes above have shown, many of the students have expressed focusing on being observant of cultural differences during the collaborations. A majority of the written reflections and evaluations testify to the students making a conscious effort to keep an open mind seasoned with goodwill and to accommodate the culture-related communicative differences, as expressed in this quote:

"The seminar of cultural differences in the beginning of the week I really liked. That gave all of us something to think about when we interacted with each other and planned the project. We learned more each day how to interact and adjust to our differences."

As noted above, the reflections on cultural differences reveal that most of the students have little or no prior formal knowledge of intercultural competence when they start the global collaboration courses in our learning environment. Some of them have prior intercultural experience, for example from having studied abroad, and some may have worked in multicultural teams in their local environment at the university, or spent a semester studying abroad. However, very few students claim to have studied or addressed the various dimensions of intercultural interactions before, and for many, this is the first time they become aware of the potential challenges at hand. This is depicted in the following quotes:

"Thanks to the presentation by Helena, a good portion of understanding my Swedish counterparts has already been taken care of. That was quite possibly one of my favorite days of the week in Sweden and by far the most eye opening experience. I had never considered many of the different aspects of their culture that Helena brought up."

and

"The Swedes have an entirely different cultural perspective and in some cases an entirely different world view driving their perception of this problem."

5. CONCLUSION

The paper is based on experiences from close to ten years experiences of running international student collaborations with an aim to improve intercultural competence among the students. Our conclusion from this is that there are no "silver bullets", but also that it can be done. We believe that the above given definition of intercultural competence is clearly beneficial in arranging a learning environment where students develop intercultural competence.

The intercultural competence definition has, for instance, been useful in finding a balance between scaffolding and openendedness in the educational setting. Scaffolding like the introduction of lectures by an expert in cultural awareness has been highly successful, in particular with regard to both the knowledge and the attitude aspects of the competence. One difficulty with providing scaffolding is that it complicates assessment of intercultural competence in that the students learn what to say to appear having the competence rather than acquiring it. This view is based on having a social constructivist (Vygotsky 1978) view of learning and that intercultural competence is achieved through real involvement in such a community.

The use of assignments where the students reflect on their collaboration is another example of scaffolding in that it provides aid to observing the mechanisms of the intercultural collaboration. These assignments have also been introduced to raise understanding of potential consequences of cultural differences and especially lead to a higher appreciation of the contribution of their peers. The aspect of creating an environment where it is easier for the students to appreciate each other is an interesting consequence of striving to achieve intercultural competence in an educational setting. To value the contributions of peer-students is a central learning activity in the Contributing Student Pedagogy (Hamer et al 2008) and our approach fits smoothly with the philosophy of this pedagogy.

REFERENCES

Byram, M. 1997. *Teaching and Assessing Intercultural Communicative Competence.* Multilingual Matters Ltd.

Cajander, Å., T. Clear, and M Daniels, et al. 2009a. *Students analyzing their collaboration in an International Open Ended Group Project*, ASEE/IEEE Frontiers in Education conference, San Antonio, USA, pp M1D 1-6

Cajander, Å., T. Clear, and Daniels, M. 2009b. *Introducing an External Mentor in an International Open Ended Group Project.* in 39th ASEE/IEEE Frontiers in Education Conference., (San Antonio, Texas 2009), IEEE, T1A1-T1A6.

Cajander, Å., M. Daniels, R. McDermott, and B. von Konsky, 2011. Assessing Professional Skills in Engineering Education, *Australian Computer Science Communications*, vol 33, no 2, pp 145-154

Chidanandan A., L. Russell-Dag, C. Laxer, and R. Ayfer R., 2010. *In their words: student feedback on an international project collaboration*, ACM technical symposium on Computer Science Education, Milwaukee, USA

Daniels, M. 2010. The Contribution of Open Ended Group Projects to International Student Collaborations, *ACM Inroads*, vol 1, no 3, pp 79-84

Daniels, M., Å. Cajander, T. Clear, and A. Pears, 2010. Engineering Education Research in Practice: Evolving Use of Open Ended Group Projects as a Pedagogical Strategy for Developing Skills in Global Collaboration, *International Journal of Engineering Education*, vol 26, no 4, pp 795-806

Daniels, M. 2011. Developing and Assessing Professional Competencies: a Pipe Dream? Experiences from an Open-Ended Group Project Learning Environment, *Digital Comprehensive Summaries of Uppsala Dissertations from the Faculty of Science and Technology 808*, Uppsala, Sweden.

Deardorff, D. 2009. *The SAGE Handbook of Intercultural Competence*, SAGE Publications, Inc., Thousand Oaks, CA, USA.

Hamer, J., Q. Cutts, and J. Jackova, et al.. 2008. Contributing student pedagogy, *SIGCSE Bulletin*, vol 40, pp 194–212

Hofstede, G. and G. Hofstede, 2005. *Cultures and Organizations – Software of the Mind*, McGraw-Hill, New York, USA.

Howard-Hamilton, M., B. Richardson, and B. Shuford, 1998. Promoting multicultural education: A holistic approach. *College Student Affairs Journal*,vol 18, no 1, pp. 5-17.

Laxer, C., M. Daniels, Å. Cajander, and M. Wollowski, 2009. Evolution of an International Collaborative Student Project, *Australian Computer Science Communications*, vol 31, no 5, pp 111-118

Marra, R. and T. Wheeler, 2000. The impact of a student-centered engineering
design project on student motivation, *ASEE/IEEE Frontiers in Education Conference*, Kansas City, USA, pp F2C 8-13.

OECD 2005. *The Definition and Selection of Key Competencies: Executive Summary*, https://www.pisa.oecd.org/dataoecd/47/61/35070367.pdf

Schwartz, D., L. Xiaodong, and J. Holmes, 2003. Technologies for learning from Intercultural Reflections. *Intercultural Education*, Vol. 14, No. 3. Carfax Publishing, Taylor & Francis Group

Schön, D. 1983. *The Reflective Practitioner - How Professionals Think in Action*, Basic Books.

Trampenaars, F. and P. Wooliams, 2003. *Business Across Cultures, Capstone* Publishing Ltd., Chichester, UK.

Vygotsky, L. 1978. *Mind in Society: The Development of Higher Psychological Processes*, Harvard University Press, Cambridge, USA.

CHAPTER 36

Scope and Scale: A Framework for Social Analysis

David L. Sallach, Michael J. North and W.A. Rivera

University of Chicago
Chicago, USA
sallach@uchicago.edu

ABSTRACT

Prior theoretical work has identified categories of socio-cultural structures that provide an analytical context supporting comparative analysis of diverse settings and time periods. The present analysis supplements these dimensions with two exogenous categories, scope and scale and, thereby, extends the basis for analysis of comparative social models.

Keywords: Eco-demographic niche, geo-strategic niche, multiscale social actors, analytical context

1 OVERVIEW

The social sciences have taken important strides in recent decades through the utilization of agent-based modeling and simulation (ABMS). Agent models allow fine-grain simulation of individual actors, localized propagation effects, and direct and network-mediated interaction. However, notwithstanding their contributions and promise, with rare exceptions, social agent modelers have struggled to represent the rich details of structural and cultural forces and, thus, have been challenged to address deeper historical and policy-oriented issues.

Prior analysis has suggested several interrelated strategies for advancing social modeling including the representation and integration of social theory in various forms (Sallach 2003; 2008), and the definition of socio-cultural structures (Sallach 2002; 2007; 2012), that can facilitate a comparative analysis of diverse settings and time periods. In the latter work, three interleaved dimensions are proposed: 1) bio-functional differentiation, 2) resource stratification, and 3) geo-cultural collectivities.

The present paper, drawing upon the Modeling Strategic Contexts project (Sallach 2011), seeks to characterize broad contexts that may have the potential to reshape, or disrupt, strategic initiatives, social structures and even historical trends. These ultimate contexts thereby underlie and supplement the dimensions of socio-cultural structure with an exogenous architectonic framework. Clear conceptualization of the categories of scope and scale thereby supports and extends the basis of comparison and analysis of social processes.

Scope combines ecological and demographic characteristics. Demography, of course, is a relatively mature social science emphasizing birth and death rates, immigration and emigration, the population effects of epidemics and plagues, refugee dynamics and other population characteristics and trends (Kertzer 1997; Keyfitz & Caswell 2005; Bacaër 2011).

Another aspect of scope is used to place population dynamics in the context of the socio-ecological niches that constrain and shape regional demographics. By identifying and analyzing eco-demographic niches, a context for the analysis of historical conjunctures is established.

The second exogenous category, *scale*, addresses the fact that social actors have diverse sizes and distinct histories. Although the historical details depend upon multi-vocal semantics (Levine 1988), informally: nation-states are larger than the political parties that seek to govern them, those parties are larger than the movements and factions that vie for their leadership, and both are larger than the groups that compose them. Social scale is often formalized, so that nations are larger, by definition, than the states or provinces that compose them, which are larger than their successive, counties, cities and wards. Similarly, alliances are, by definition larger in scale than the nations that compose them. Identifying social actors of varying types as constituting a single dimension, (with an initial focus on their relative scale) allows modelers to isolate attributes that social actors may share, and/or how social actors may differ in other attributes and dynamics, across scale. In addition, each specific actor, of whatever scale, has a distinct history that shapes the interpretation of situations and events.

The two exogenous categories form a natural pairing in that they ultimately build from the biological individual. That is, both the eco-demographic component of scope and the scale of social actors combine to define an encompassing framework for the emergent social structures they contain. Accordingly, a given model can be located at specific scale and scope location, which defines the context within which socio-cultural structures arise and take form. Taken together, these categories provide a generic framework within which social structures, and the dynamic processes to which they give rise, emerge.

2 THE SCOPE OF ECO-DEMOGRAPHY

Ecological niches and the populations and practices they support are woven together (cf., Diamond 1997). Considering this relationship explicitly provides a foundation for a wide range of social models.

2.1 Demographic Processes

Demographic shifts can redefine power balances and introduce unexpected generational challenges. It is not uncommon for radical political movements to draw upon demographic strategies as a part of their larger plan. The role of eugenics within Nazi grand strategies, from increasing the German birth rate to the Holocaust, is well known (Weikart 2006).

Demographic factors influence historical trends in less direct ways as well. Pareto (1968), for example, asserts that, as families move into the societal elite, their birthrates decline (cf., Johansson 1987). Turchin and Nefedov (2009) supplement this insight by suggesting that the onset of periods of political instability tends to be preceded by 'elite overproduction', i.e., a dramatic expansion in the size of the dominant elite. In this case, the historical logic appears to be, not only that the expanded elite weighs more heavily on the supporting population, but also that such growth is associated with larger and more frequent fissures within the elite that generate conflicts which often ripple throughout the body politic.

2.2 Niches and Population Distributions

For demographic processes to be adequately contextualized, they must be located in their ecological settings (Alihan 1964; McNeill 1976; 1980). Fortunately, ecological modeling is undergoing rapid progress, and many of the associated innovations can be used within social modeling as well. Peterson, *et al.* (2011) provide an excellent overview and synthesis of this progress. First, niche modeling explicitly incorporates both Grinellian and Eltonian definitions. The former focuses on geographic scales and requirements and emphasizes non-interacting attributes of the niche, while the latter attends to the functional role of the inhabitants, and includes resource consumption and impacts. Both are vital to an integral concept of the niche.

Second, an association between niches and population distributions is established. In ecological models, the Eltonian definition focuses upon the population distribution of species, of course, whereas social models will need to (equivalently) emphasize geo-cultural collectivities. This forms an interface between scope, as defined here, and socio-cultural structures (Sallach 2012).

Another distinction between ecological and social models is the type of resources that need to be represented. In ecological models, resources are basically limited to food, water and natural habitats (or resources used in their construction). A social niche must address a far wider range of resources in order to determine the types of social, economic and cultural activities that can be supported within that region. Collins (1978) holds that access to strategic resources is one advantage that differentiates between dominant and subordinate states. However, as summarized below, other niche characteristics are important as well.

2.3 Geo-Strategic Niches

In representing the relations among nation-states, niches are defined by additional properties that have played a historically significant role in national and international conflicts (Collins 1978). One of the most important is a geo-strategic configuration that provides the basis for a natural defensive posture. Collins (1978:7) calls this a 'heartland', and describes it as a territory with some geographical unity that makes it "more easily and uniformly accessible to military control from within than from any point outside." Heartlands, he asserts are the basic units of geopolitics. Countries that are enveloped in mountains, or surrounded by seas, or other natural barriers, are easier to defend and, thus, are better able to retain their autonomy in the face of various possible threats.

A second configuration of interest is the position of being on the periphery of a cluster of similarly situated states. Collins refers to this as a 'marchland' position, and observes, "Marchland states are the first to break free from an empire, the longest to hold out against an empire, and most importantly, the principle creators of conquest states." Why is being on the periphery an advantageous position? Collins explains that the benefit lies in having potential enemies aligned along a single front (p. 13) although, of course, such alignments take many forms.

Other configurations are less geo-strategic and more situational, such as various balances of power that may emerge. Periods of stalemate and overextension are part of the same process. Each suggests that consequences of various geo-strategic niches emerge over time. It is to this issue that we now turn.

2.4 The Dynamics of Eco-Demographic Strategic Processes

From Pareto to Turchin and his colleagues, political models of instability are becoming ever stronger. However, even when demographic factors are incorporated (Goldstone 1991; Turchin & Nefedov 2009), the context provided by geological and ecological factors is ignored. Conversely, when innovative agent-based models incorporate geologies and ecologies, it is typically done in either a stylized way (Cederman 1997; 2008), or else specialized to a specific case (Dean *et al.* 2000). Therefore, a generalized theory of eco-demographic niches, including their dynamic strategic implications, will make a generalized contribution.

Social niches and their associated population distributions form the context for aspects of social structure. The ecology of niches will influence the division of labor, and eco-demographic niches shape and influence geo-cultural collectivities. The result is models that are more deeply grounded.

3 THE SCALE OF SOCIAL ACTORS

Following the example of the physical processes, social actors are often modeled in an exogenous way, whether determinate or stochastic. However, social actors arise through social, meaning-oriented interaction and, arguably, computational models need to be able to address fine-grain, endogenous interactions in order to represent emergent social and historical issues such as those raised above. The following sections consider this question.

3.1 Social Actors

Few scholars have contributed more to the concept of multi-scale social actors than James Coleman (1986). Especially in his later work, Coleman (1990; Coleman & Hao 1989), extends economic equilibrium processes to more generic social equilibria, and, thereby, implicitly moves beyond the limits of a rational choice framework (Adams 2010). Coleman (1990:667-688) recognizes that the concept of economic actors can be extended to generalized social actors. For Coleman, diverse social actors exist at various scales and in many forms. Such actors seek to control resources and events, however, the outcome may be quite different from those intended by the agents.

The broader implications of models that represent multiscale social actors have yet to be fully pursued within the social sciences. One of the primary purposes of the present paper is to consider, at a much more extensive scale, how best to take multiscale interactions seriously.

3.2 Multiscale Social Actors

Over the course and breadth of history, countless types of social actors have been identified. Prior to conceptualization (while they are still in the tacit realm of social practice), social actors begin to identify one another and distinguish themselves (Luhmann 2002), albeit in a somewhat arbitrary way. As issues arise regarding the attribution of social consequences, actor types become distinguished, are labeled and, frequently, congeal into conscious sources of contention.

Table 1 Types of Social Actors

Social Actor	Description	Elaboration
Nation-state	Geographically based institution concerned with governance and the control of coercion.	Focusing outward, analysis of the nation-state incorporates the breadth of international relations. Looking inward, the nation-state addresses domestic institutions, and their interaction.
Institution	Imperatively coordinated	Family, economy, state and religion (or equivalent organized belief systems).

	associations adapted to address basic social needs.	Their interrelationships and level of coherence contribute to social stability, or instability.
Party	Subtype of organization, focusing on the acquisition and use of political power.	Widely divergent types of parties have arisen in human history, from the totalitarian to the evanescent, with a wide range of ideologies.
Movement	Diffuse associations that combine in diffuse forms in order to achieve broad objectives or express common sensibilities.	Some movements exist as currents of common practice with little organizational form (e.g., jazz, expressionism). In other cases, there may be dozens of overlapping organizations, each with its own focus or emphasis. In all such cases, a movement is broader than any component organizations, and each has its own distinctive lifecycle.
Organization	Pervasive formation that typically involves an intentional focus on collective goals, and rationalized planning.	In general, organizations and movements should be considered at the same level of analysis. Movements may contain many organizations.
Faction	Instrumental group forged to achieve diverse goals, often ideological. It may operate through networks, organizations and/or institutions.	Factions are small enough to achieve strong cohesion, and large enough to control much larger organizations and institutions.
Network	Assemblage of relations (links) among persons or other social actors.	Networks may be constructed from diverse sources. Stable relations lend themselves to formal analysis. The relations that serve as links may evolve over time, or be repurposed. Groups may sometimes serve as nodes in the network.
Group	Small and relatively intense associations formed through frequent and multi-faceted interaction.	The broadest and most intense forms of group interaction involve face-to-face meeting and communication. Persistent groups typically do not exceed twelve members.
Social Self	Defined by the higher social actors with which it is affectively engaged.	Each of the higher actors forms a social collectivity with an orientation and identity that influences and potentially shapes correlative orientations and identities down to the smallest scale unit.

Such social actors exist at various scales, assuming diverse forms in particular locations and times. As scholars seek to generalize and rationalize these constructs, they often become more abstract, conceptual and broadly applicable. Examples of such named concepts plus suggestive descriptions are illustrated in Table 1. In practice, however, each instance remains socially and historically distinct and, common labels notwithstanding, actors situated within that time and place recognize and respond to the uniqueness of the social actors that they encounter.

The social actors described in Table 1 are neither mutually exclusive nor exhaustive. Here they are intended as social scientific concepts, but many of them are also commonly used in broader public discourse, albeit with a range of overlapping definitions.

Given multiple levels of social scale (as depicted in Table 1), there are two strategies by which they might reasonably be addressed and ultimately modeled. The first and most common is to identify relevant social entity types and model each individually. However, this is a strategy that has multiple limitations. Social entities proliferate while each historical instance manifests on a slightly different scale and in idiosyncratic forms combining distinct features with general characteristics. Nor are such labeled patterns exhaustive, suggesting that effective scientific concepts have yet to be identified. To be applied to particular problems or policy issues, the resulting abstractions must then be translated into the historically specific forms under investigation. The more complex the mapping, the more limited is the explanatory power of the putative generalization.

The second strategy is to develop an integrated model of how intertwined social actors interact across levels. This may seem more difficult than the first strategy but, by avoiding artificial distinctions, it may be better able to represent social processes as they actually arise. The analysis presented in the next several sections undertakes to illustrate the second strategy.

3.3 Groups, Networks, Movements and Organizations

In the sociological and social-psychological literature, groups, networks movements and organizations are each treated as a distinct social formation. And, of course, each has a unique form and distinctive characteristics and these must be understood in their own terms. However, none of them exists in isolation and, in historical settings, specific instances of each type of sociality form a context for, and influence the others, in situated ways.

For those of us involved in historical and/or policy-oriented research, we may benefit from developing an integrated configuration of social forms. We then regard any isolated form as degenerate (i.e., the other social formations, that are always influential in some way, have been ignored).

3.4 Factions, Movements, Parties and States

The same can be said for larger social forms although, in this case, the result is less generic. Factions, movements, parties and states are integral parts of political sociology. Coherent groups or factions may persist over decades while moving coherently across multiple scales. Factions can assert themselves within movements, parties and, ultimately, states. However, the actual dynamics vary widely.

In some cases, a hard-core faction holds to a rigid ideology and/or identity and imposes itself on successively larger actors (Apter & Saich 1994). Under this strategy, the decision mechanisms of higher-level actors need to be restructured in order for the factional identity to continue to assert itself.

In other cases with a comparable ascent, the personal relations among the underlying group create a sense of loyalty and trust, but the decision mechanisms of the higher-level actors are essentially unchanged (cf., Mattson 2010). Emerging political issues are given group scrutiny and, in most cases, a common response is formulated. Differences within the group are minimized and/or downplayed.

3.5 Dynamics of Multi-scale Interaction

The specific dynamics of multi-scale interaction are widely diverse. However, Figure 1 shows a stylized view of how a (non-exhaustive) hierarchy of social actors of varying scales each influence outcomes within the larger social process. The important characteristic of the described process is that it is causally dual. The left side depicts voluntary social actors from the smallest to the largest, while the right traces structural affects from the largest to the smallest. In this way, it expresses two dualities: small/large scale, and under/over socialization.

More specifically, Figure 1 depicts a micro-macro process (Collins 1988) that begins (left side) moving up from the smallest to the largest (voluntary) actor scale, and returns (right side) moving down from the largest structure scale to the smallest. The structural influence is more fully expressed in the socio-cultural structures model.

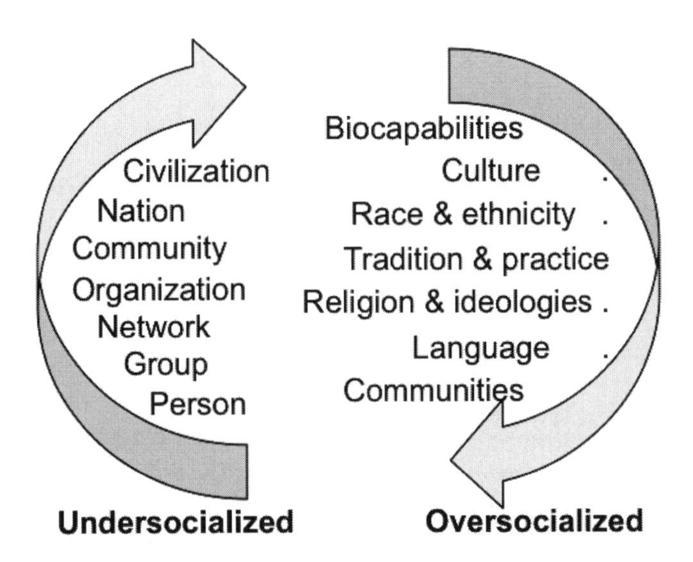

Figure 1: Agency and Structure as Causally Dual.

Undersocialization versus oversocialization is one way that the social sciences express the tension between the choice of the actor and the constraint of structure (Turk 1965; Wrong 1999; Raub, Buskens & van Assen 2011). In disciplinary terms, economics has tended to favor agency, while sociology has been more oriented toward structure (although, of course, this is only a tendency). However, neither alone can be causally adequate. In social life, their relative efficacy fluctuates, and that is what cross-cutting models must effectively capture. Thus, it is a tension best expressed as a duality, which is what Figure 1 is designed to convey.

4 THE INTERACTION OF SCOPE AND SCALE

Scope and scale, as considered here, are quite different in form, yet their historical and prospective interactions are why it is useful to consider them in conjunction. Together, they represent an inherent frame for social models of all types. Regarding scope as a horizontal axis and scale as a vertical axis (and addressing also the time span over which they interact), scope and scale can be considered as an exogenous container for social models of quite variegated foci and duration, an architectonic framework for all the situated social processes that they contain.

ACKNOWLEDGEMENT

The authors gratefully acknowledge support for this project from the Office of Naval Research, Award No. N00014-09-1-0766.

REFERENCES

Adams, Julia. 2010. "The unknown James Coleman: Culture and history in *Foundations of Social Theory*." *Contemporary Sociology* 39:253-258.

Alihan, Milla Aïssa. 1964. *Social Ecology: A Critical Analysis.* New York: Cooper Square Publishers.

Apter, David E. & Tony Saich. 1994. *Revolutionary Discourse in Mao's Republic.* Cambridge, MA: Harvard University Press.

Bacaër, Nicolas. 2011. *A Short History of Mathematical Population Dynamics.* London: Springer-Verlag.

Cederman, Lars-Erik. 1997. *Emergent Actors in World Politics: How States and Nations Develop and Dissolve.* Princeton, NJ: Princeton University Press.

_____. 2008. Articulating the geo-cultural logic of nationalist insurgency. Pp. 242-270 in S.N. Kalyvas, I. Shapiro & T. Masoud, eds., *Order Conflict and Violence.* New York: Cambridge University Press.

Coleman, James S. 1986. "Social theory, social research and a theory of action." *American Journal of Sociology* 91:1309-1335.

_____. 1990. *Foundations of Social Theory.* Cambridge, MA: Harvard University Press.

Coleman, James S. and Lingxin Hao. 1989. "Linear systems analysis: Macrolevel analysis with microlevel data." *Sociological Methodology* 19:395-422.

Collins, Randall. 1978. Some principles of long-term social change: The territorial power of states. *Research in Social Movements, Conflict and Change* 1:1-34.

_____. 1988. The micro contribution to macro sociology. *Sociological Theory* 6(2), pp. 242-253.

Dean, J.S., G.J. Gumerman, J.M. Epstein, R.L. Axtell, A.C. Swedlund, M.T. Parker & S. McCarroll. 2000. Understanding Anasazi culture change through agent-based modeling. Pp. 179-207 in T.A. Kohler & G.J. Gumerman, eds., *Dynamics in Human and Primate Societies: Agent-based Modeling of Social and Spatial Processes.* New York: Oxford University Press.

Diamond, Jared. 1997. *Guns, Germs and Steel: The Fates of Human Societies.* New York: W.W. Norton.

Goldstone, Jack A. 1991. *Revolution and Rebellion in the Early Modern World.* Berkeley, CA: University of California Press.

Johansson, S. Ryan. 1987. Status anxiety and demographic contraction of privileged populations. *Population and Development Review* 13 (September):439-470.

Kertzer, David I., ed. 1997. *Anthropological Demography: Toward a New Synthesis.* Chicago: University of Chicago Press.

Keyfitz, Nathan & Hal Caswell. 2005. *Applied Mathematical Demography.* New York: Springer.

Levine, Donald N. 1988. *The Flight from Ambiguity: Essays in Social and Cultural Theory.* Chicago: University of Chicago Press.

Luhmann, Niklas. 2002. *Theories of Distinction.* Stanford, CA: Stanford University Press.

Mattson, Kevin. 2010. 'What the Heck Are You Up To, Mr. President?': Jimmy Carter, America's 'Malaise,' and the Speech That Should Have Changed the Country. London: Bloomsbury Publishing.

McNeill, William H. 1976. *Plagues and Peoples.* Garden City, NY: Anchor Books.

_____. 1980. *The Human Condition: An Ecological and Historical View.* Princeton, NJ: Princeton University Press.

Pareto, Vilfredo. 1968. *The Rise and Fall of the Elites.* Somerville, NJ: Bedminster Press.

Peterson, A. Townsend, Jorge Soberòn, Richard G. Pearson, Robert P. Anderson, Enrique Martìnez-Meyer, Miguel Nakamura & Miguel Bastos Araùjo. 2011. *Ecological Niches and Geographic Distributions.* Princeton, NJ: Princeton University Press.

Raub, Werner, Vincent Buskens & Marcel A.L.M. van Assen. 2011. Micro-macro links and microfoundations in sociology. *Journal of Mathematical Sociology* 35:1-25.

Sallach, David L. 2002. Toward a synthetic theory of social structure. Paper presented to the Midwest Sociological Society. Milwaukee.

_____. 2003. "Social theory and agent architectures: Prospective issues in rapid-discovery social science." *Social Science Computer Review* 21:179-195.

_____. 2007. Logic for situated action. Pp. 13-21 in *Advancing Social Simulation: The First World Congress*, edited by S. Takahashi, D. Sallach, and J. Rouchier. Tokyo: Springer.

_____. 2008. Modeling emotional dynamics: Currency versus field. *Rationality and Society* 20:343-365.

_____. 2011. Modeling Strategic Contexts: A Theory-Grounded Modeling Language. University of Chicago: Computation Institute.

_____. 2012. Socio-cultural structure: A categorial Synthesis. Paper presented to the Midwest Sociology Society, Minneapolis.

Turchin, Peter & Sergey A. Nefedov. 2009. *Secular Cycles.* Princeton, NJ: Princeton University Press.

Turk, Herman. 1965. An inquiry into the undersocialized conception of man. *Social Forces* 43 (May):518-521.

Weikart, Richard. 2006. *From Darwin to Hitler: Evolutionary Ethics, Eugenics and Racism in Germany.* New York: Palgrave Macmillan.

Wrong, Dennis H. 1999. *The Oversocialized Concept of Man.* Piscataway, NJ: Transaction Publishers.

Section VI

Commercial Research and Applications of Social-Cultural Science

CHAPTER 37

Military Application of Non-Defense Related Social Science Research Methods

Theodore Stump

Strategic Analysis, Incorporated
4075 Wilson Blvd, Suite 200
Arlington, VA. 22205
tstump@sainc.com

ABSTRACT

There exists a significant body of social-cultural research and related technologies and methods that have evolved over the years which have been traditionally focused on commercial or non-defense related applications. Applications ranging from consumer behavior, public relations media strategies, financial market models, epidemiology, and gaming, are just a few areas where the social sciences have been utilized in an attempt to understand and model human dynamics. More recently, with the advent of social media applications and mobile devices, significant information is now available with regards to attitudes and behaviors that may provide businesses and other organizations with a wealth of information. We plan to review those domains that have employed social science for non-defense related purposes and discuss where and how they could be used with regards to national security.

DESCRIPTION

The application of social-cultural research and methods within commercial industry is hardly a new phenomenon having started in earnest back in the 1920's at the dawn of an earlier communications revolution, the introduction of the radio into American homes. Market research began to be conceptualized and put into formal practice during this period as a derivative of the advertising boom of the Golden Age of radio. Businesses who advertised on the radio began to realize the impact of demographics as revealed through the sponsorship of specific radio programs. Though earlier efforts with opinion research using mailed questionnaires date back to the 1890's, it wasn't until the advent of the radio and public broadcasting was this new field recognized as a valuable business activity. In addition to trying to regulate and control the radio frequency spectrum, the Commerce Department also took an interest in the developing field of market research holding a conference in Washington in 1926 to discuss and adopt a priority list of fundamental research projects on the topic of market research.

During the 1930's, the use of market research began to spread beyond just the commercial sector and into other domains where there was a similar need to understand personal choice and individual behaviors. While working for a New York ad agency, George Gallup recognized the strong ties between selling products and politics. Subsequently, in 1935, the Gallup Poll was born with a focus on government policies and voter tendencies. Previous methods of obtaining this information relied on mailed surveys sent to a large numbers of households. In contrast, Gallup would carry out biweekly polls – interviewing people in person – and the samples were scientifically selected so as to represent a larger group, including all classes, races and regions.

During the 1950's, large corporations and business schools conducted research in individual purchasing behavior including motivation research, social determinants of consumer behavior, factors in brand loyalty, and household decision making. Research became more scientifically driven and included both public surveys as well as focused laboratory research with controlled experiments using human subjects. In the 1960's and 1970's, advances in technology allowed for more accurate and scientifically valid research on non-verbal measures of personal preferences and attitudes including brain wave activities, galvanic skin response, voice pitch analysis, and eye-tracking. Growing computational power provided ground-breaking capability in large scale data mining giving businesses with a detailed understanding of individual and group patterns of activity.

Today's interconnected and highly mobile world of smart phones and social media platforms have created a treasure of information on personal preferences, purchasing behaviors, and attitudes in near real time. Text analytics and sentiment analysis methods allow the financial community to continuously evaluate attitudes and preferences among consumers and use that information to make better informed investment decisions. Social network analysis and influence metrics provide Madison Avenue with invaluable insight on how to influence brand awareness and conduct targeted, and more successful marketing campaigns. Similar techniques have been used to measure the effectiveness of political campaign advertisements and provide insight into what messages best influence certain segments of voters.

Social media has recently sparked interest within the federal government and specifically within the national security community. Social media played a key role in the Arab Spring by helping to coordinate protests and to communicate important information to fellow activists and news services. "Twitter terrorism" is part of an emerging trend and several al Qaeda franchises are increasingly using social media like Facebook, MySpace, YouTube and Twitter to broadcast their messages. Social media has helped terrorist groups recruit individuals, fund-raise and distribute propaganda much more efficiently – and arguably more effectively - than they have in the past. The al Qaeda branch in Yemen, for example, has proved especially skilled at distributing propaganda and commentary through several different social media networks. Instability like that experienced in North Africa and terrorist communication networks are just two examples of where social media is highly relevant to U.S. strategic interests. Social media also has the ability to provide the U.S. military and Intelligence Community with the platform and data to:

- Forecast behaviors of groups or perhaps key individuals in foreign operational contexts.
- Develop capabilities to better access and produce knowledge on complex social communication systems and on the perceptions, attitudes, and beliefs of populations and stakeholders.
- Better understand the direct and indirect effects of potential actions and signals on perceptions, attitudes and beliefs, and formulate and deliver timely and culturally attuned messages.

Commercial industry has already begun exploiting social media to more efficiently collect consumer data and provide important insight into market trends. A number of the technologies and methods employed by commercial firms are also relevant to the Department of Defense (DoD).

Social-behavioral science research within the DoD can be grouped into four general, operationally-oriented categories. This conceptual framework is relevant to both the military and commercial domains as both communities need to understand, detect, forecast, and mitigate behaviors and trends whether it be in either the marketplace or the area where military operations are conducted.

- **Understand:** Both government and commercial organizations seek to understand what factors influence the behavior whether they are consumers or foreign populations.
- **Detect:** Understanding sentiments and attitudes and when they are at inflection points is important in detecting potential hot spots in the world or in understanding when a product or brand has gotten traction or started to experience bad press with consumers.
- **Forecast:** An important objective is to model behaviors or sentiment and be able to project that into the future in order to determine the best course of action given key indicators. Commercial industry is interested as well in forecasting the demand for products and services and has started to utilize social media analytic tools to capture this information. The National Security Community would find the capability useful in predicting state and regional stability.
- **Mitigate:** Understanding what specific actions will have the desired effects is of high interest to commercial firms that need to effectively market their products and services as well as to the U.S. Government who needs to better understand the effects the various course of actions available.

To that end, the DoD is looking to develop the capabilities of a "social radar," which when fully realized would provide our military leaders with a global and persistent indications and warning capability consisting of technologies to detect sociocultural behavior signatures with operational or strategic relevance. This ambitious initiative, funded under the Human Social Culture Behavior Modeling (HSCB) Program, seeks to provide the DoD with the capability to correlate social "signatures" with indicators across a variety of domains (economic, political, social, health and welfare, etc.). These signatures of change detection are also important to commercial industry as well as the financial community as they seek early indicators of expressed perceptions, sentiments and attitudes that may in turn materialize into real life events.

There exists a significant body of social-cultural research and related technologies and methods that have evolved over the years which have been primarily focused on commercial or non-defense related applications. Several of these domains that are relevant to the DoD include:

- **Consumer psychology / behavior**: The study of consumer behavior helps firms and organizations improve their marketing strategies by understanding issues such as:
  - The psychology of how consumers think, feel, reason, and select between different alternatives (e.g., brands, products, and retailers);
  - The psychology of how the consumer is influenced by his or her environment (e.g., culture, family, signs, media);
  - How markets can adapt and improve their marketing campaigns and marketing strategies to more effectively reach the consumer (i.e. Brands are the associations that people make with a product name. They are formed through each and every interaction people have with a product line and those who stand behind it).
- **Public relations media strategies**: This includes communications primarily directed toward gaining public understanding and acceptance – usually deals with issues rather than products or services, and is used to build goodwill. In today's global media environment, messages are spread to audiences broader than originally intended, with potentially negative consequences. Culturally based perceptions can compound these negative effects as audiences perceive messages and actions in ways not intended.
- **Financial markets models:** Financial markets invest significant money in emerging markets (i.e. developing countries) and need to understand the risks associated with those investments. Not unlike the DoD and IC, investment banks employ state stability models to improve their analysis and decrease the risk with their investment portfolios. Social-behavioral models are also employed domestically to anticipate market swings and predict broader macroeconomic trends.

- **Healthcare – Epidemiology**: Public health officials faced with recent natural and man-made threats, ranging from Hurricane Katrina to the recent H1N1 pandemic influenza outbreak, understand the critical role the public's emergency preparedness and response (EP&R) behaviors play in the successful management and response to a disaster. Models incorporating social behavioral sciences provide the capability to evaluate multiple notional communication strategies and select those that will have the greatest impact on disease spread.
- **Gaming Industry:** Societal games (i.e. SimCITY) and first person games (Call of Duty, SIMS, Medal of Honor, etc.) employ simple behavioral models within the game environment. These models can be relatively simple AI behavioral models as well as more sophisticated agent based models of social behavior and governance.

Though considerable overlap exists with social science research applications between the DoD and commercial industry, there has been limited interaction between the two groups. The DoD has funded some cutting-edge research in a variety of different technologies that have significant commercial potential, but paths to commercialization have rarely been realized. On the other hand, commercial industry has funded research and created technologies that have already overcome some of the issues still being worked by government-funded research. Some of the challenges in bridging the gap between these two worlds include:

- Limited interactions between the research communities (cultural, organizational). Few companies participate in both markets.
- Commercial industry may not have the cleared personnel / facilities to gain access to some DoD opportunities.
- Limited rights on the part of the government for commercially procured tools / software. License fees and lack of technology transparency may be a strong deterrent to government program managers.
- Concerns over IP and data rights, export controls, and limitations on international partnering
- Government contracting processes; non-standard cost accounting, profit and overhead policies, documentation, dispute resolution, etc.
- General contracting processes; poorly written or inflexible requirements, delayed billing on change orders

ACKNOWLEDGEMENTS

Dr. Ivy Estbrooke, Program Officer, Expeditionary Maneuver Warfare & Combating Terrorism Department, Office of Naval Research

Dr. Laura Leets, Booz-Allen-Hamilton

BIOGRAPHY

Mr. Stump is the Vice President, Advanced Concepts Division at Strategic Analysis Incorporated and has responsibility for corporate internal research and development projects. He provides technical support to the Human, Social, Cultural, and Behavioral (HSCB) modeling program managed through the Office of Naval Research and recently coordinated a conference on *Social Media: Current and Emerging Technology and Applications in Government and Industry*. He has been a technical consultant on a broad range of research and engineering programs at the Defense Research Projects Agency (DARPA), Office of Naval Research (ONR), Naval Research Laboratory (NRL), and the Department of Homeland Security Science and Technology Directorate (DHS S&T).

He has specific expertise in computational social science modeling technologies as applied to Counterinsurgency Operations (COIN) and Stability, Security, Transition, and Reconstruction (SSTR) efforts. He also has significant experience with technical and economic assessments of emerging technologies as well as national security science and technology policy. Mr. Stump has B.S. in Engineering from The Pennsylvania State University and M.A. in Economics from George Mason University.

REFERENCES

[1] Todd C. Helmus, Christopher Paul, Russell W. Glennrand, Enlisting Madison Avenue: The Marketing Approach to Earning Popular Support in Theaters of Operation, RAND Corporation.

[2] Carrick Mollenkamp, Serena Ng, Liam Pleven and Randall Smith, "Behind AIG's Fall, Risk Models Failed to Pass Real-World Test," *Wall Street Journal*, November 03, 2008.

[3] Jill D. Egeth, Dr. Jennifer Mathieu, Meredith A. Keybl, Paula J. Mahoney, Maeve C. Kluchnik, and John H. James, Modeling the Impact of H1N1 Cognitions on Vaccination Rates and Disease Spread, MITRE Corporation.

[4] Barry G. Silverman, Gnana Bharathy, Kevin O'Brien, Jason Cornwell. Human Behavior Models for Agents in Simulators and Games: Part II: Gamebot Engineering with PMFserv. *Presence: Teleoperators and Virtual Environments* 15:2, 163-185 (2006).

[5] Sociocultural Behavior Research and Engineering in the Department of Defense Context, Human Performance, Training and BioSystems Research Directorate, Office of the Assistant Secretary of Defense (Research and Engineering), September 2011.

CHAPTER 38

Measuring and Identifying Culture

Michael D. Young

Social Science Automation
Hilliard Ohio, USA
michael@socialscience.net

ABSTRACT

Culture is one of the most important factors that influence and predict human behavior. Many of the current approaches to measuring culture in the Intelligence Community focus on capturing the propositions that are prevalent within a defined group of individuals. Social Science Automation applied the propositional approach to a set of advocacy documents by opponents and proponents of gay marriage and abortion rights. Although noise in the texts rendered overall difference measures ineffective, recent progress in noise reduction combined with new culture indicators allowed us to correctly classify 95% of the documents as either abortion or same-sex marriage documents. The most promising of these indicators are, concepts shared across documents, propositions shared across documents, concept valence, and threat/victim.

Keywords: culture, proposition, valence, threat, victim

1 THE PROPOSITIONAL APPROACH TO CULTURE

Culture is widely regarded within academia and within the Intelligence Community (IC) as one of the most important factors that influence and predict human behavior. Agencies such as the Defense Intelligence Agency (DIA) are searching for ways to enable their analysts to identify, compare, and contrast the cultures of various populations in order to support planners and warfighters. For example, DIA Information Operations planners would like to know what the beliefs are that lead to support for the Taliban in Afghanistan, who holds those beliefs, and how those beliefs can be changed.

Most of the current approaches to measuring culture in the IC (see references) are compatible with a propositional approach to culture that defines culture as:

> shared cognition (beliefs, values, attributes) that is transmittable between individuals and therefore across time and space.

The assumptions of this approach are that members of a culture will share and express the same beliefs, desires, and intentions and may also use their own vocabulary of concepts when they express those beliefs. Therefore, a culture can be identified as a set of individuals whose expressed beliefs and ways of expressing those beliefs are similar to one another's. For example, although we know that abortion is salient in both pro-life documents and in pro-choice documents, we expect the propositions about abortion to be quite different in each set of documents. All propositional approaches to culture require (A) the extraction of propositions from culture repositories such as speeches, newspapers, books, expert judgments, field reports, internet media, and other sources, and (B) subsequent analysis using network analysis techniques variously referred to as causal loop, influence, or cultural network analysis.

In previous work, Social Science Automation has used the Text Mapping coding scheme, Profiler Plus, and WorldView to apply the propositional approach to a set of advocacy documents (N = 78) written by opponents and proponents of gay marriage and abortion rights. The primary finding of those efforts was that overall document difference measures (transformation cost and incongruence) are insufficient to group documents and isolate cognitions that reflect a culture. The overwhelming "noise" that occurs naturally in unstructured texts produces difference scores in excess of 0.95 on a scale of 0.0 to 1.0. However, despite the low signal to noise ratio in unstructured texts from the "wild", several possibilities were identified to greatly reduce the noise. We have subsequently made progress on noise reduction, added DIA's Critical Network Analysis Tool (CNAT) to our toolkit, and identified several promising indicators that may, in combination, identify documents from a single culture.

2. PROMISING INDICATORS

Although the culture signal in our test documents is evident to human readers, the low signal to noise ratio in unstructured text appears to preclude any prospect of using transformation cost or other gross measures of difference to discriminate between cultures. However, the promising indicators explored in the current work are all intended to isolate or amplify aspects of the signal including:

- Shared concepts.
- Shared propositions.
- Concept valence.
- Threat/Victim.

2.1 Shared Concepts

A standard method of document clustering is to group documents based on the terms they share and on the co-occurrence of those terms (n-gram analysis, see also latent semantic analysis). However, this procedure does not retain the propositional content of the documents. Shared concept analysis is in some ways less powerful than n-gram analysis, but it can be performed within WorldView while retaining the propositional content of the documents.

WorldView does not provide any way to evaluate whether shared concepts analysis discriminates between groups. However, the shared concept and relation reports from WorldView can be transferred to CNAT which has routines for identifying cohesive groups. In CNAT, documents from the same culture (abortion versus same-sex marriage) are expected to form a strong cohesive subgroup that is distinguishable from the opposing culture with a classification accuracy significantly greater than 0.50 (chance). Although a plot in CNAT of the network of the 15 most shared concepts and their containing documents produces no useful discrimination between the two sets of documents, the discriminating power of each shared concept can be evaluated by examining changes in the non-directional cohesive strength for two groups as concepts are added and removed from the network and it should be possible to determine a minimum set of shared concepts which maximizes cohesive strength. If shared concepts do discriminate between the two cultures, the classification accuracy of groupings selected by CNAT should be greater than 0.5 (chance) on scale of 0.0 to 1.0

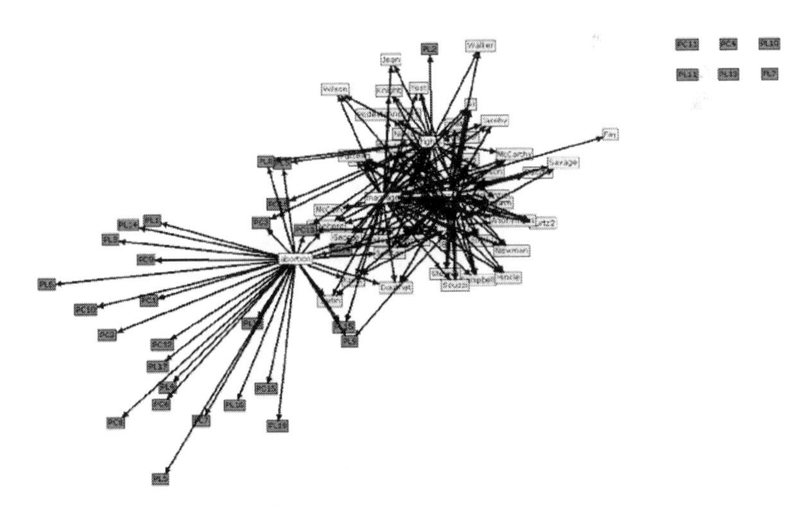

Figure 1. Connections between the six best most shared concepts and the same-sex marriage and abortion documents (N=78).

As a test of this approach, the cohesive strength measure was used to select the six concepts (marriage, rights, same-sex, abortion, couples, gay) that provided the best discrimination between the groups. When CNAT is allowed to select the two groups, an overall cohesive strength of 0.821 is achieved with a classification accuracy of 0.81. However, if documents sitting between the two groups are classified by hand, the cohesive strength increases slightly to 0.823 and classification accuracy increases to 0.91 (Figure 1.). This suggests that it is possible to construct an optimization algorithm that is blind to the meaning of the concepts and yet can still achieve high classification accuracy.

As a further test, a similar optimization methodology was repeated for each of the two document subsets (abortion, same-sex marriage). For the abortion document subset, two groups are obtained with 0.75 non-directional cohesive strength and a classification accuracy of 0.64 (Figure 2.). However, for the same-sex marriage document subset, only one group is obtained and the classification accuracy is barely above chance at 0.52.

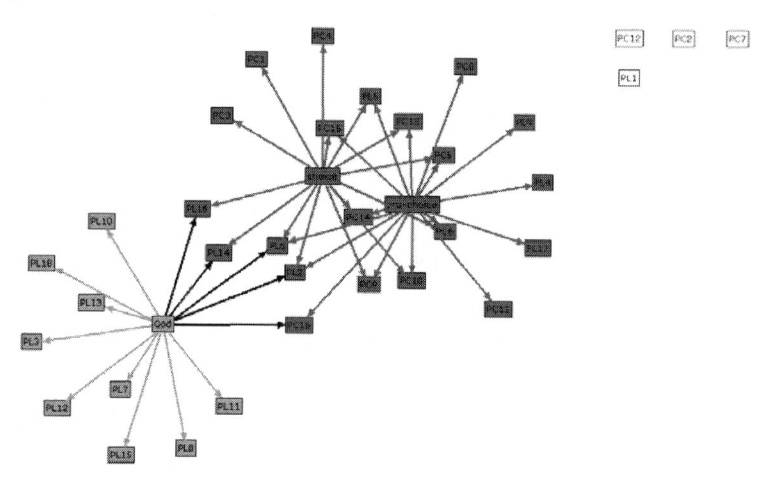

Figure 2. Optimized groups for the abortion document set. Non-directional cohesive strength = 0.75, accuracy = 0.64

2.2 Shared Propositions

Although we expect cultures to differ in their use of concepts, some distinct cultures may use the same concepts, but in different ways. For example, abortion is salient in both the pro-life documents and in the pro-choice documents. However, the propositions about abortion should be quite different. Pro-life documents are likely to describe abortion as *wrong*. On the other hand, Pro-choice advocates may describe abortion as *a medical procedure*. Thus, although the concept abortion may be shared across the two cultures, the cognitions about that concept in each culture may be very different.

Repeating the analysis process used for shared concepts, a shared proposition analysis for the entire document set nicely separates the abortion documents from same-sex marriage documents, but unfortunately, it only yields one group and numerous singletons and both the cohesive strength and classification accuracy are undefined in this case. Applying the analysis process to the same-sex marriage and abortion documents separately did not yield any result with a classification accuracy greater than chance.

2.3 Concept Valence

In the absence of shared concepts or propositions that distinguish documents and cultures, it may be possible to distinguish documents and cultures by how they evaluate shared concepts. For example, using the nonsense word deetchzeeb, consider the following propositions taken from one of six imaginary documents: deetchzeeb is wrong, deetchzeeb is good, deetchzeeb is evil, deetchzeeb is honorable, deetchzeeb is criminal, and deetchzeeb is pleasant. All six propositions and imaginary documents share the concept deetchzeeb but none of them share a proposition. However, it is quite easy to group the six propositions (and their documents) into two groups, one where deetchzeeb is positive and one where deetchzeeb is negative.

To explore the usefulness of concept valence, a very rough concept valence prototype coding scheme was created and used to generate data for both sets of documents. The concept valence coding scheme simply tags words with an evaluation. For example:

Abortion is wrong.
->
(abortion valence1 true factual present bad1)

The initial results of a shared proposition analysis using valence data are promising (Figure 5.), but the classification accuracy of 0.58 is only a little better than chance. Although substantial work remains to improve the performance of the concept valence coding scheme, the initial results justify additional work.

2.4 Threat/Victim

One additional aspect of concept valence that can be explored is the directionality of the valence. For example, if we use a simple non-directional good/bad valence indicator, "A attacks B" is "bad" for both A and B and this becomes a shared valence relation, tending to group A and B together. However, if the "threat" and "victim" of the proposition are distinguished they may provide greater discrimination. In both the same-sex marriage and abortion documents there is clear disagreement between the sides about what the threats are and who the victims are. For example, there is a prevalent belief among those opposed to same-sex marriage that "same-sex marriage will undercut marriage" even as some proponents believe that gay and lesbian Americans are discriminated against under current laws. Distinctions between threats and victims such as these are also likely to be valid and relevant in cultures of interest to the IC and may provide insight into cultural perceptions of constraints and threats. Complementary Benefit/Friend indicators may also prove useful but were not explored. For example, in many of the Anti same-sex marriage documents, the people, as voters in referenda, are seen as trustworthy.

In the initial exploration of Threat/Victim, a Victim concept (typically an actor) is any concept that is under attack or otherwise threatened; a Threat concept (also typically an actor) is a concept that is wielding illegitimate power or actively threatening or displaying a threatening posture. For example, in the hand-coded sentences below, examples of concepts that are coded threat are **bold** and concepts coded victim are *italic*. The terms that indicate these relationships are underlined.

- An even more substantive danger lies in the consequences of **gay marriage** on *the next generation.*
- There are valid -- and secular -- reasons to believe that **same-sex marriage** will undercut *marriage* itself.
- *Gay men and lesbians* suffer **discrimination**.

A prototype Threat/Victim coding scheme was developed that identifies threat and victim noun phrases. Illustrative Profiler Plus output for the sample sentences above is given below:

(gay power true attribute na (marriage power true na na marriage))
(next victim true attribute na (generation victim true na na generation))
((both same sex) power true attribute na (marriage power true na na marriage))
(marriage victim true na na marriage)
(gay victim True attribute na (man victim true na na man))
(lesbian victim true na na lesbian)

A shared proposition analysis of the Threat/Victim data produces two groups with an overall non-directional cohesive strength of 0.822 and a classification accuracy of 0.68 (Figure 6.) providing grounds for further analysis.

A subsequent shared proposition analysis applied to the 34 same-sex marriage documents with clear positions produced two groups with non-directional cohesive strength of 0.888 and a classification accuracy of 0.71. These results provide further evidence both that creating a culture identification routine is possible, and that combining data from more than one indicator may produce increased classification accuracy. However, applying the same procedure to the abortion documents proved less successful with a classification accuracy less than chance (0.41).

2.5 Combined Indicators

Three of the indicators examined provide some evidence of discriminatory power and, although there is substantial overlap between document groups identified for each indicator the overlap is not complete. This partial overlap suggests that the indicators may perform even better in combination. To assess this possibility, a shared item analysis was conducted for all 78 documents using both the 9 best concepts and 7 best valenced concepts producing two groups with a non-directional cohesive strength of 0.73 and a classification accuracy of 0.95 (Figure 3.) Unfortunately this success was not repeated with the addition of the Threat/Victim data.

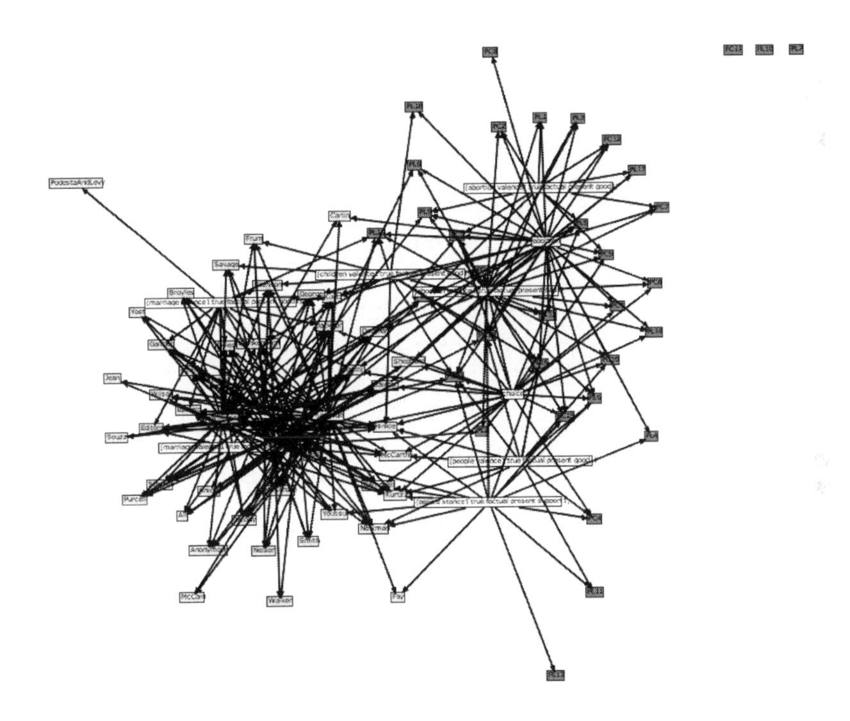

Figure 3. Best 9 Concepts and 7 valenced concepts for all documents. Non-directional cohesive strength = 0.73; classification accuracy = 0.95.

3 CONCLUSION

Despite a discouraging start to our investigations, our work to date suggests that, with some refinement, the propositional approach will lead to a useful methodology for identifying cultural groups and measuring cultural content.

REFERENCES

U.S. Army TRADOC Analysis Center—Monterey "Modeling Populations in Stability Operations Modeling Populations in Stability Operations using Cultural Geography." Paper presented at Focus 2010 Conference, 5-7 August, Chantilly Virginia, 2009.

Silverman, B. "UPenn's Leader Modeling Capability: An Holistic Socio-Cognitive Agent Approach." Paper presented at Focus 2010 Conference, 5-7 August, Chantilly Virginia, 2009.

Miller, C.A. and P. Wu. "An Architecture for Interpersonal Interactions; Politeness Modeling and Where it Can Take You." Paper presented at Focus 2010 Conference, 5-7 August, Chantilly Virginia, 2009.

Wright, W., D. Schroh, and D. Jonker. "Web-Based Geospatial Visualization Web Based Geospatial Visualization for HSCB Behaviors." Paper presented at Focus 2010 Conference, 5-7 August, Chantilly Virginia, 2009.

Sims, E. and K. Knudsen. "Plug and Play Avatars for Training and Mission Rehearsal." Paper presented at Focus 2010 Conference, 5-7 August, Chantilly Virginia, 2009.

University of Texas at Dallas "The First-Person Cultural Trainer." Paper presented at Focus 2010 Conference, 5-7 August, Chantilly Virginia, 2009.

Ahmad, A. and C. Phillips. "Sociocultural Communication Skills Training – A Serious Game Approach." Paper presented at Focus 2010 Conference, 5-7 August, Chantilly Virginia, 2009.

Ntuen, C. A. "A Cognigraphic Training Tool For Human Terrain Understanding." Paper presented at Focus 2010 Conference, 5-7 August, Chantilly Virginia, 2009.

Taylor, G. "PSTK: A Toolkit for Modeling and Simulation of Power Structure Networks." Paper presented at Focus 2010 Conference, 5-7 August, Chantilly Virginia, 2009.

Parunak, V. "The Environment - Attitudes - Actions Theory of Human Behavior." Paper presented at Focus 2010 Conference, 5-7 August, Chantilly Virginia, 2009.

Sallach, D. L. "Modeling Strategic Contexts: The Axes of Uncertainty." Paper presented at Focus 2010 Conference, 5-7 August, Chantilly Virginia, 2009.

Kruger, M. "Lessons Learned from the Field." Paper presented at Focus 2010 Conference, 5-7 August, Chantilly Virginia, 2009.

Sieck, W. "Cultural Network Analysis: Method and Application." Paper presented at Focus 2010 Conference, 5-7 August, Chantilly Virginia, 2009.

Nicholson, D., E. Salas, R. Shumaker, S. Burke, D. Barbar, and M Salazar. "Conceptualizing our Nations Together: A Cultural Taxonomy (CONTACT).". Paper presented at Focus 2010 Conference, 5-7 August, Chantilly Virginia, 2009.

Abbe A. "Training and Mission Rehearsal." Paper presented at Focus 2010 Conference, 5-7 August, Chantilly Virginia, 2009.

Sae S., D. Nicholson and M. L. Shuffler. "Temperament as a Lens for Culture." Paper presented at Focus 2010 Conference, 5-7 August, Chantilly Virginia, 2009.

Fidopiastis, C., S. Lackey, D. Nicholson. "Cross-cultural Analysis Methods Using Advertising and Psychophysiological Measures." aper presented at Focus 2010 Conference, 5-7 August, Chantilly Virginia, 2009.

CHAPTER 39

Improve Your Global Vision with Social Media Analytics

Karl Sanchack

Lockheed Martin Corporation
Suffolk, VA
karl.h.sanchack@lmco.com

ABSTRACT

"We can't know with absolute certainty what the future of warfare will hold, but we do know it will be exceedingly complex, unpredictable and – as they say in the staff colleges – "unstructured." Just think about the range of security challenges we face right now beyond Iraq and Afghanistan: terrorism and terrorists in search of weapons of mass destruction, Iran, North Korea, military modernization programs in Russia and China, failed and failing states, revolution in the Middle East, cyber, piracy, proliferation, natural and man-made disasters, and more. And I must tell you, when it comes to predicting the nature and location of our next military engagements, since Vietnam, our record has been perfect. We have never once gotten it right, from the Mayaguez to Grenada, Panama, Somalia, the Balkans, Haiti, Kuwait, Iraq, and more – we had no idea a year before any of these missions that we would be so engaged" - Defense Secretary Robert Gates (February 2011). To address the challenges of this complex international security environment, Lockheed Martin is developing a World Desk concept to collect, process, exploit and disseminate advanced analyses and indications and warnings of significant events reflecting potential shifts in population dynamics and sentiment. This creates a future state where national security missions are enabled by social and behavioral sciences to effect a lasting positive change.

In essence, the world desk concept is a 24-hour, rotation-based, open-source reporting environment with both commercial and proprietary software solutions, in which population trends, events, and sentiments are fused to provide actionable intelligence. Lockheed Martin has developed a model for a world desk view of such social and behavioral dynamics through which open media analytics help leaders understand, identify, and even forecast sentiments before critical inflection points. Lockheed Martin is leveraging data from forums, news posts, blogs, microblogs such as Twitter, and social media collaboration environments in order to lay a mission-critical foundation for cultural insight. Knowledge of target populations of interest, organizations, and teams would be derived and exploited from baseline content available through open source intelligence channels in real-time. That information could then be augmented and validated via appropriate defense intelligence services to aid decisions and create plans for appropriate mitigations.

To achieve that end, there is an initial foundation of human-centered sciences at multiple geographic locations in a net-centric structure that offers collection through dissemination of socio-cultural information fused with other types of intelligence sources including: cyber, economic, environmental, geographic and political to produce an in-depth situational understanding. As a critical differentiation from today's intelligence centers, the world desk concept incorporates seeks to leverage emerging capabilities in mobility to create a totally different vision for the content provider and analyst. Essentially, in the net-centric and mobile world, the analyst will be part of a larger "wisdom of the crowd" source set in the field producing content real-time and collaborating with associates around the world from remote regions in addition to fixed centers. This paradigm is similar to a news organization with field reporting but leverages analytical leadership in social and behavioral sciences to provide a core of verifiable information sources to baseline the larger super-set of content produced by the open market of discussion. At this time, progress has been largely limited to the analytics of the open source medium but efforts are underway to move these social media analytics through the combatant commands for additional information means.

Meanwhile, illustrated through a real-world social media analytics research project, described below, we have explored collection, processing and exploitation of open data using emerging technologies from partner, corporate, and business unit perspectives. Business initiatives already underway offer opportunities to answer the defense challenges of effective population engagement and shaping of strategic communications in a soft power system.

Keywords: social media, business innovation, open source intelligence (OSINT)

1 WORLD DESK CONCEPT AND INVESTIGATION

As concept background, consider a vision of World Desk Operations that encompasses a real-time, collaborative network for exchanges between analysts and field operatives, which fuses current and proven intelligence services with expanding open information sources such as: cyber worlds, social media, social networks, and crowd sourcing. Lockheed Martin achieved this objective through a multi-year, whitespace investigation called the Human Terrain Pathfinder, which launched in December of 2008 at the Center for Innovation in Suffolk, VA. The vision of the pathfinder was to enhance national security missions through social and behavioral analytics. This pathfinder effort forms the basis for an umbrella of research & development across the corporation and with national laboratory partners to explore ways and means by which the social and behavioral sciences could achieve demonstrable changes in understanding of populations and groups.

Figure 1. Vision for the World Desk center with data and analysis integrated from open source, field operatives and traditional intelligence sources to drive understanding of populations.

The categories of investigation in this pathfinder included data & modeling management, open source analytics, visualization and exploitation, cultural readiness training, and mobile edge collection & collaboration. This work was derived in part from formal research underway within a Defense Department program from which the data and modeling management concepts benefited. Integration of internal research & development (R&D) across these investigative categories resulted in a discovery experiment in 2010 intended to demonstrate the ability to develop a rich, contextual awareness of foreign populations and groups of interest around the world. A socio-cultural suite of systems addressed published customer gaps and needs across a broad spectrum of operations by supporting tactical collections in austere environments through open source data collects within cyberspace, which enabled analyses, visualizations and modeling based on national to local events and information. The experiment evaluated the capabilities of the system against a series of scenarios based on both civilian and military missions in a notional, disruptive and unstable locale in Africa.

The selected scenarios demonstrated an end-to-end approach taken in designing a socio-cultural system, as the experiment staff grappled with a complex local culture in the process of executing their mission objectives. System capabilities utilized in the experimentation included deployable expeditionary networks; mobile data collection; and data, modeling and visualization. Measures of performance, effects and outcomes were gathered through 35 questions from multiple surveys where participants rated such things as the system's ease of use, provision of additional useful information, and ability to improve shared understanding. Results of this discovery level experiment were fed back to the system's integrated product team for further development and improved integration. US Combatant Commands (COCOM), Armed Services and other customers were invited to observe the experiment, use the systems, and provide feedback through their own surveys. Follow-on with one customer enabled a contract for 6.3 research and development funds to mature part of the suite of software to Technology Readiness Level 9 for field operations at the COCOMs.

More specifically, the R&D investigation results of this discovery experiment shaped a cooperative effort with GE Global Research Center in the development of the Web Information Spread Data Operations Module (WISDOM). The WISDOM software suite capitalizes on the cyber domain as a near-real-time source of open source intelligence (OSINT) to meet a range of customer requirements while addressing capability gaps, such as automated collection and cyber-specific analytics. The growth of social media offers a profound wealth of changes to strategy and operations down to tactics, techniques and procedures. The pathfinder was a key environment for experimentation, testing, and product hardening of technologies like WISDOM through the Lockheed Martin innovation management process, which takes a rough concept from idea generation through fielded transition.

2 MATURITY PATH FOR THE INNOVATION OF WORLD DESK ANALYTICS

Lockheed Martin's Corporate Engineering & Technology (CE&T) office used three phases from its innovation management practice to explore the previously described pathfinder and, ultimately, created a baseline social media analytic suite of applications. Each phase engaged different types of innovators and tested a range of topics in this domain. Hypotheses for mission applications were tested and validated and growth of each analytic tool occurred within relatively low investment thresholds.

The three phases involved in this innovation management practice were as follows:

1. **Ideation**: Focuses heavily on collaboration across innovators with research and proof-of-concept development in order to establish hypotheses around new products and services.
2. **Investigation**: Explores and experiments with a variety of use cases in order to validate mission responsiveness for each proof-of-concept or prototype. Proof-of-concepts are further refined to meet a range of customer requirements. Business model development throughout this phase

prepares the product for transition from a laboratory environment to operational test and evaluation.

3. **Implementation**: Demonstrates mission relevancy and outcomes, closing the final step of the product hardening process targeted to commercial and/or military application.

Each phase of this innovation management process involved development of a new business case or refinement to an existing one, as well as the technical steps necessary to transition from vision through deployment. What follows is a description of the use cases and outcomes for WISDOM under the focus of the Social Media Analysis Mission, which has value to both commercial and military domains.

3 SOCIAL MEDIA ANALYTIC MISSIONS – COMMERCIAL AND DEFENSE

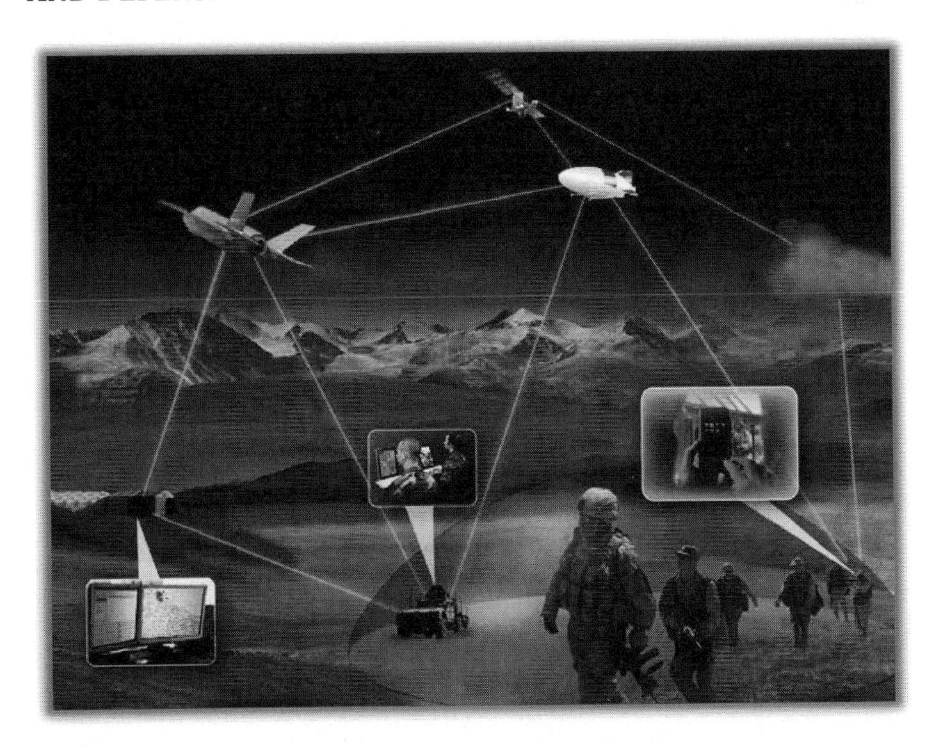

Figure 2. The Human Terrain Pathfinder Experiment's Concept of Operations demonstrates the connection between analysts, field operative and intelligence sources.

As has been reported in cases like Egypt and Libya, real world events have been followed by a series of social media dialogues moving from Facebook to Twitter to blogs/forums and ultimately into the mass news medium. The rapid gestation of events within a few weeks poses challenges to the established intelligence paradigm but offers opportunities in this approach for gaining population insights and sentiment prior to movement from words to action. Social media, in many cases, breaks the news before traditional news sources have had a chance to report it formally, so information gleaned from social media can readily be used as a tip off to unfolding events of interest. The recent political uprisings in key Middle Eastern countries have highlighted the value of harnessing the power and speed of information spread through social media networks as a vehicle to understanding population sentiment and intentions. The US and our allies see the importance of assessing social media content in the course of strategic and tactical planning, but are aware of the challenges to success (FBI, 2012; Mayfield, 2010; NATO, 2012; Schoen, 2011). These disruptive moments illustrate the critical need for indications and warnings of potential insurgencies and instability events. In that context, WISDOM's maturation process paralleled the evolution of the social medium for creating events of deep impact.

3.1 Ideate

In 2008, Lockheed Martin conceived an idea for a software product to monitor online media in support of such evolving national security missions. Given General Electric (GE) Research Laboratory's prior experience in web monitoring for commercial products, Lockheed Martin partnered with GE under its Shared Vision research program, which was a cooperative R&D arrangement between the entities. Beginning with the Lockheed Martin Human Terrain Pathfinder vision, the organizations sought to develop a reliable system for monitoring and characterizing online discussion with geographic focus, as well as identifying important player roles within cyber discussions. That system became known as the Web Information Spread Data Operations Module (WISDOM). At the outset of the Ideation Phase, the Shared Vision team identified the most fundamental differences between OSINT from the Internet and other types of OSINT like printed foreign media:

1. The Internet propagates information faster than print media;
2. Blog discussions may provide a means to measure population perceptions of, and responses to, media-reported events; and
3. Cyber data is layered with Uniform Resource Locator (URL) linkages that may provide a basis for identifying and analyzing online communities of interest.

The WISDOM concept sought to capitalize on these differences through software-enabled automation of online blog and news, data discovery and harvesting, text scraping and trending (e.g., quotes, NGrams, entities), and URL in-link processing and community mapping. With WISDOM automation, the concept was to enhance and accelerate the daily activities of World Desk collection officers, analysts, and desk officers. The Shared Vision Team conducted historic and innovative research to support this concept, from which the WISDOM proof-of-concept software suite was developed. Lockheed Martin continued its partnership with GE under the Corporate Shared Vision program throughout 2010, and then fully transitioned the WISDOM development and intellectual property rights to a leading business area within Lockheed Martin called Information Systems & Global Solutions in 2011 to finalize the product hardening and proceed to field deployment.

3.2 Investigate

As the World Desk system concept's technical maturity evolved, the investigation team worked horizontally across business areas at Lockheed Martin to establish a WISDOM beta instance with COCOM users. This pilot investigation not only shaped WISDOM's technical development and refinement to better meet our traditional Department of Defense (DoD) customers' requirements, but it also aided socializing the product's capabilities with the defense and intelligence communities.

A team of intelligence analysts and subject matter experts conducted case studies against the following areas to determine the utility of WISDOM collection and analysis capabilities:

1. United States political and diplomatic figures involved in Afghanistan and Pakistan security;
2. Hacker communities of interest to cyber security specialists;
3. Jihadist propaganda used by adversary communities to influence populations;
4. Government and population reactions to natural disasters and subsequent relief efforts.

In our World Desk intelligence paradigm, analysts would achieve superior situational awareness by aggregating information from WISDOM, prosumers (users of the system who also provide content similar to Wikipedia®), RSS feeds, traditional news, and crowd-sourcing feeds. This social radar's ebb and flow of information between the tactical edge and the analysis center allows the World Desk community to:

- Identify, study and understand potential audiences and their needs & motivations
- Produce and coordinate signals (information and actions) designed to affect the perceptions, attitudes, beliefs, and behaviors of selected audiences in ways that support the accomplishment of the mission
- Monitor and assess the effects of these signals
- Iteratively modify actions & information products as a result of assessment process

Using innovative analysis algorithms, as well as existing technologies for alerts, tagging, and visualization, frees up analyst time from collection, allowing them to concentrate on analysis. Through small group studies and metrics-based testing by operational users, WISDOM was evaluated to enhance and accelerate OSINT collection and analysis activities to meet capability gaps for a range of intelligence and defense customers. Based upon feedback from these studies, the Shared Vision team scaled the system to meet multiple user needs, and developed additional capabilities to further enhance the WISDOM software, such as geo-visualization, sentiment and volume trending, and individual online roles and influence.

Growing from the positive outcomes of these military investigations, WISDOM was launched at tradeshows, conferences, and at customer site locations around the globe. Through these engagements and further business analysis (Frost & Sullivan, 2010; Frost & Sullivan, 2011), WISDOM was evaluated for its broad applicability within many industries and services beyond law enforcement and traditional government applications, to include retail, healthcare, and financial services. With this feedback, a secondary business model was created to target product, brand and service assessments in commercial markets. This evaluation was further expanded and corroborated by a field study executed by the Wharton Business School as a graduate project in 2011.

3.3 Implement

As WISDOM moved forward in the implementation phase, the WISDOM team completed the final steps needed to support the first WISDOM licensing and analytic service offering in the last quarter of 2011. From 2010 to present, the Lockheed Martin WISDOM team has been hardening the WISDOM code to provide a licensable product to its current and potential customers. Because requirements have expanded to include requirements for multiple simultaneous users over extended periods of time, scenario templates to meet customer collection requirements, and WISDOM interface enhancements for improved usability, code hardening has included scalability and sustainability development.

To maintain relevance in our traditional military market space, Lockheed Martin has also executed and demonstrated studies of current interest to DoD customers. For example, the corporate technology office sponsored a WISDOM study to identify early indicators and warnings of social and political unrest in the Middle East North Africa region based on online social media tracking and analysis. The intent was to demonstrate the accuracy and precision of forecasting based on the real-time development of dynamic organizations and groups in the region. The results suggested that social media dynamic patterns could be established with thresholds for levels of response and activity in these groups.

We also have continued our pursuit of commercial markets, as corporations and other enterprises continue to look to understanding sentiment in order to respond to their customers and predict and prepare for future events (Gannes, 2012). Lockheed Martin has held pilot projects to monitor such public sentiment around events or analyze social media for market development at other corporations, as well as monitored our own market. In the case of working with other corporations, we collected metrics which demonstrated WISDOM's capabilities in these areas, resulting in follow-on work. As illustrated in this Social Media Analysis Mission use case, WISDOM has progressed successfully through the three phases of innovation management, from ideation to implementation, and supports ongoing commercial and defense missions.

4 INTEGRATE ONLINE COLLABORATION AND ANALYTICS

Further, social media offers collaborative capability for idea development and disruptive change within the industrial marketplace. Several Fortune 500 entities have adopted off-the-shelf products for ideation and brainstorming as part of their innovation practices. Tools range from internet-hosted systems like inno360™ through intranet deployments like Brainstorm™ by Intuit, with varying success in capturing the game-changers within a corporation. Along those lines, Lockheed Martin has explored these techniques to engage its engineers and scientists.

For example, nearly a year before a critical cyber crime center contract award made headlines, the seeds of its success were sown by idea contributors within the business' online collaboration and idea-sharing program. A commercial-off-the-shelf product was used to bring the power of the entire organization together to shape information for whitepaper drafts, validate customer needs, and develop sharpened discriminators by collaborating with internal experts across the entire organization. By the time the contract proposal was submitted, more than 150 people had contributed to its proposal and more than 400 people had viewed the relevant topics within the system. Shaping the future and tracking these outcomes with such analytics offers practical change to multiple industries and missions.

All of this is in line with our customers, who are expanding their use of social media to communicate and to understand the needs of their population. For example, the US Army notes the DoD policy allows access to social media with guidelines. Because of this, the US Army is using blogs to improve training, and using Twitter and YouTube™ as outreach tools. They believe these tools will enhance not only unit performance, but the soldiers' professional performance and in their family life. (US Army, 2011).

5 CONCLUSION

Understanding the Human Terrain -- thoughtful gauging of the local population, culture and customers, sociological make-up and history – is the current state of the art for missions dealing with instability from economic to humanitarian/disaster-related to military or political. Lockheed Martin's World Desk with innovative capabilities such as WISDOM, takes advantage of processing and analyzing all the new data out there, which is so abundant that the analysts have difficulty keeping pace with it manually, to enable more successful population-centric missions. The World Desk vision is one in which end users have access to the Forward Operating Base (FOB), to the regional command, to the Non-governmental Organizations (NGOs), and all the way to companies, labs and institutions that can provide them the means to accomplish their mission, in real-time or near real-time. The World Desk system utilizes capabilities used commercially, such as an app store, social media analysis, and cloud computing, to provide the individualized support required right at their hand-held device. It also incorporates the tools and data needed by the analyst to forecast instability early and plan the best use of resources at the tactical through strategic levels to proactively address the challenges faced in order to address regional needs and objectives. With an integrated suite of capabilities, an analyst can discover events, identify sentiment shifts, and understand more about real world concerns through these cyber societies.

As part of it, we presented here how unique tools and techniques have been created and tested from a general concept, and developed into technologies that mine information on the Internet to provide near real-time monitoring and socio-cultural insights. This innovation process culminates with the implementation phase, but each phase contains its own business case to provide a comprehensive and attainable transition from ideation through deployment.

ACKNOWLEDGMENTS

The author would like to acknowledge the work of the HT Pathfinder team; in particular the work of Tina Chau as lead WISDOM analyst, Linda Foster for leadership of the 2010 experiment and ongoing Pathfinder, and Dianna Julian for her assistance in authoring the documentation of efforts throughout the Pathfinder.

REFERENCES

Frost & Sullivan, Inc. 2010. Social Networking for Customer Contact: Market Insight. Mountain View, CA: Frost & Sullivan.

Frost & Sullivan, Inc. 2011. Social Media Use Among the U.S. Healthcare Provider Institutions. Mountain View, CA: Frost & Sullivan.

Ganes, L, 2012. "Facebook Gives Politico Deep Access to Users' Political Sentiments." Accessed February 24, 2011, http://allthingsd.com/20120112/facebook-gives-politico-deep-access-to-users-political-sentiments/

Gates, R. 2011. Speech presented at the United States Military Academy, West Point, NY. February 25, 2011. http://www.stripes.com/news/text-of-secretary-of-defense-robert-gates-feb-25-2011-speech-at-west-point-1.136145.

Mayfield, T. 2010. "The Impact of Social Media on the Nature of Conflict, and a Commander's Strategy for Social Media." Carlisle, PA: U.S. Army War College. http://www.dtic.mil/cgi-bin/GetTRDoc?AD=ADA545261&Location=U2&doc=GetTRDoc.pdf.

North Atlantic Treaty Organization (NATO). 2011. HFM-201 Specialists Meeting on Social Media: Risks and Opportunities in Military Applications, April, 16th-18th, 2012, Tallinn, Estonia. www.rto.nato.int/ACTIVITY_META.asp?ACT=HFM-201.

Schoen, R. 2011. "Social Media: Valuable Tools in Today's Operational Environment." Newport, RI: U.S. Naval War College.

United States Army. Office of the Chief of Public Affairs. 2011. Army Social Media: harnessing the power of networked communication.s http://www.dtic.mil/cgi-bin/GetTRDoc?AD=ADA549448&Location=U2&doc=GetTRDoc.pdf.

United States. Federal Bureau of Investigations (FBI). 2012. Request for Information: Social Media Application. www.fbo.gov: Solicitation Number: SocialMediaApplication.

CHAPTER 40

State of the Practice and Art in Sentiment Analysis

Stephen M. Shellman

Strategic Analysis Enterprises
Williamsburg, Virginia, USA
steve@strategicanalysisenterprises.com

Michael A. Covington

Institute for Artificial Intelligence
The University of Georgia
Athens, Georgia, USA
mc@uga.edu

Marcia Zangrilli

Strategic Analysis Enterprises
Williamsburg, Virginia, USA

ABSTRACT

Sentiment analysis (opinion mining) is the automatic extraction of opinions, feelings, or likes and dislikes from text. This paper is an overview of sentiment analysis with its current technical and theoretical challenges, with particular reference to the SAE Text Analysis system (SAEtext/Pathos). We give special attention to the nature of sentiment, its relation to the psychology of emotion, and the requisite semantic representations and language understanding techniques.

Keywords: sentiment analysis, opinion mining, text mining, text analytics, natural language processing, emotion

1 ABOUT SENTIMENT ANALYSIS

When newspapers report that people on Twitter are happier about President Obama this week than they were last week, they are reporting the results of

sentiment analysis, also known as **opinion mining.** That is the automatic extraction of feelings, likes and dislikes, or opinions from text. It is one of two types of **text mining** (**text analytics**; when applied to social media, **social media analytics**). The other type, known as **idea mining** or **information extraction**, aims to extract recurrent themes in the information content of the text (Evangelopoulos and Visinescu 2012).

This paper is an overview of sentiment analysis with its current technical and theoretical challenges, with particular reference to the SAE Text Analysis system and its sentiment analysis component, Pathos (hereinafter SAEtext/Pathos).[1] We do not attempt a complete literature review. Good surveys of this fast-developing field, up to their publication dates, are given by Pang and Lee (2008) and Liu (2010). To these one can add Liu (2011) for broader context and Shanahan, Qu, and Wiebe (2006) for research background.

One of the first domains of sentiment analysis was online product reviews, particularly movie reviews, starting with Pang, Lee, and Vaithyanathan (2002), who have generously shared their corpora. In a product review, we already know what is being evaluated, and the text is often accompanied by a numerical rating of the product by the same author, thus providing a criterion to test against.

Nowadays, social media are a major focus, and Twitter "tweets" (**microblogging**) are especially suitable for analysis because they are short, rely on little or no context, and express the sender's spontaneous thoughts and feelings. They are often explicitly tagged for subject and are always accurately time-stamped. Often written just seconds before publication, they provide an especially fast-moving indicator and can even be used for real-time monitoring of unfolding terrorist incidents (Cheong and Lee 2011).

Newspapers, blogs, and forums are harder to analyze, though in many cases more valuable. Many newspaper articles express little or no opinion or feeling (they are of low **subjectivity**). News events with good or bad consequences can be recognized, of course, but doing so is not exactly sentiment analysis. Editorials and blogs are more sentiment-laden, but also more complex, often relying on context and expressing sentiment about more than one thing in a single text, making it necessary for the sentiment analyzer to divide the text up into subtopics (O'Hare et al. 2009) or identify specific entities within it (Godbole et al. 2007). Some blogs are accompanied by mood icons selected by the user, making it easy to correlate moods with their textual expressions (Keshktar and Inkpen 2011). Before jumping into the technical details of sentiment analysis, we begin with some definitions.

---

1 *Pathos* is ancient Greek for 'sentiment.' The other components of SAEtext are the *Taxis* text classifier (Greek for 'classification') and the *Xenophon* event coder (named after an ancient Greek writer with a journalistic style). SAEtext is a product of Strategic Analysis Enterprises, Inc., Williamsburg, Virginia.

2 WHAT IS SENTIMENT?

All of the discussion above and to follow assumes we know what we mean by "sentiment." In much work on sentiment analysis, that has not been made clear; sentiment has been assumed to be whatever correlates with positive or negative evaluation, inclination to buy a product, or inclination to vote for a candidate. Sentiment might be an emotion, a behavioral disposition, or a rational judgment.

The simplest formal model of sentiment is a scale from "bad" to "good" or "dislike" to "like." This is often called a **valence** scale, and it presumes nothing about the nature of sentiment. Multiple valence scales are possible, such as moral approval vs. emotional liking.

If sentiment is emotion, it becomes possible to link sentiment analysis to psychology. Psychologists have several ways of classifying emotions.

Osgood, May, and Miron (1975) classify the emotional associations of words on a three-axis scale, Evaluation, Potency, and Activity (E, P, A), where E corresponds to sentiment or valence as more commonly understood. A similar system is used in the ANEW database (Affective Norms for English Words, Bradley and Lang 2010, earlier version 1999). This is a collection of words, most of them seemingly emotionally neutral, rated for "valence, arousal, and dominance" by a large population of human raters.

Plutchik (1958, 2003) arranged eight emotions in a wheel like primary colors (Fig. 1, left). In his theory, like primary colors, emotions can be mixed. There is also an intensity dimension along the length of each arrow.

Some of the mixtures predicted by the wheel make sense (such as fear+surprise), but others are harder to understand (fear+trust), and one commonly observed mixture is forbidden by the wheel layout (fear+anger). Plutchik himself (2003: 80-83) cites the results of several other researchers showing that fear and anger are not opposites but are both fairly close to disgust. On this basis we suggest the rearrangement shown in Fig. 1 (right), which is far from definitive.

Many psychologists use a system with five discrete emotions, happiness (joy),

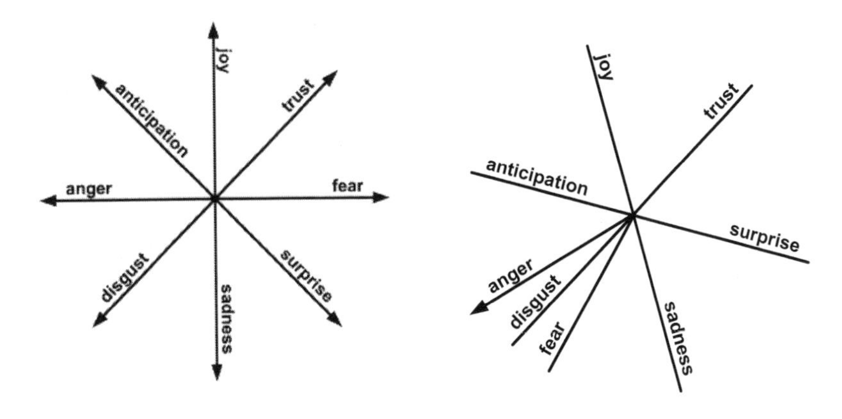

Fig. 1. Plutchik's Wheel of Emotions (left) and a suggested rearrangement (right).

anger, sadness, fear, and disgust. That is Plutchik's system without anticipation, surprise, and trust (which arguably are cognitive states, not feelings) and without the wheel. Stevenson et al. (2007) reworked the original ANEW database with a new set of human raters using this system. In both politics and advertising, though, we see a need for Plutchik's "trust" axis.

Insofar as it is based on counting words with emotional associations, sentiment analysis resembles a much older line of psychological text characterization called **content analysis** and implemented, for example, in the General Inquirer software package (Stone et al. 1966), still supported (http://www.wjh.harvard.edu/~inquirer/). In its current incarnation, General Inquirer rates text on several emotional scales as well as scales indicating the prominence of content areas.

A related content analysis system is Linguistic Inquiry and Word Count (LIWC, pronounced "Luke", Pennebaker et al. 2007). LIWC has been applied largely to psychology in a narrower sense – personality types, behavior, response to trauma, and minor mental illness. It rates texts along dozens of dimensions.

In our own work, we use both the positive to negative valence scale and the discrete emotion categories depicted in Figure 1(right).

3 WHERE TO GET THE DICTIONARY

Sentiment analysis requires, at minimum, a dictionary (lexicon) of words rated for the sentiment they express. Where to get this dictionary is an important question.

Given a collection of product reviews accompanied by numerical or "star" ratings, or blog entries marked with mood icons, it is straightforward to use machine learning to extract the words that indicate positive or negative valence (Pang et al. 2002, Keshktar and Inkpen 2011). This reduces sentiment analysis to text classification, perhaps with some measure of goodness of fit (as in the bag-of-words classifier in SAEtext), and requires no theory of sentiment or emotion beyond what is already given in the data set.

From there, the dictionary can be extended automatically, to some extent, by looking for patterns of usage that indicate synonymy or antonymy. For example, if something is described as *A and B,* then *A* and *B* probably convey similar sentiment; if the text says *A but B*, then *A* and *B* are probably opposites (Hatzivassiloglou and McKeown 1997).

However, many texts don't come pre-tagged with ratings or moods, nor do they sufficiently resemble other available texts that do. Accordingly, one must use other sources for the sentiment dictionary, such as the ANEW data set (Bradley and Lang 2010; Stevenson et al. 2007) or other psychological corpora.

A third way of building or expanding the dictionary is to perform *ad hoc* experiments with human raters. Collect words that occur regularly in the texts of interest and are likely to indicate sentiment; use a thesaurus to expand the collection; and ask a panel of human raters to judge their sentiment values, even crowdsourcing the work via Amazon Mechanical Turk (Taboada et al. 2011). Human raters are used internally at SAE, with careful consistency checking of results.

Once such a dictionary is started, statistical and machine-learning methods can be used to refine and expand it (Grefenstette et al. 2006, Pitel and Grefenstette 2008, Agarwal et al. 2009, Lu et al. 2011).

4 SEMANTICS OF SENTIMENT

4.1 Structures to represent sentiment expressions

The information content of a sentiment expression is more than just a number; sentiments are always *about* something. In product reviews, Liu (2010, 2011) represents every sentiment as a quintuple, *<object, feature, orientation, holder, time>*, tying it not only to a product but also to a feature of the product. Clearly, the first two elements, object and feature, are actually part of the ontology of product reviews. The third element is the sentiment *sensu stricto.* The author and time tags are properties of the text.

In theoretical linguistics, thematic relation theory (Gruber 1976) assigns noun phrases to roles such as *theme, agent, experiencer,* and *instrument.* Of these, only *theme* and *experiencer* apply to sentiment. Accordingly, in SAEtext/Pathos, the theme and experiencer are called by the more memorable names **evoker** and **reactor**. The reactor is the one who feels and evinces the sentiment evoked by the evoker. For example, if A praises B, A is the reactor and B is the evoker (which does not imply any action on B's part). The reactor is often the author of the text. Other attributes of the sentiment expression, such as date and time of utterance, are handled separately.

4.2 "Valence shifters" – intensification and negation

"Valence shifters" (Polayni and Zaenen 2006) are words such as *not, slightly,* or *could* that modify the effect of a sentiment expression. Obviously, *not good* is not the same as *good,* which is not the same as *very good.* A pure word-counting approach will not detect this; somehow *not* or *very* needs to be combined with the adjective to alter its sentiment value. How to do so is not necessarily obvious.

Fitting numerical values to valence shifters is, on the surface, easy. If *bad* is –1 and *good* is +1, then maybe *very good* is about +1.6, and *very* is a multiplier with a value of 1.6. Taking this tack, Cliff (1955), Lilly (1968), and Taboada et al. (2011) surveyed human raters and assigned numerical values to a wide range of intensifiers.

Negation interacts with intensifiers in a poorly understood way. Clearly, if *good* and *very good* are positive, then *not good* and *not very good* are negative. However, *very good* is stronger than *good,* but *not very good* is weaker than *not good.* Negation not only reverses the polarity but also, in some way, reverses or shifts the scale. Taboada et al. (2011) model this by saying that negation subtracts a constant, rather than multiplying by a negative number.

5 FROM TEXT CLASSIFICATION TO TEXT UNDERSTANDING

The first generation of sentiment analyzers all performed **document-level sentiment analysis**, assigning a sentiment to an entire text, generally using a **bag-of-words** approach. That is, words were counted but not related to each other in any other way. To modernize an old joke, when a dog bites a man, it's not news; when a man bites a dog, it's news; and bag-of-word techniques can't tell the difference.

In product reviews, editorials, and blogs, finer-grained analysis is necessary. A sentiment analyzer that looks at units smaller than the whole text must recognize the named entities to which sentiments pertain (**named entity recognition**). This can be done with a dictionary or semi-automatically. Pronouns must also be coreferenced; otherwise, every occurrence of *he* or *it* will be an unknown entity.

The Holy Grail of natural language analysis is deep understanding, which requires **parsing** the entire text to recover its sentence structure, then constructing a full representation of the meaning. Doing this reliably on a large scale is not a solved problem. Qiu et al. (2009) use parsing to connect sentiment words to the entities they describe, but they use it only for dictionary development.

6 A SPEECH-ACT-BASED APPROACH

6.1 Sentiment expressions as speech acts

The key idea behind the new approach to sentiment analysis taken by SAEtext/Pathos is that *expressing a sentiment is an act, and hence an event.* This follows from speech act theory (Austin 1962, Searle 1979) and particularly the *F*(*P*) hypothesis, which states that people do not communicate by transmitting ideas directly to other people – every proposition that they express is wrapped in a speech act intention, or to put it more technically, an illocutionary force. The triad of **locution** (act of speaking), **illocution** (intent), and **perlocution** (effect) is essential to all communication.

The act of speaking is performed by a specific speaker at a specific time (the last two elements of Liu's semantic quintuple). This immediately takes us beyond just measuring the author's sentiment; we can also measure *reported* sentiment. If a newspaper reports that A endorsed B and C denounced D and E, it has reported two speech acts, expressing the sentiment of A toward B and of C toward D and E respectively. We call these **dyadic** (A to B) or **multi-adic** (C to D and E) **sentiments** since each has an overtly expressed reactor and evoker, not just a single entity. The author's own sentiment (**monadic sentiment**) is a special case (as when the author writes *A is good*); the author is the reactor. Authors can also report sentiment while naming only the reactor and not the evoker, as in *A is happy.*

If sentiment expressions are events, they can be studied quantitatively like other events. We connect sentiment analysis to the well-established technology of

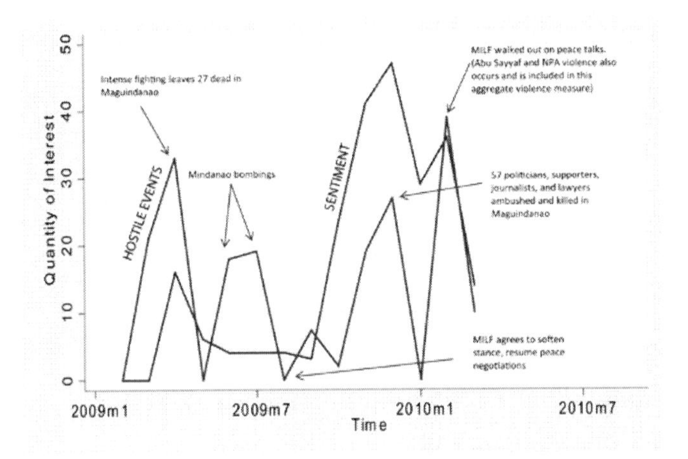

Fig. 2. Sentiment toward Moro Islamic Liberation Front (MILF) in Philippine blogs compared with reported actions of the MILF.

quantitative analysis of political and military events (Schrodt 1994, 2001; Shellman 2006, 2008). Figure 2 depicts sentiment and event trends towards the Moro Islamic Liberation Front (MILF). One can see that attitudes and opinions towards the MILF on a simple valence scale trend well with reported MILF actions. Below we describe in more detail how Pathos works.

6.2 SAEtext/Pathos infrastructure

The SAE Text Analysis System (SAEtext) comprises several tools: a general-purpose bag-of-words text classifier; a bag-of-word sentiment analyzer that handles some valence shifters; a political event coder and, most importantly, a dyadic sentiment (sentiment speech act) coding engine. The last of these, SAEtext/Pathos, is what we shall describe here.

On inputting an English-language text, SAEtext/Pathos breaks the string of characters into words; labels the parts of speech with a hand-optimized tagger; reduces each word to its dictionary form for easy recognition; performs chunk-style parsing; finds clause boundaries; and coreferences shortened names (e.g., *Jones* for *Colonel John Jones*) and pronouns (*he, she, they,* etc.) using heuristics based on those of Mitkov (2002).

Then, sentiment expressions (mostly verbs and adjectives) are connected to their respective reactors and evokers using a combination of template matching and syntactic heuristics. Finally, **discourse contexts** are classified; that is, sentences are marked as such if they are negative, conditional or hypothetical, future, or historical (referring to events long before the time of authorship). In that way, non-real and non-current sentiment expressions can be excluded from analysis.

Coding requires a dictionary of named entities (to which SAEtext/Pathos will suggest additions) and a dictionary of sentiment expressions (verbs, adjectives, etc.).

The dictionary format essentially matches words by syntactic category Sentiment codes can have multiple components, such as dimension or emotion plus value, so that multiple emotions/sentiments can be coded in a single run.

6.3 Examples of SAEtext/Pathos in use

Figures 2 and 3 show two examples of using SAEtext/Pathos to track sentiment changes associated with events reported in the news media. Fig. 2 tracks sentiment of Philippine bloggers (from GoogleReader) toward the Moro Islamic Liberation Front alongside acts of that organization reported in news stories available through Lexis-Nexis. Fig. 3 tracks Philippine sentiment toward the United States alongside actions that are perceived as (subtly) hostile or cooperative, and the sentiment is factored into emotions. From this we see that cooperative actions increase trust, and hostile actions lead to sadness, then fear, then anger, very much as predicted by a wealth of social psychology literature.

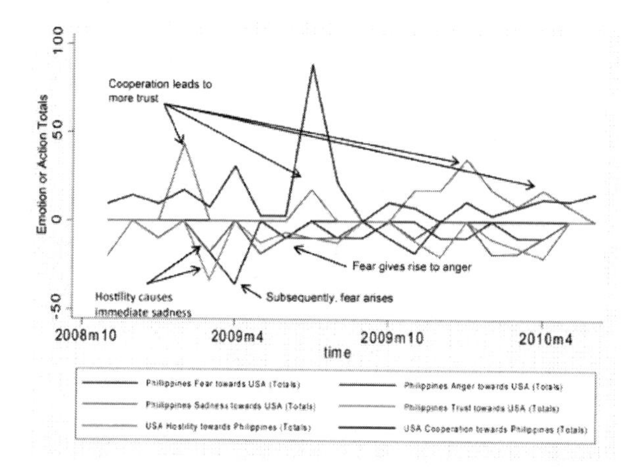

Fig. 3. Philippine sentiment toward the U.S. separated into emotions, correlated with U.S. actions perceived as cooperative or hostile.

ACKNOWLEDGMENTS

This research was supported by the U.S. Office of Naval Research.

REFERENCES

Agarwal, Apoorv; Fadi Biadsy; and Kathleen R. McKeown. 2009. Contextual phrase-level polarity analysis using lexical affect scoring and syntactic N-grams. *Proceedings of the 12th Conference of the European Chapter of the ACL (EACL 2009)*, 24—32.

Austin, J. L. 1962. *How to do things with words.* Oxford: Oxford University Press.

Bradley, Margaret M., and Peter J. Lang. 2010. Affective norms for English words (ANEW): Stimuli, instruction manual and affective ratings. Technical report C-2, The Center for Research in Psychophysiology, University of Florida. (Earlier edition, C-1, 1999.)

Cheong, Marc, and Vincent C. S. Lee. 2010. A microblogging-based approach to terrorism informatics: Exploration and chronicling civilian sentiment and response to terrorism events via Twitter. *Information Systems Frontiers* 13:45—59.

Cliff, Norman. 1959. Adverbs as multipliers. *Psychological Review* 66 (1) 27—44.

Das, Sanjiv R., and Mike Y. Chen. 2001. Yahoo! for Amazon: Extracting market sentiment from stock message boards. *Proceedings of the Asia Pacific Finance Association (APFA) Annual Conference.*

Das, Sanjiv R., and Mike Y. Chen. 2007. Yahoo! for Amazon: Extracting market sentiment from small talk on the Web. *Management Science* 53:1375—1388.

Evangelopoulos, Nicholas, and Lucian Visinescu. 2012. Text-mining the voice of the people. *Communications of the ACM* 55 (2) 62—69.

Godbole, Namrata; Manjunath Srinivasaiah; and Steven Skiena. 2007. Large-scale sentiment analysis for news and blogs. *Proceedings, International Conference on Weblogs and Social Media (ICWSM 2007).* Retrieved February 19, 2012, http://www.cs.sunysb.edu/~skiena/lydia/sentiment.pdf.

Grefenstette, Gregory; Yan Qu; David Evans; and James Shanahan. 2006. Validating the coverage of lexical resources for affect analysis and automatically classifying new words along semantic axes. In Shanahan et al. (eds.) (2006), pp. 93—107.

Gruber, Jeffrey. 1976. *Lexical relations in syntax and semantics.* Amsterdam: North-Holland.

Hatzivassiloglou, Vasileios, and Kathleen R. McKeown. 1997. Predicting the semantic orientation of adjectives. *Proceedings of the Eighth Conference of the European Chapter of the Association for Computational Linguistics (EACL'97),* 174—181.

Keshtkar, Fazel, and Diana Inkpen. 2011. A hierarchical approach to mood classification in blogs. *Natural Language Engineering* 18:61—81.

Lilly, Roy S. 1968. Multiplying values of intensive, probabilistic, and frequency adverbs when combined with potency adjectives. Journal of Verbal Learning and Verbal Behavior 7:854—858.

Liu, Bing. 2010. Sentiment analysis and subjectivity. Chapter 26 of Nitin Indurkhya and Fred J. Damerau, eds., *Handbook of Natural Language Processing,* 2nd ed., pp. 627—666.

Liu, Bing. 2011. *Web Data Mining.* 2nd ed. Heidelberg: Springer.

Lu, Yue; Malu Castellanos; Umeshwar Dayal; and Cheng Xiang Zhai. 2011. Automatic construction of a context-aware sentiment lexicon: an optimization approach. *Proceedings, WWW 2011.* Retrieved February 19, 2011, http://dl.acm.org/citation.cfm?id=1963456.

Mitkov, Ruslan. 2002. *Anaphora resolution.* London: Longman.

O'Hare, Neil; Michael Davy; Adam Bermingham; Paul Ferguson; Páraic Sheridan; Cathal Gurrin; and Alan F. Smeaton. 2009. Topic-dependent sentiment analysis of financial blogs. *Proceedings, 1st International CIKM Workshop on Topic-Sentiment Analysis for Mass Opinion Measurement (TSA '09)*, Hong Kong. Accessed February 15, 2012, http://doras.dcu.ie/14830/1/tsa20-ohare.pdf.

Osgood, C. H.; W. H. May; and M. S. Miron. 1975. *Cross-cultural universals of affective meaning.* Urbana: University of Illinois Press.

Pang, Bo, and Lillian Lee. 2008. Opinion mining and sentiment analysis. *Foundations and Trends in Information Retrieval,* vol. 2, nos. 1-2, pp. 1-135. Accessed February 14, 2012, http://www.cs.cornell.edu/home/llee/omsa/omsa-published.pdf.

Pang, Bo; Lillian Lee; and Shivakumar Vaithyanathan. 2002. Thumbs up? Sentiment classification using machine learning techniques. *Proceedings of the Conference on Empirical Methods in Natural Language Processing (EMNLP)*, pp. 79—86.

Pennebaker, James W.; Roger J. Booth, and Martha E. Francis. 2007. *Linguistic Inquiry and Word Count* (software). http://www.liwc.net.

Pitel, Guillaume, and Gregory Grefenstette. 2008. Semi-automatic building method for a multidimensional affect dictionary for a new language. *Proceedings, Language Resource and Evaluation Conference (LREC 2008).* Retrieved February 19, 2012, http://www.lrec-conf.org/proceedings/lrec2008/pdf/264_paper.pdf.

Polanyi, Livia, and Annie Zaenen. 2006. Contextual valence shifters. In Shanahan et al. (eds.) (2006), pp. 1—10.

Plutchik, Robert. 1958. Outlines of a new theory of emotion. *Transactions of the New York Academy of Sciences* 20:394–403.

Plutchik, Robert. 2003. *Emotions and life.* Washington: American Psychological Association.

Qiu, Guang; Bing Liu; Jiajun Bu; and Chun Chen. 2009. Expanding domain sentiment lexicon through double propagation. *Proceedings, International Joint Conference on Artificial Intelligence (IJCAI-09).*

Schrodt, Philip A. 1994. Event data in foreign policy analysis. In Laura Neack, Jeanne A.K. Hey and Patrick J. Haney, eds., *Foreign policy analysis: Continuity and change.* New York: Prentice-Hall, pp. 145—166.

Schrodt, Philip A. 2001. Automated coding of international event data using sparse parsing techniques. Presented at the International Studies Association, Chicago. Accessed February 15, 2012, http://polmeth.wustl.edu/media/Paper/schro01b.pdf.

Searle, John R. *Expression and meaning: studies in the theory of speech acts.* Cambridge: Cambridge University Press, 1979.

Shanahan, James G., Yan Qu, and Janyce Wiebe (eds.). 2006. *Computing attitude and affect in text.* Dordrecht: Springer.

Shellman, Stephen M. 2006. "Process Matters: Conflict & Cooperation in Sequential Government-Dissident Interactions." *Security Studies* 15(4): 563-99.

Shellman, Stephen M. 2008. "Coding Disaggregated Intrastate Conflict: Machine Processing the Behavior of Substate Actors Over Time and Space." *Political Analysis* 16(4), pp 464-477.

Stevenson, Ryan A.; Joseph A. Mikels; and Thomas W. James. 2007. Characterization of the Affective Norms for English Words by discrete emotional categories. *Behavior Research Methods* 39:1020—1024. Archived data sets retrieved February 15, 2012, http://www.springerlink.com/content/a16t327670463155/MediaObjects/Stevenson-BRM-2007.zip.

Stone, Philip J.; Dexter C. Dunphy; Marshall S. Smith; and Daniel M. Ogilvie. 1966. *The General Inquirer: a computer approach to content analysis.* Cambridge, Mass.: MIT Press.

Taboada, Maite; Julian Brooke; Milan Tofiloski; Kimberly Voll; and Manfred Stede. 2011. Lexicon-based methods for sentiment analysis. *Computational Linguistics* 37:267—307.

CHAPTER 41

Using Social Analytics to Discover Emerging Issues

Kevin D. McCarty, Anthony Sardella

Kevin D. McCarty, LLC
kevin@kevindmccarty.com

ABSTRACT

The ability to discover emerging issues has been a Holy Grail for analysts across many disciplines, from the commercial, financial and political sectors to practitioners in the realm of national security. Renewed focus on the need for a more consistently predictive methodology to discover emerging issues was brought about by the almost complete surprise faced by the national security communities during the Arab Spring of 2011. Much like the proverbial blind men describing the elephant, many analysts saw differing signs of what was to come. However, none saw the whole picture – the timing, speed, intensity, drivers and ultimate effects. Even in retrospect it is difficult with today's commonly accepted analytic methods to tell why the Arab spring occurred when and as it did. Yet these events did not occur in a vacuum and were not random. Nor is the basic problem of predicting issues different here than it is in other disciplines. Somehow the people involved in the events or issues were driven to common action and timing. Many analysts are looking to the field of social analytics to see if it can provide the necessary data and rigor to provide a more consistently predictive methodology. The relatively new field of social analytics, which combines technology to harvest the vast, new digital information environment, structure and quantify it, and then analyze it with social science in near real time, has provided a rich, new, actionable group audience understanding tool. It is important to understand, though, that not all social analytic capabilities are the same, and that the level of application and process necessary to be predictive is of an advanced nature that goes beyond most available methods today. This paper discusses the necessary capabilities and practices needed to make social analytics reliably predictive. From the basic concept of understanding how perception and fact interrelate and are discovered, building the right information environment and extracting the relevant information and drivers from it, measuring impact, and monitoring baselines to predicting the growth and adoption of a new idea are covered.

Keywords: emerging issues, social analytics, predictive behavior analysis

1 WHAT IS AN EMERGING ISSUE?

Discovering emerging issues is a valuable capability. Everyone from politicians, businesses and national security practitioners try to find tools, experts, disciplines and technology that would allow them to do this. The ability to find an emerging issue, know what is causing it, and be able to shape it either for benefit or to prevent crisis could mean the difference between success or failure, opportunity or loss. In this age of digital communication and advanced computing technology, the ability to collect and analyze vast amounts of communications content has given rise to a whole new business sector often referred to as Social Analytics. Most within this new sector promise vast insights to any problem set, including discovering emerging issues. Yet none of the promises or capabilities quite seem to provide what the user needs. As one esteemed friend with great access to this new sector said to me, "many promise to be able to do this, many show how they could have done it after the issue had arisen, but none have credibly shown me they predicted it before it happened." So, how do you evaluate if social analytics can provide you what you need?

The first step in being able to detect and understand an emerging issue is to define what it is:

An emerging issue is a perception change in an audience that has potential to cause change in that audience's behavior.

Within this definition are several key points that require clarification, as they will play an important part in understanding how to discover and understand emerging issues.

1.1 Perception

It is often argued about as to what is more important – perception or reality. What is perception? What is reality? Or is perception reality? Which should I study to determine or predict behavior? The answer is that perception is what the audience believes reality is, and it is their beliefs that they act on. For discovering emerging issues you need to know what audiences believe and will act on – regardless of whether it reflects reality or not. Knowing how their perception is formed is not important to discover an issue, only that it exists and it has enough potential impact to cause behavior change. The likelihood and intensity of potential behavior resulting from these beliefs is based on both the amount of exposure and the intensity and type of emotional drivers behind the beliefs. The more intense the emotional drivers, the more likelihood it will be both talked about and acted on. The more volume it has, the more people it reaches, the more likelihood it will become an issue. The communication medium that you would most likely find unfiltered audience beliefs is social media. Blogs, Twitter, texting, Facebook, etc., unlike traditional media venues where reputation for balance, fairness and facts is more important, are more known for stating opinions and venting emotion. Social media is like "the word on the street".

To know how to influence the potential behavior resulting from audience belief (once discovered), you have to understand what forms their perceptions. What forms their perception is an ever-changing interplay between basically three things: reality, the information environment and the audience's perspective. Reality is the factual, data driven experience of an issue. It answers questions like "what is the actual performance; does an economic depression actually exist; what is the actual impact on me?" The information environment is what information and opinion the audience is being exposed to – it may or may not reflect reality. The audience perspective is the cultural, historical, and belief-based lens through which they filter both experience with reality and the information they encounter. It is how they expect the world to work and what is important to them. These three things are constantly shifting and varying in impact as circumstances shift around an audience's life. One may be more dominant than another at a given time. Another may come in and change the other two. Being able to understand these three, how they shift and interplay, is the key to influencing behavior.

1.2 Change

The fact that an issue is emerging denotes a change from "normal". In order to detect change, one must be able to persistently monitor "normal" and know what it is. Random, or periodical samplings are not sufficient to pick up and recognize change. Change could occur and disappear between sample cycles. Issues caught during a sample could be overinflated in value, as they might be short-lived anomalies. A new issue may not be recognized as new. Polls and surveys only provide limited points of reference on what that they ask about. They cannot discover new issues, nor identify normal. In fact, by the mere fact that they are introduced into the environment they change it. Persistent, non-invasive and non-biased monitoring is key to proper evaluation.

The clearest analogy to this is monitoring the electronic spectrum to know what signals exist and what their normal levels are (noise). What is normal depends totally on the location and what is active there. When a receiver picks up either a new or stronger existing signal (relative to the existing noise level), what you now have is a break in the "noise" that shows something new is in the environment. By looking at the relative strength (signal to noise ratio), volume and type of signal you can determine what it means to you. You cannot just analyze the signal independent of the environment, or you will mischaracterize its impact. A signal that has significance in one environment (great signal to noise) may be totally lost in another (below the noise level). So it is with emerging issues. You must be able to baseline and monitor the "noise" of issues and their related volume and emotional drivers. Issues and drivers that might cause behavior change in one environment may not even break the noise level in another. You also must have the ability to detect new issues, as well as changes in existing ones. This requires persistent, large data collection and near real time analysis in order to detect change in a timely and useful manner.

1.3 Audience

Defining an audience is another key aspect. In the marketing world it is known as segmenting. You may segment by demographics, behavior or a combination thereof. Knowing how to segment properly for the issues you care about is essential. Gross generalizations will not have the sensitivity to pick up either perceptions or change properly. Pulling in large amounts of information without knowing what parts of it affect your key audiences makes it impossible to analyze properly for impact. In detecting emerging issues, you must be able to behaviorally segment audiences in the area you are concerned with. From this segmentation, it is necessary to build and monitor an appropriate information environment that is used by the segment. Only if you do this are you able to build the correct "normal noise environment" that allows you to detect change.

2 WHY SOCIAL ANALYTICS?

Technology has forever changed how people communicate. It has erased borders and made vast amounts of data and information available to almost anyone. The speed by which information can flow is unprecedented, and is unhindered by geographic borders. Old, demographic based social orders have broken down and multiple new, behaviorally based ones have arisen. The ability to manually monitor and analyze information flow has all but become obsolete. The world is reordering itself. On the other hand, technology has enabled the ability to collect and analyze the vast new databases of information and opinion. Where an analyst might once have monitored a few radio and TV stations and a handful of newspapers and magazines, analysts now have access to terabytes, petabytes or even zeta bytes of data or more a day. Social analytics, which combines technology to harvest the vast, new digital information environment, structure and quantify it, and then analyze it with social science in near real time, has provided a rich, new, actionable group audience understanding tool. It is perhaps the only tool that can collect, organize, monitor and analyze the new digital data streams in a fast and accurate enough manner to provide the audience insights needed today. As in all tools, all social analytics, or listening platforms, are not equal. In order to have the sophistication needed to detect and evaluate the impact of rising issues, a baseline of capabilities is needed.

2.1 The information environment

It all starts with the information environment. To be able to detect and evaluate emerging issues, a platform must be able to collect the proper data. It is important to note that I define this as the "proper" data, which is different from quantity of data. Quantity alone is a poor measure of required data. Properly conditioned "right" data is the better measure. What capabilities are needed?

Persistent monitoring: Data must be collected and monitored persistently in order to denote change. Typically a minimum of six to twelve months of data should be on hand to develop a baseline of normal conversation. Without this baseline, or normal "noise level", you cannot ever truly evaluate either change or impact. Just like any scientific measurement or experiment, you have to have a control, or norm, to evaluate the impact and change of anything. Change is measured in deviation, not absolute terms.

Persistent monitoring is also critical to finding measures of effectiveness. Once you have a baseline and have detected an issue, you may want to do something to affect it. By continued monitoring you can see how the emerging issue changes and what causes that change. This is perhaps the most advanced means of measurement of effectiveness available today – but it cannot be done with comprehensive monitoring for other factors and baselines to compare change to.

Find the "unknown unknown": An emerging issue comes in two types – the famous Donald Rumsfeld "unknown unknown" and the known issue that takes on new significance. To find the "unknown unknown" you must be able to collect all available data. Because you don't know what you don't know, you cannot directly search for it. This is where manual methods, polls and surveys cannot compete. If you don't know what it is, you can't ask about it. Instead, you must collect available data and see what it tells you. This is something you can only reliably accomplish with an automated tool such as social analytics. Even within social analytics platforms, there are varying capabilities in this area. Most platforms will use search strings to reach into various data aggregators to find "relevant" information. Like polling and surveys, this method lacks when it comes to the unknown because you are tied to the search string and hope you get everything you need. A few platforms have the advanced data storage and transfer capability to actually pull in all available data from their feeds, and then do analytics internally – unrestrained by the search strings. They can truly see what the data tells them, rather than ask the data questions. This is a key capability needed for finding emerging issues.

Condition the data: Just pulling in large amounts of data is not sufficient. The data must be reviewed for accuracy and completeness. Large amounts of data pulled in can have large numbers of duplicate and redundant sources. These must be cleaned and adjusted or else they will sway your analytics. Automated analytics will rely on machine translations. Although machine translations are seldom accurate enough for verbatim translation accuracy, they are sufficient for large data analysis. Machine translators often, however, are not complete enough to recognize name variations, colloquialisms and slang. They need to be monitored and adapted as necessary. Platforms should do this or their results will be suspect.

Information segments: A good platform should have the ability to define information environments around audience segments of interest. The ability to define an audience segment and then determine what information the segment normally looks at is essential. This type of audience analysis must be done before data collection is started so that the data can be properly tagged as it is collected and then used for accurate analysis. Assuming that all people see all data will only provide inaccurate analysis. Understanding the audience segment information environment can also help to ensure that you are collecting all the right data – even to the point of manually collecting and inserting key data if it is not digitally available. Missing key data is another danger to useful, accurate analysis. If a data source is key to the audience but is not digitally available, it must still be gathered and entered into your database to have accurate analysis.

2.2 Analytics

Analytics is a confusing area for many users. So much data can be sorted and presented that it becomes daunting to determine what will work best. There are a wide variety of dashboards, or data display capabilities, which users tend to focus on rather than what is behind them. Within these dashboards there are several layers of analytics that can be accomplished. These include basic sorting by volume, key word, source and other Meta tags identifiers. Many platforms offer a further level of sentiment analysis – is the topic being portrayed in a positive or negative manner? Few platforms can go the next crucial step for emerging issues – emotional driver analysis. Understanding and quantifying over time the relation between volumes, emotional impact and sentiment is the key to detecting emerging issues. Let's review what analytic capabilities are needed to accomplish emerging issue detection.

Data sorting: The platform must have a robust data sorting capability. It must be able to sort data by topic, author, source (publication, web, twitter, etc.), target audience, sentiment, emotional driver and time.

Display: The platform must be able to display relevant data in a timeline. The display needs the ability to show the information environment in a three-dimensional form that clearly shows volume, sentiment and emotional driver impact over time. These are the key indicators to detecting emerging issues.

Sentiment: The analytics need to be able to determine complex, contextual sentiment analysis. Simple key word sentiment analysis is not sufficient to detect and recognize the granularity and understanding needed for emerging issue detection.

Emotional Drivers: Social research is available and is applied by some platforms that will detect and quantify emotional drivers. These are the basic human emotions that cause people to react and behave. Being able to detect and track emotional drivers over time is essential to predicting perception and behavior change. Social norming has been done on emotional drivers in the United States, however little has been done on non-U.S. audiences. Because of this, absolute values are not the best guideline to predict change on these factors. Instead, looking for relative values and deltas in in impact against a baseline is the necessary approach for detecting emerging issues.

2.3 Predicting growth of an idea

The ability to understand the spread of an idea has become one of the most critical challenges for modern business, society and government. The ability to identify why some ideas spread to impact people's views and behaviors while others among a vast constellation of ideas remain dormant, is a capability greatly valued but has been difficult to attain.

There are several key factors that have demonstrated an ability to reveal the likelihood for an idea to spread and the rate to which it can create a change in behavior. Historically, the ability to identify and understand the presence or structure of these factors was limited. However, social analytics opens a new door to introducing the ability to gain deeper level of insights.

The determinants found to provide an enhanced understanding for the potential of an idea to grow and spread include emotion, network, and vehicle.

Emotion: Is the idea supported by emotional constructs with the potency to alter ones perception and behavior? Significant amounts of research have determined that not all emotion is equal. Certain emotional constructs demonstrate greater impact on changing our perceptions. Moreover, their potency can greatly affect the degree to which it changes our behavior. Behavioral change can include actions to further spread the idea through communications of other forms of action that facilitate a change. The spread of an idea requires action. These potent emotions drive action.

Network: Potent emotional drivers increase the likelihood of an idea spreading but the spread needs a network of people, organizations and interests to sustain its growth. Network analysis has been redefined with the emergence of social and digital media. It enables the identification and segmentation of stakeholders as nodes within a network that facilitate the spread of ideas among others. What has been revealed with social analytics is the various roles that are required to spread an idea. Spread requires members who act as authoritative sources of information. Others members act as central nodes to facilitate information flow while others are instrumental in connecting two disparate communities enabling the spread of an idea outside the main community from which the idea originated. The spread of an idea requires messengers. Networks and their structures drive messaging.

Vehicle: While networks act to spread messaging, they require an organization instrument. An instrument is needed to overcome the collective action problem. The collection action problem is where you have many people, organizations with the same interests of ideas. However, they are dispersed across vast disparate geographies or networks. Organizing these interests to action as a network is a collective action problem. Social and digital media has provided a highly efficient means for overcoming the collective action problem. It enables greater coordination of networks acting as an organizing tool for the networks. Furthermore, it provides the vehicle to not only organize the network but to disseminate the message to the vast communities of interests. Networks require a dissemination vehicle to spread ideas and to organize. Social and digital media provides an efficient vehicle.

Social analytics provides a powerful new way to understand the major drivers behind the successful spread of an idea. It provides unprecedented ability to evaluate emotion, networks and the vehicle that are the determinants of an idea spreading versus remaining dormant.

3 CONCLUSIONS

Social Analytics is the best approach currently available to discover emerging issues. It's ability to combine technology to harvest the vast, new digital information environment, structure and quantify it, and then analyze it with social science in near real time is unmatched by any other methods. Although other methods may be arguably more precise in any one or two areas, none combine all the needed capabilities together. Social Analytics ability to track and compare in near-real time conversation volume, sentiment and emotional impact – particularly in the unfiltered "word on the street" world of social media - provides the right insights. Simple aggregation and basic analytics will not work. Occasional sampling will not work. Social Analytics must persistently monitor the information environment to baseline it, and must be able to search widely enough to find the "unknown unknown" of emerging issues. Social Analytics, employed by users with the right framework and advanced capabilities, can do the job.

Section VII

ICEWS and SAA: Early Warnings

CHAPTER 42

Lessons Learned in Instability Modeling, Forecasting and Mitigation from the DARPA Integrated Crisis Early Warning System (ICEWS) Program

Brian Kettler and Mark Hoffman

Lockheed Martin Advanced Technology Laboratories
Arlington, VA, USA
brian.p.kettler@lmco.com, mark.hoffman@lmco.com

ABSTRACT

The DARPA Integrated Crisis Early Warning System (ICEWS) was a four-year flagship effort to apply computational social science models to dynamic data sources to produce actionable forecasts of foreign country stability. Lockheed Martin Advanced Technology Laboratories led a multidisciplinary team of computer and social scientists from several universities and small businesses to achieve this objective. This paper summarizes some of the key products of that effort, which has resulted in deployed operational prototypes and ICEWS event monitoring capabilities transitioned into a major DOD program-of-record. Major products include a mixed-methods suite of statistical and agent-based models for more than 50 countries that produce highly accurate forecasts; an extensible model integration framework (ADAMS) that provisions, executes, and manages models; innovative forecast aggregation methods; and tools for the ingest, processing, and analysis of dynamic data feeds including more than 16 million news stories. Several contributions to social science theory were made along with advances in data processing (event coding) technologies. Key lessons learned will be discussed including the data processing and quality assurance effort; the need for transparency in model forecasts to foster trust among operators; and the desire for actionable forecasts that can be addressed by Diplomatic, Military, Information, and Economic (DIME) actions. Some directions for future work will be presented.

Keywords: modeling, forecasting, instability, computational social science

1 ICEWS OVERVIEW

The DARPA Integrated Crisis Early Warning System (ICEWS) project (2007-2012) was a flagship effort in computational social science to predict instability in foreign countries of interest. In Phase 1 (2008), the Lockheed Martin Advanced Technology Laboratories' (LM ATL) Team demonstrated that it was possible to forecast select events of interest (EOIs) at high accuracy using a mixed methods approach (O'Brien, 2010). The resulting Phase 2 and 3 efforts by the LM ATL team built on this success and resulted in the research; development; operational test and evaluation; and transition of ICEWS capabilities to the USSTRATCOM Integrated Strategic Planning and Analysis Network (ISPAN) program of record.

ICEWS has been described by one senior commander as a key, new capability for Intelligence, Surveillance, and Reconnaissance (ISR) by providing near-real time exploitation of dynamic data sources such as news and social media. ICEWS complements other more expensive, less widely available ISR assets. The LM ATL Team combined insights from social scientists, computer scientists, operational users, and subject matter experts from large and small businesses, universities, and government for a highly interdisciplinary approach.

ICEWS has four primary sets of interoperating capabilities, shown in Figure 1: (1) iTRACE—event trend analytics from news media and other sources; (2) iSENT—sentiment analysis from social media; (3) iCAST—stability EOI forecasting; and (4) iDIME—stability prevention and mitigation via Diplomatic, Information, Military, and Economic (DIME) actions. These capabilities are integrated through a web-based portal and a shared data repository. This paper focuses on lessons learned in the more mature capabilities: iTRACE, iCAST, and the supporting ICEWS data environment. iTRACE capabilities have been transitioned to the ISPAN program to support joint course-of-action planning by providing global situation awareness to Combatant Commands (COCOMs) worldwide. iCAST capabilities are currently being expanded for transition to ISPAN under the new Worldwide ICEWS (W-ICEWS) project sponsored by the Office of Naval Research (ONR). The following sections describe these capabilities, summarize the results, and discuss key lessons learned.

The more exploratory iDIME and iSENT capabilities are not described in detail here. iDIME supported the selection of DIME actions to address forecasted

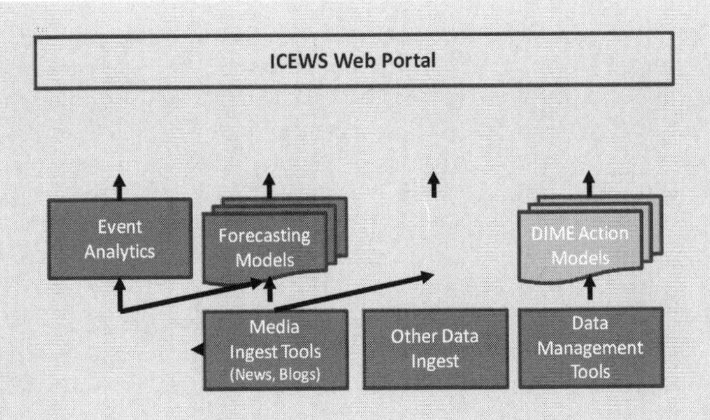

Figure 1 ICEWS Functional Capabilities

instability. A taxonomy (formal ontology) of several hundred DIME actions was developed from multiple sources. Several DIME actions were modeled in detail using inputs from subject matter experts and historical cases (Shellman et al., 2011). These efforts were hampered by lack of data; expense of SMEs; and the effects of DIME actions being highly conditional on their context, difficult to isolate, and sometimes lagging and varying over time. Thus iDIME work was deemphasized in favor of maturing the iTRACE and iCAST capabilities.

iSENT capabilities include sentiment analysis of social media (e.g., blogs, Twitter, etc.) and news sources as other potential inputs to iCAST models and iTRACE analytics. Early efforts demonstrated feasibility but were superseded by the incorporation of LM's Web Information Spread Data Operations Module (WISDOM) product, developed under prior efforts. This was fielded as a prototype capability under ICEWS (Starz et al., 2012) Under the new ONR efforts, the WISDOM-based iSENT capabilities are being enhanced for transition to ISPAN.

2 ICEWS CAPABILITES, RESULTS, AND LESSONS LEARNED

2.1 Forecasting Events of Interest (iCAST)

iCAST (ICEWS Forecasting) capabilities include an extensible, mixed-methods suite of computational social science models for forecasting events of interest (EOIs). The focus EOIs defined by DARPA are Domestic Political Crisis, Rebellion, Insurgency, Ethnic/Religious Violence, and International Crisis. Models produce a monthly EOI probability for a given country for 6-12 months into the future. Countries include 53 focus countries with populations over 500,000 in the PACOM and SOUTHCOM areas of responsibility (AORs). Under the new ONR W-ICEWS effort, models are being expanded to cover 177 countries worldwide for transition, along with the iCAST infrastructure, to ISPAN in 2012.

Our mixed-methods approach integrates a number of statistical and agent-based models for each EOI using a flexible model integration, data-provisioning, and execution framework (ADAMS) and Bayesian methods for aggregating forecasts from those models for each country and EOI (Mahoney et al., 2011; Montgomery et al., 2011). These methods generate an aggregate forecast primarily by weighting each model's current forecast by its prior performance as compared to a ground truth data set (GTDS). The GTDS was based on SME assessment of whether or not an EOI occurred in a given country in a given month using as a resource the coded event data described in Section 2.2. This GTDS set covering 1998-2012 was used to estimate the statistical models and validate them using a split sample methodology.

The key Phase 1 result showed that the aggregate forecast outperformed that of any of the individual models (O'Brien, 2010). Furthermore, models often complemented one another with some performing better on some EOIs and countries than others. In Phases 2-3, the performance of the models was improved from around 80% accuracy (number of correct predictions) to over 95%, with less than 5% false positives.

Statistical models include logistic regression (logit) models that combine traditional, more static, country-level indicators (e.g., level of democracy, GDP, ethnic fractionalization) with more dynamic aggregations of dynamic event data coded from news stories, described in Section 2.2. These models capture structural conditions in countries that may predispose them towards instability and behavioral factors (e.g., recent interactions between government, insurgent, and other types of actors) that can trigger specific EOI occurrences.

Other statistical (autologistic) models exploited networks of relationships among countries such as spatial proximity, trade ties, people flows, etc. (Ward et al., 2012). Instability can "spread" over such networks. Thus the occurrences of an EOI in two countries related by a network may not be truly independent events. These models were also successfully demonstrated on disaggregated data to model Afghanistan at the provincial and district levels and to predict the level of anti-government conflict in Thailand (Metternich and Ward, 2011; Metternich et al., 2011).

A key challenge for statistical models is the relatively rare occurrence of the focus EOIs in the GTDS. Operators are particularly interested in the onset and cessations of EOIs (e.g., when a rebellion starts and ends). These have been hard to forecast with accuracy. DARPA stated a strong preference for statistical models that used as input more innovative indicators, grounded in social science theory, rather than more traditional, less actionable indicators such as whether the country was mountainous. Although these latter indicators have predictive value (e.g., Fearon and Laitin, 2003), the ability of the US to affect levers suggested by such indicators is very limited. Modelers also avoided the use of lagged dependent variables as the utility of predicting future rebellion, as a function of recent rebellion is also limited.

A model accurately predicting multiple countries was preferred over a model forecasting just one country given the goals of predictive generality and minimizing the number of models that would have to be run and maintained on an ongoing basis, particularly post-deployment. Initially each statistical modeler created one model per EOI per set of countries in an AOR (PACOM or SOUTHCOM). This approach averages the effects of a model's independent variables on the EOI across all of the countries. But not every variable in these models will have such an effect in every country, especially given the variety of countries across an AOR such as PACOM (e.g., Australia, China, Thailand, Fiji, etc.). When looking at a given forecast for a country, some variables would show minimal impact and thus be less amenable to manipulation through DIME actions. This led to the development of mixed effect models in which the coefficients of some of the variables were allowed to vary by country. Models were also developed for sets of countries grouped by polity type (e.g., democracy, autocracy, etc). These techniques improved model performance and actionability. With the expansion to 177 countries under W-ICEWS, new groupings of countries that cross AOR boundaries will be explored.

The agent-based models (ABMs) model select countries as time-driven simulations at high levels of detail using the Political Science-Identity (PS-I) agent-based modeling framework (e.g., Alcorn et al., 2011). Seven virtualized country models were developed for analysis and forecasting (Bangladesh, Sri Lanka, Thailand, Philippines, Malaysia, Vietnam, and Indonesia). Sub-country models were developed for Kandahar and Kunduz provinces of Afghanistan. These models are grounded in social science theory (e.g., constructivist identity theory) and deeply rooted in geography. These models combine elements of cellular automata (for local behavior) and social networks that model authority structures (for more "global" behavior). In each PS-I model, dozens to hundreds of agents represent population elements (i.e. groups of collocated people or, in some cases, specific elite/powerful individuals) organized geographically. Each agent has a repertoire of identities they can express that impacts their behavior. Identities include salient government affiliations; ethnic, religious, and political group membership; social class; etc.

Models are configured to represent a country at a specific point in time. During a model run, agents make decisions at each time step (e.g., which identity in their repertoires to express) based on their environmental configuration, which includes neighboring agents and agents they are connected to over social networks. As agents interact, emergent behaviors (e.g., protests, mobilizations, etc.) occur. These behaviors are aggregated to determine whether a specific EOI has occurred. Multiple runs are made with perturbations of the input parameters to control for chance. Each run or trajectory, representing an alternative future, can be analyzed, compared, and grouped. For example, in some trajectories one might observe the government, military, or other identities gaining ascendancy in particular parts of the country, while other identities are suppressed.

Visualizations of agents as changing cells on a map-based view (area cartogram) typically required significant training and expertise to interpret. To help make sense of the large amount of ABM output, several modeling constructs (e.g., Lustick et al., 2010) and visualizations (e.g., Crouser et al., 2012) were created though more work is needed to qualitatively and quantitatively cluster various futures. This area is ripe for future research that could benefit all agent-based models of this size and complexity. For example, data mining techniques could be applied to ABM output.

Statistical and agent-based models complement one another. Although the ABM models proved less accurate than the statistical models in forecasting specific EOIs, they can provide a wealth of information about how EOIs could potentially occur as a result of the interactions among population elements in a country. This level of detail can suggest DIME actions to mitigate potential instability. The statistical models provided the most accurate forecasts using models estimated over a variety of large, historical data sources. They are computationally inexpensive to run. The ABMs require significant computational cycles to execute (e.g., many runs are made) and generate lots of data (e.g., data on every agent for every time step of every run). Currently statistical models are estimated from actual historical data, but ABMs can provide a rich variety of alternate histories to exploit. One could also estimate statistical models from time-series captured during one or more runs. Finally, ABMs require significant effort (days to weeks) by subject matter experts to initially configure and maintain from sources such as census reports, election results, NGO reports, case studies, other monographs, and geographic data. Over time, the models must be updated to incorporate recent real-world changes in the country. Towards this, some initial work was done to build templates and automation to help create new ABMs from SMEs inputs and various data sources.

Finally, a major challenge for all models is fostering an operator's trust in the results by providing actionable detail from the forecasts. In ICEWS, the use of multiple, heterogeneous models make this even more important. To this end, we developed a framework for model forecast transparency (Wedgwood et al., 2012). Transparency is generally provided by interactive visualizations, such as the ones in Figure 2, for end users (operators, analysts) and model developers. These views include heat maps showing forecasted EOIs, detail on aggregate forecasts (from the aggregation methods), and detail on individual model forecasts, including which input variables (and values) had what kind of impact in each model's forecast.

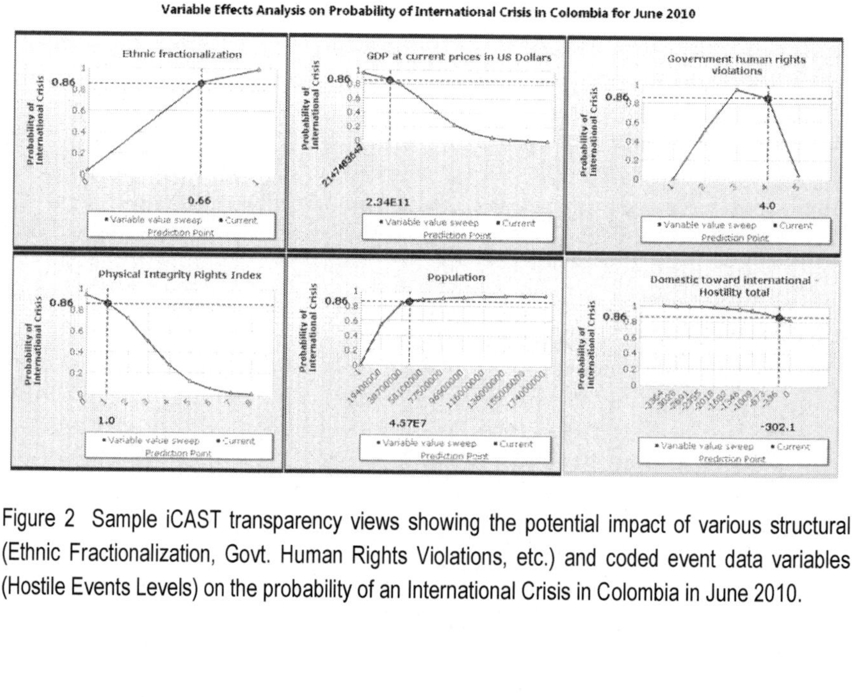

Figure 2 Sample iCAST transparency views showing the potential impact of various structural (Ethnic Fractionalization, Govt. Human Rights Violations, etc.) and coded event data variables (Hostile Events Levels) on the probability of an International Crisis in Colombia in June 2010.

Some of these visualizations support a "what if" exploration capability: users can alter the values of model input variables and see the resulting impact to the forecast. This exploration enables a user to gain an intuitive feel for the relationships among the input variables to a model and its output: e.g., correlations, sensitivities, etc. A user can also simulate a DIME action and its EOI probabilities by altering these "levers." Additional transparency views display a model's pedigree, track record of accurate predictions, etc. An ontology of epistemological elements ranging from the social science theory in a model to its operationalization in software and specific data sources has been developed under another project to improve model verification and validation (Ruvinsky et al., 2012). Such an ontology could be used to organize various detailed transparency views and provide additional information on how a model has been verified and validated.

Transparency views have proven useful, but have not been fully vetted because iCAST models have not yet been used in a production mode. Current views are more oriented towards users with some expertise in computational models. Ultimately one can envision different paths through the transparency views for different kinds of users who want different amounts of detail. An unexpected customer for these views has been the ICEWS model developers, who want to get a feel for which variables might be the most useful. These views have shed light on some modeling issues such as the need for mixed effect models, discussed above.

2.2 Monitoring and Analyzing Events (iTRACE)

The primary goal of the iTRACE (ICEWS Trend Recognition and Assessment of Current Events) capabilities in ICEWS is to provide a fully automated capability to monitor and view current and historical political activity around the globe. These capabilities have been integrated into ISPAN and include an extensible suite of analytics and interactive (web-based) views for operators (e.g., COCOM Country Desk Officers) of geopolitical events for 177 countries since 2001. For example, users can quickly view which regions are increasing in instability; which countries are interacting and how; and what specific events are happening where—all without having to individually monitor hundreds of dynamic data feeds.

Coded event data is automatically extracted from news stories using JabariNLP, a shallow event coder (Van Brackle, 2012). Based on the TABARI tool from the Univ. of Kansas, JabariNLP uses patterns defined in verb and actor dictionaries and syntax trees generated by the OpenNLP parser to rapidly extract events from text. Events specify a source actor, an event type, and a target actor: i.e., who did what to whom. Actors can be individuals, groups, organizations, and countries. Event types are specified by the geopolitical CAMEO event taxonomy with more than 300 types ranging from conflicting (attacks, etc.) to more neutral (meetings, etc.) to cooperative (positive statements of support, financial aid, etc.) (Schrodt et al., 2008). These extracted events, much broader than the five EOIs forecast by iCAST, are scored by type on the Goldstein scale that ranges from -10 (maximally hostile) to +10 (maximally cooperative). This scoring of events enables their numeric aggregation in addition to aggregation by type, actor, and timeframe.

The largest fully automated event coding effort to date was done during ICEWS to process more than 16M news stories (mostly in English) from more than 250 publishers, ranging from international sources (e.g., BBC) to more local ones (e.g., the Jakarta Post). Stories covered 177 countries from 2001-2012. JabariNLP coding precision of 78% approaches peak levels of human manual coding performance (King and Lowe, 2003) that is several orders of magnitude slower on much lower data volumes. Semi-automated methods and a variety of SMEs were used to develop actor dictionaries with more than 50K unique, specific actors and their group affiliations.

As early users of the iTRACE prototype began to explore the coded event data and the source news stories, shortfalls in the precision and recall of the event coding became apparent. In Phase 1, this event data was only used by the models that, tolerant of some noise, still produced accurate forecasts. Users were less tolerant and a new target of 80% precision was established (the TABARI coder in use was baselined at 58% precision). This led to a large, sustained effort to improve the coder, dictionaries, and related tools (Shilliday and Lautenschlager, 2012). Many iterative evaluations of coder results were done by trained humans, including independent, operational evaluations by COCOM users. Future such evaluations could benefit from a broader set of metrics and additional process automation.

Key natural language processing (NLP) challenges included a wide variety of publishers and journalistic standards and styles found in the news stories; coding the relatively fine granularity of CAMEO event types; and coding foreign language translations. In general, the development and application of data acquisition, configuration, processing, evaluation, and improvement tools and methods was a huge, critical path, and significantly underestimated effort. Maintenance of actor dictionaries, story filters, etc. remains a significant cost. Under W-ICEWS, additional coding improvements are expected from leveraging statistical NLP tools. W-ICEWS will also code other text sources and foreign language texts.

The coded event data is consumed by iCAST statistical models (in various aggregations) and iTRACE analytics and visualizations. The latter include heat map, time series (from daily to annually), time line, and other views of aggregated events, such as the ones shown in Figure 3. Users can drilldown to relevant groups of events from these views and ultimately to the original news stories from which the events were extracted. Views can be configured by users depending on their job.

The development and transition of iTRACE capabilities to ISPAN presented a variety of challenges (Starz et al., 2012). Valuable feedback was obtained by PACOM and SOUTHCOM users of the prototypes. iTRACE was used to support several operational exercises during ICEWS. It will be available as a production system via ISPAN in the Spring of 2012 to COCOMs and other users worldwide.

Future iTRACE enhancements planned under W-ICEWS include integrating other data sources including social media sentiment analysis via iSENT as a leading indicator of unrest in advance of news reports. Other analytics and visualizations could also be incorporated such as the tracking transnational and coalition activity; tracking activity by issue (e.g., counter-drug); and social networks.

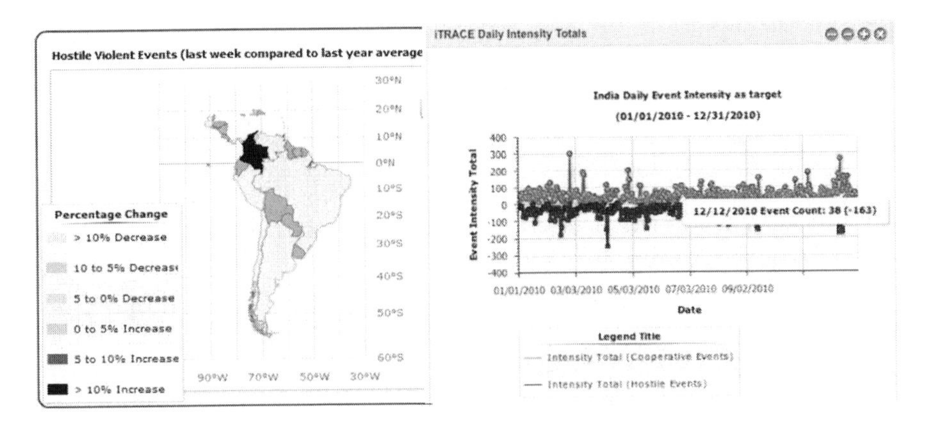

Figure 3 Sample iTRACE views: heat map for SOUTHCOM (left) showing changes in hostile violent events (compared to the prior year) and daily intensity totals of events directed at India in 2010 (right) with cooperative events (blue/above) and hostile (red/below).

3 CONCLUSION

The DARPA ICEWS program was a successful, multi-disciplinary, research and development effort that demonstrated the value of innovative computational social science models and data sources in monitoring of event trends and producing accurate, actionable, and transparent forecasts of EOIs. Initial ICEWS (iTRACE) capabilities were successfully transitioned to ISPAN with additional (iCAST and iSENT) capabilities scheduled to transition under the ONR W-ICEWS effort in 2012-2013. ICEWS can provide commanders with lower cost, nontraditional methods of ISR that can exploit the vast array of dynamic information now available in open media, as well as more traditional sources, to provide timely situational awareness and ultimately DIME action course-of-action generation, monitoring, and assessment to prevent or mitigate instability.

ACKNOWLEDGMENTS

Core team members included Duke University (led by Dr. Michael Ward); Innovative Decisions, Inc. (led by Drs. Suzanne Mahoney and Dennis Buede), Lustick Consulting, Inc. (led by Dr. Ian Lustick), the Pennsylvania State University (led by Prof. Phil Schrodt), and Strategic Analysis Enterprises, Inc. (led by Dr. Steve Shellman). ICEWS was supported by the Defense Advanced Research Projects Agency (DARPA). We especially thank Drs. Sean O'Brien and Philippe Loustaunau for their vision in creating and guiding the ICEWS Program and the ICEWS operational champions at SOUTHCOM, PACOM, and STRATCOM. Future work is being supported by Worldwide ICEWS contract through the Office of Naval Research.

REFERENCES

Alcorn, B., Garces, M., and Hicken, A. 2011. VirThai: A PS-I Implemented Agent-Based Model of Thailand as a Predictive and Analytic Tool. *International Studies Association.*

Crouser, R.J., Kee, D.E., Jeong, D.H., and Remco, C. 2012. Two Visualization Tools for Analysis of Agent-Based Simulations in Political Science. *IEEE Computer Graphics and Applications*: 67-77.

Fearon, J. and Laitin, D. 2003. Ethnicity, Insurgency and Civil War. *American Political Science Review*. 97(1).

King, G, and Lowe W. 2003. An Automated Information Extraction Tool For International Conflict Data with Performance as Good as Human Coders: A Rare Events Evaluation Design. *International Organization*. 57:617-642.

Lustick, I., Alcorn, B., Garces, M., and Ruvinsky, A. 2010. From Theory to Simulation: The Dynamic Political Hierarchy in Country Virtualization Models. *American Political Science Association Annual Meeting*, Sept. 2010.

Mahoney, S., Comstock, E., deBlois, B., and Darcy, S. 2011. Aggregating Forecasts Using a Learned Bayesian Network. *Proceedings of the Twenty-Fourth Florida Artificial Intelligence Research Society Conference*, May 18–20, 2011.

Metternich, N.W. and Ward, M.D., 2011. Now and Later: Predicting the Risk of Violence in Afghanistan. Research Report.

Metternich, N.W., Dorff, C., Gallop M., Weschle, S., and Ward, M.D. 2011. Anti-Government Networks in Civil Conflicts: How Network Structures Affect Conflictual Behavior. *Workshop on Theory and Methods in the Study of Civil* War.

Montgomery, J.M., Hollenbach, F. and Ward, M.D. 2011. Improving Predictions Using Ensemble Bayesian Model Averaging. Submitted and under revision.

O'Brien, S. 2010. Crisis Early Warning and Decision Support: Contemporary Approaches and Thoughts on Future Research. *International Studies Review*. 12:1, 87-104.

Ruvinsky, A., Wedgwood, J.E., Welsh, J.J. 2012. Establishing Bounds of Responsible Operational Use of Social Science. *2nd International Conference on Cross-Cultural Decision Making: Focus 2012.*

Schrodt, P.A., Yilmaz, O., Gerner, D.J., and Hermreck, D. 2008. The CAMEO (Conflict and Mediation Event Observations) Actor Coding Framework. *Annual Meeting of the International Studies Association*, 26 - 29 March 2008.

Shellman, S.M., Levey, B., and Leonard, H. 2011. Countering the Adversary: Effective Policies or a DIME a Dozen? *American Political Science Association Meetings.*

Shilliday, A., and Lautenschlager, J. 2012. Data for a Global ICEWS and Ongoing Research. *2nd International Conference on Cross-Cultural Decision Making: Focus 2012.*

Starz, J., Hoffman, M., Roberts, J., Losco, J., Spivey, K., and Lautenschlager, J. 2012. Supporting situation understanding (past, present, and implications on the future) in the STRATCOM ISPAN program of record. *2nd International Conference on Cross-Cultural Decision Making: Focus 2012.*

Van Brackle, D. 2012. Improvements in the Jabari event coder. *2nd International Conference on Cross-Cultural Decision Making: Focus 2012.*

Ward, M.D., Metternich, N.W., Carrington, C., Dorff C., Gallop, M., Hollenbach, F., Schultz, A. and Weschle, S. 2012. Geographical Models of Crises: Evidence from ICEWS. *2nd International Conference on Cross-Cultural Decision Making*, June 2012.

Wedgwood, J., Ruvinsky, A., and Siedlecki, T. 2012. What lies beneath: Forecast transparency to foster understanding and trust in forecast models. *2nd International Conference on Cross-Cultural Decision Making: Focus 2012.*

CHAPTER 43

Geographical Models of Crises: Evidence from ICEWS

Michael D. Ward, Nils W. Metternich, Christopher Carrington, Cassy Dorff, Max Gallop, Florian M. Hollenbach, Anna Schultz, and Simon Weschle

Duke University
Durham, USA
michael.d.ward@duke.edu

ABSTRACT

Developing political forecasting models is not only relevant for scientific advancement, but also increases the ability of political scientists to inform public policy decisions. Taking this perspective seriously, the International Crisis Early Warning System (ICEWS) was developed under a DARPA initiative to provide predictions of international crisis, domestic crisis, rebellion, insurgency, and ethnic violence in about two-dozen countries in the US PACOM Area of Responsibility. As part of a larger project coordinated by Lockheed Martin Advanced Technology Labs, a team at Duke University created a series of geographically informed statistical models for these events. The generated predictions have been highly accurate, with few false negative and false positive categorizations. Predictions are made at the monthly level for three months periods into the future.

Keywords: prediction, international crisis, security

1 INTRODUCTION

Predicting crises has been a research priority of the US intelligence and warning community for decades. For the past several years under DARPA funding, a large, multidisciplinary team of computer and social scientists from universities and small businesses developed the Integrated Crisis Early Warning System (ICEWS). ICEWS provides Combatant Command staffs (COCOMs) with highly accurate and timely forecasting of instability events of interest (EOIs) using an innovative combination of state-of-the-art computational social science models (O'Brien, 2010). ICEWS exploits dynamic, high-volume, heterogeneous data sources to estimate these models and provide operators with situational awareness of past and current events in countries of interest. Many components of this system are being transitioned into the Integrated Strategic Planning and Analysis Network (ISPAN) program of record at the United States Strategic Command by Lockheed Martin. ICEWS has also been deployed for user testing and evaluation at the Pacific and Southern Combatant commands in 2010 and 2011.

In political science, prediction is typically conceptualized as a conditional exercise, in which values on a dependent variable are calculated based on some estimated, or conditional, statistical model, and then compared with the actual observed values (Hildebrand, Laing and Rosenthal, 1976). But there is also a recent tradition of attempting to make political predictions about things that have not yet occurred. An early proponent of using statistical models for making such predictions in the realm of international relations was Stephen Andriole, a research director at ARPA in the late 1970s (Andriole and Young, 1977). In 1978, a volume edited by Nazli Choucri and Thomas Robinson provided an overview of the then current work in forecasting in international relations, much of which was done in the context of policy oriented research for the U.S. government during the Vietnam War. There were a variety of efforts to forecast or evaluate forecasting efforts, including Freeman and Job (1979), Singer and Wallace (1979), & Vincent (1980), and a few projects began to forecast internal conflict (Gurr and Lichbach, 1986). However, the median empirical article in political science (as well as sociology and economics) used predictions only in the sense of in-sample observational studies. Doran (1999) and others provided some criticism but most scholars avoided making predictions, perhaps because their models had enough difficulty in describing accurately what had happened. Still a few scholars continued to make predictions, including Gurr and Harff (1998), Pevehouse and Goldstein (1999), Schrodt and Gerner (2000), King and Zeng (2001), O'Brien (2002), Bueno de Mesquita (2002), de Marchi, Gelpi and Grynaviski (2004), Enders and Sandler (2005), Ward, Siverson and Cao (2007), Brandt, Colaresi and Freeman (2008), Bennett and Stam (2009), and Gleditsch and Ward (2010), among others. A summary of classified efforts was declassified and reported in Feder (1995) and a nice overview of the historical efforts along with a description of current thinking about forecasting and decision-support is given by O'Brien (2010).

The basic task of the ICEWS project is to produce predictions for five dependent variables, for 29 countries[1] in PACOM, for every month from 1997 through the present plus three months into the future. The variables in question are rebellion, insurgency, ethnic violence, domestic political crises, and international political crises. Each month we receive a drop of two sets of data. The first of these comprises the five dependent variables in this study, known as the ground truth

---

[1] The twenty-nine countries are Australia, Bangladesh, Bhutan, Cambodia, China, Comoros, Fiji, India, Indonesia, Japan, Laos, Madagascar, Malaysia, Mauritius, Mongolia, Myanmar, Nepal, New Zealand, North Korea, Papua New Guinea, Philippines, Russia, Singapore, Solomon Islands, South Korea, Sri Lanka, Taiwan, Thailand, & Vietnam.

data. In addition, we receive data for each event that transpires within or involving each of the 29 countries in the sample. These event data are gleaned from natural language processing of a continuously updated harvest of news stories, primarily taken from Factiva™ (Dow Jones), an open source, proprietary repository of news stories from over 200 sources around the world. The baseline event coder is called JabariNLP, and builds on the TABARI/KEDS software developed by the Philip Schrodt and colleagues (see http://eventdata.psu.edu/). It combines a "shallow parsing" technology of prior coders with a richer exploitation of syntactic structure. This has increased accuracy (precision) from 50% to over 70%, as demonstrated in a series of ongoing (informal) evaluations of its output by human graders (peak human coding performance is around 80% (King and Lowe, 2003)).

These data are augmented with a variety of other attribute and network data. In particular we use attributes, coded on a monthly or yearly basis from the Polity, MAR, and World Bank data set. We also include information about the election cycles (if any) in each of the countries. In addition, we use information about relations among the 29 countries, including geography, the length of shared borders, the amount of trade, the movement of people across borders, the number of refugees, as well as the number and types of events between each pair of the 29 countries (plus the US).

The next section provides a brief review of the theoretical motivations for each of our models, summarizes relevant variables and literature, and informs the reader of the type of model used to generate the results. A full explanation of mixed effects models is provided in the second section, followed by an overview of the model specifications for each dependent variable. Finally, we demonstrate the predictive power of all the models and explore the cumulative probabilities of predicting any event of interest.

2 MODELING EVENTS OF INTEREST

ICEWS focuses on five EOIs: Rebellion, Insurgency, Ethnic Violence, Domestic Crisis, and International Crisis. To model each EOI, we draw from relevant literature in political science to suggest a set of conditions under which the specific event is likely to occur.

Rebellion. Our model for predicting rebellion uses proxies for the level of latent conflict between the government and the opposition, and then models the circumstances under which this latent conflict will lead to rebellions. The proxies are directional measures of the number of conflictual words ("demand", "disapprove", "reject", "threaten") stated from the government towards opposition groups and vice versa. We suggest that the effect of conflict on the probability of rebellion depends on the number of ethnically relevant groups that are excluded from power. When there are no excluded ethnic groups, rebellion should be very unlikely, as disagreements can be solved in the political arena. However, if a large number of excluded groups exist, coordination problems arise, which also mitigate rebellions. Hence, rebellion becomes most likely when few excluded groups exist.

We also include proximity to elections, which can bring about an increase in violence. A recent example is the case of Kenya, where following the victory of incumbent President Mwai Kibaki, the opposition denounced the results and widespread protests led to violence. As Snyder argues, while elections and democracy are often seen as important mechanisms in the peace building process, they can actually increase the likelihood of violence (Snyder, 2000). Additional predictive factors are detailed in Table 1.

Insurgency. Access to power is a key variable to understanding the causes of insurgencies. Insurgencies involve groups attempting to wrest political power from the sitting government, and so groups without access to political power are especially of interest. The larger this excluded population, the more likely violence will be used to change the political landscape. Furthermore, evidence has shown that violence designed to undermine the government is faced with a collective action problem (Kalyvas and Kocher 2007). However, if anti-government groups observe attacks against the government, they may change their calculus. Thus, we include a measure of dissident groups' actions against the government because such actions can be used as a rallying force and recruiting tool, increasing the probability of insurgency. Similarly, it follows that insurgencies in nearby countries may update individuals' beliefs about who else will act against the government of their own country. For this reason we include a measure of insurgencies in nearby countries, lagged by three months. We also suggest that nearby insurgencies could potentially disrupt effective government repression, liberating sources of weapons, money, and information for would-be insurgents in the target country.

Ethnic Violence. While most quantitative studies focus on the effect ethnicity has on conflicts between rebels and the government, we are interested in inter-ethnic and inter-religious violence. Thus, our concept of ethnic violence matches ideas of non-state war (Sarkees and Wayman, 2010), non-state conflicts (Eck, Kreutz and Sundberg, 2010), or subnational wars (Chojnacki, 2006), where the primary actors are non-state. In line with recent work on ethnic conflicts (Cederman, Wimmer and Min, 2010), we argue that government policies play an important role in explaining these dynamics. Thus, in our models, we include the number of politically excluded ethnic groups in a country and the overall proportion of the excluded ethnic population. The existing literature also points to a polarization effect of political exclusion, which suggests including the squared term of the proportion excluded. In addition, we argue that periods of political transition increase incentives to lock in political power in future institutions. Hence, we include Polity and its squared term to model political transition periods (Hegre et al., 2001). Finally, we model the spatial component of ethnic conflict. An increasing number of scholars not only highlight the transnational dimensions of civil conflict (Gleditsch, 2007; Salehyan, 2009), but also its ethnic component (Cederman, Buhaug and Rød, 2009). Thus, our model takes into account possible spillover effects from neighboring countries.

Domestic Crises. Domestic violence and protests are frequently triggered by elections that were perceived to be unfair. We include proximity of elections in our model, with different effects depending on the level of executive constraints. We propose this approach because in countries with moderate levels of executive constraints, elections have meaningful implications regarding who holds office, but governments have the latitude to manipulate the elections and therefore domestic crises are more likely to center around elections. A second major factor that we believe affects the propensity of domestic crises onsets is a country's ethnic composition. When ethnic groups are excluded from political processes grievances are likely to arise. In authoritarian systems this effect is likely to differ from democracies, so in our model, the effect of the number of excluded groups varies by executive constraints. Hence, the likelihood of domestic conflict is conditional on different levels of executive constraints, with the coefficients for the proximity to elections and the number of excluded groups also varying by executive constraints. In addition to the random effects for proximity to election and number of excluded ethnic groups, we control for GDP per capita, population size, and a spatial lag of domestic crises.

International Crises. Our model of international crises tries to capture those situations when a leader is unable or unwilling to make the necessary concessions to avoid a crisis. A leader's incentives to avoid international crises will be conditional on domestic political institutions. Leaders in more democratic regimes may be less able to make concessions internationally due to threat of domestic audience costs. The costs of a crisis might be lower for leaders of more autocratic regimes since their constituency will not bear the brunt of any potential fighting (Bueno de Mesquita et al., 2003; Schultz, 2001). To account for systematic differences between the prevalence of crises under different regime types, the model includes a random intercept based on a country's democracy polity score. In addition, homogeneous populations impose few constraints on the bargaining of leaders in democracy. So, the model also includes a random effect for the number of politically relevant ethnic groups conditional on level of democracy. We also control for population size, international crises in politically similar states and include measures for both domestic political pressure and domestic conflict.

3 METHODS: PREDICTING EVENTS OF INTEREST USING MIXED EFFECTS MODELS

We model rebellion, domestic conflict, and international conflict using hierarchical models in which both the intercept and slope vary. This means that we group the data along an indicator, such as level of executive constraints, creating a different intercept for each group. Thus, the varying intercepts correspond to group indicators and the varying slopes represent an interaction between predictor variables x and group indicators:

$$\Pr(y_{it} = 1) = \text{logit}^{-1}(\alpha_{j[it]} + \beta^{G}_{j[it]} x^{G}_{it} + \beta^{O}_{j[it]} x^{O}_{it} + Z_{it}\gamma)$$

$$\begin{pmatrix} \alpha_j \\ \beta^{G}_j \\ \beta^{O}_j \end{pmatrix} \sim N\begin{pmatrix} \mu_\alpha \\ \mu^{G}_\beta \\ \mu^{O}_\beta \end{pmatrix}, \Sigma)$$

where i denotes the countries, t the month and j the grouping variable, α_j are the grouping variable's random intercepts. x_{it}^G and x_{it}^O are predictor variables; β_j^G and β_j^O are the associated random coefficients; γ is a vector of fixed effects associated with Z_{it}. Table 1 provides an overview over all model specifications.

For an illustrative example, this equation accurately presents our model of rebellion, where j denotes the tercile in which a country falls with respect to the number of excluded ethnic groups, α_j are the grouping-specific random intercepts, x_{it}^G is the number of conflictual words from the government against the opposition in country i at time t, and x_{it}^O is the number of conflictual words from the opposition against the government; β_j^G and β_j^O are the associated random coefficients. All models except ethnic violence take this form, which does not include grouping variables.

Table 1 Overview of variables in each EOI prediction model

	Grouping Variables	Controlled Effects
Rebellion	Excluded groups grouped by: Conflictual words opposition → government Conflictual words gov → opposition	Proximity to election Competitiveness of executive recruitment Executive constraints GDP per capita (log) Rebellions in surrounding countries
Insurgency	Country	Proportion of population excluded Number of excluded ethnic groups Exclusion interaction: population × number of groups High intensity actions: dissidents → government Insurgencies in nearby countries
Ethnic Violence	None	Number of excluded groups Number included groups Proportion of population excluded Squared proportion of population excluded High intensity actions: ethnic groups → government Polity score Squared polity score Violence in neighboring countries
Domestic Conflict	Level of executive constraints grouped by: Number of excluded groups Proximity to election	Population (log) GDP per capita (log) Crises in neighboring countries
International Conflict	Number of ethnic groups grouped by: Level of democracy	Population (log) Domestic EOIs Conflictual words: any domestic group → government International crises in politically similar countries

All effects are lagged three months, except proximity to election.

4 PREDICTIVE POWER

In this section we demonstrate the predictive power of our models and show how to use the individual predictive probabilities to create a cumulative predictive probability of any crisis event occurring. We use separation plots to visualize and assess the predictive power of our models. These plots provide a summary of the fit for each model by demonstrating the range and degree of variation among the predicted probabilities and the degree to which predicted probabilities correspond to actual instances of the event (Greenhill, Ward and Sacks, 2011). Red panels represent events and non-events are left white. The line through the center of the plot represents the expected probability for each model. Thus the probability increases from left to right in the plot and a good fit would be visualized with more red panels (events occurring) stacked at the right end of the plot.

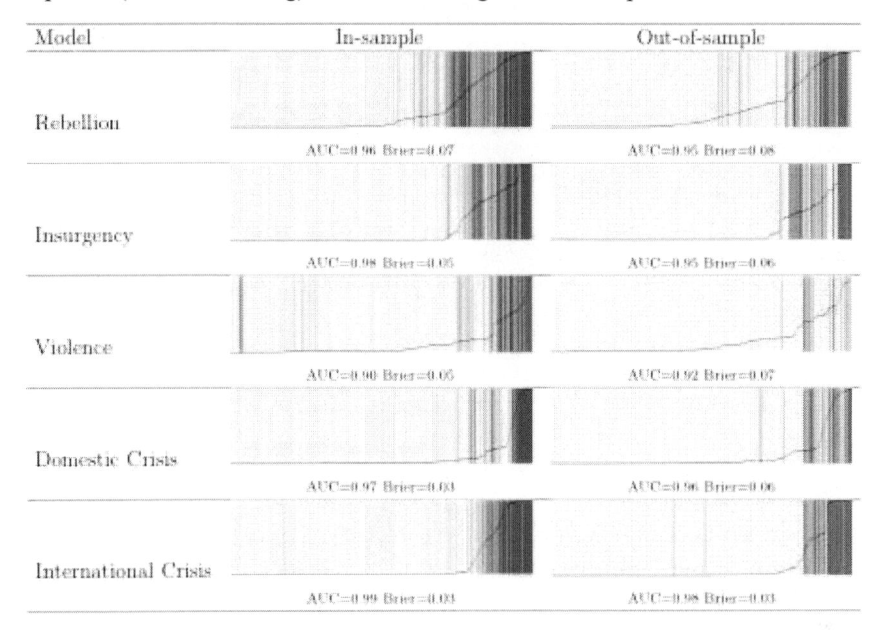

Figure 1 Predictive Capabilities of All EOI Models, including Area under the ROC curve (AUC) and Brier scores.

Another way to evaluate predictions that is employed here is the Brier score, defined as the average squared deviation of the predicted probability from the true event (Brier, 1950). It has been shown that the Brier score is one of the few strictly proper scoring rules for predictions with binary outcomes (Gneiting and Raftery, 2007). Brier scores closer to zero indicate better predictive performance. In Figure 1, we combine a visual and numeric interpretation of the models to display their predictive power. The separation plots and the provided statistics demonstrate a very good predictive performance of our models.

In addition to predicting specific crisis events, we can use the predictive probabilities derived by each of the models to create a cumulative predictive probability of any crisis event occurring. To do so we first create the new crisis variable that indicated the occurrence of any of the five crisis events above. To predict the experience of any crisis event we combine the individual predicted probabilities and calculate a cumulative probability of any crisis occurring. By simple probability theory, the cumulative probability is:

$$P(Y^C_{i,t}=1) = [1 - P(Y^C_{i,t}=0)] = 1 - \prod_{k=1}^{5}(1 - P(Y^k_{i,t}=1))$$

where k ranges from one through five, enumerating the predicted probabilities derived for country i at time t for each of the individual EOIs discussed above. In words, the probability of a country to experience any crisis event in a given month is one minus the probability of no event occurring. This can be calculated as the product of the individual probability of non-occurrence for each of the individual events. Y stands for the occurrence of a crisis event in country i at time t, where the superscript indicates the type of crisis. We then use the cumulative probability to predict if a country experiences any of the above specified crises in a given month.

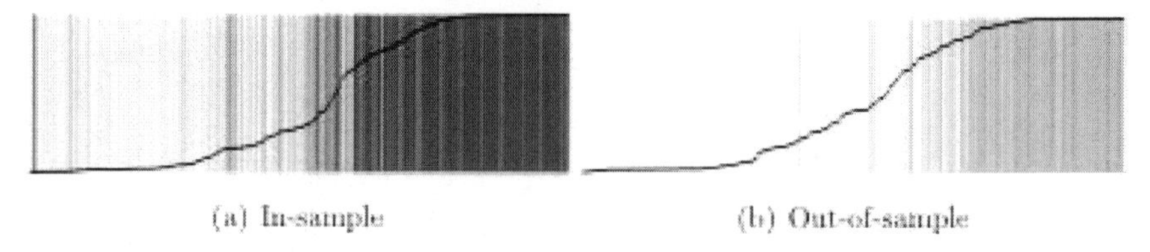

Figure 2 Separation plot for any crisis, indicating fit of the model, in and out-of-sample.

Figure 2 shows the separation plots of predicting any crisis using the cumulative predictive probabilities. There are very few actual crises that are missed, and relatively few false positives. Especially on the out-of-sample data, the cumulative prediction missed very few actual crises. A curious case is displayed on the in-sample separation plot, which shows a number of actual events on the left side with very low predicted probabilities of crisis occurrence. This is the case of the Solomon Islands in 2003, which experienced ethnic violence in five months during that year and was missed by the individual prediction models. Since the cumulative prediction is dependent on the individual models, the same case is missed in the in-sample prediction of the ethnic violence model. For the in-sample observations, the Brier score for predicting the occurrence of any crisis is 0.07, while the Brier score for the out-of-sample observations is 0.1. Thus, the model predicts quite accurate using the individual predictions to calculate a cumulative probability of any crisis occurring.

5 CONCLUSION

We have demonstrated the utility of creating forecasting models for predicting political conflicts in a diverse range of country settings, and have shown that this series of geographically informed statistical models is highly accurate, containing few false negative and positive predictions. These models can serve the public policy community and shed light on an array of critically important components of the political science literature on conflict dynamics. Moving forward, this project hopes to extend these models beyond their current geographical domains and onto a larger, worldwide data set.

ACKNOWLEDGMENTS

This project was undertaken in the framework of an initiative funded by the Information Processing Technology Office of the Defense Advanced Research Projects Agency to provide an Integrated Crisis Early Warning System (ICEWS) for decision makers in the U.S. defense community. The holding grant is to the Lockheed Martin Corporation, Contract FA8650-07-C-7749. For helpful insights we thank Scott de Marchi. All the bad ideas and mistakes are our own. We thank Mark Hoffman and Philippe Loustaunau for support and suggestions.

REFERENCES

Andriole, S.J. and R.A. Young. 1977. Toward the Development of an Integrated Crisis Warning System. *International Studies Quarterly* 21: 107–150.

Bennett, D.S. and A.C. Stam. 2009. Revisiting Predictions of War Duration. *Conflict Management and Peace Science* 26: 256-267.

Brandt, P.T., M. Colaresi and J.R. Freeman. 2008. The Dynamics of Reciprocity, Accountability, and Credibility. *Journal of Conflict Resolution* 52: 343-374.

Brier, G.W. 1950. Verification of Forecasts Expressed in Terms of Probability. *Monthly Weather Review* 78: 1-3.

Bueno de Mesquita, B. 2002. *Predicting Politics*. Columbus: Ohio State University Press.

Bueno de Mesquita, B., A. Smith, R. M. Siverson and J. D. Morrow. 2003. *The Logic of Political Survival*. Cambridge: M.I.T. Press.

Cederman, L.-E., A. Wimmer and B. Min. 2010. Why Do Ethnic Groups Rebel? New Data and Analysis. *World Politics* 62: 87-119.

Cederman, L.-E., H. Buhaug and J.K. Rød. 2009. Ethno-nationalist dyads and civil war: A GIS-based analysis. *Journal of Conflict Resolution* 53: 496-525.

Chojnacki, S. 2006. Anything New or More of the Same? Wars and Military Interventions in the International System, 1946-2003. *Global Society* 20: 25-46.

Choucri, N. and T. W. Robinson, eds. 1978. *Forecasting in International Relations: Theory, methods, problems, prospects*. San Francisco: W.H. Freeman.

de Marchi, S., C. Gelpi and J. D. Grynaviski. 2004. Untangling Neural Nets. *American Political Science Review* 98: 371-378.

Doran, C.F. 1999. Why Forecasts Fail: The Limits and Potential of Forecasting in International Relations and Economics. *International Studies Review* 1: 11-41.

Eck, K., J. Kreutz and R. Sundberg. 2010. Introducing the UCDP Non-State Conflict Dataset. Uppsala University: Unpublished manuscript.

Enders, W. and T. Sandler. 2005. After 9/11: Is it All Different Now? *Journal of Conflict Resolution* 49: 259-277.

Feder, S. 1995. Factions and Policon: New Ways to Analyze Politics. In *Inside CIA's Private World: Declassified Articles from The Agency's Internal Journal, 1955-1992*, ed. H.B. Westerfield. New Haven: Yale University Press.

Freeman, J.R. and Brian L. J. 1979. Scientific Forecasts in International Relations: Problems of Definition and Epistemology. *International Studies Quarterly* 23: 113–143.

Gleditsch, K.S. 2007. Transnational dimensions of civil war. *Journal of Peace Research* 44: 293-309.

Gleditsch, K.S. and M.D. Ward. 2010. "Contentious Issues and Forecasting Interstate Disputes." Paper presented at the 2010 Annual Meeting of the International Studies Association, New Orleans.

Gneiting, T. and A.E. Raftery. 2007. Strictly Proper Scoring Rules, Prediction, and Estimation. *Journal of the American Statistical Association* 102: 359-378.

Greenhill, B. D., M.D. Ward and A. Sacks. 2011. A New Visual Method for Evaluating the Fit of Binary Models. *American Journal of Political Science* 55: 991-1002.

Gurr, T.R. and M.I. Lichbach. 1986. Forecasting Internal Conflict: A Competitive Evaluation of Empirical Theories. *Comparative Political Studies* 19: 3-38.

Gurr, T.R. and B. Harff. 1998. Systematic Early Warning of Humanitarian Emergencies. *Journal of Peace Research* 35: 551-579.

Hegre, H., T. Ellingsen, S. Gates and N.P. Gleditsch. 2001. Toward a Democratic Civil Peace? Democracy, Political Change, and Civil War, 1816-1992. *American Political Science Review* 95: 33-48.

Hildebrand, D. K., J. D. Laing and H. Rosenthal. 1976. Prediction Analysis in Political Research. *American Political Science Review* 70: 509-535.

King, G. and L. Zeng. 2001. Improving Forecasts of State Failure. *World Politics* 53: 623-658.

King, G. and W. Lowe. 2003. An Automated Information Extraction Tool for International Conflict Data with Performance as Good as Human Coders: A Rare Events Evaluation Design. *International Organization* 57: 617-642.

O'Brien, S.P. 2002. Anticipating the Good, the Bad, and the Ugly: An Early Warning Approach to Conflict and Instability Analysis. *Journal of Conflict Resolution* 46: 791-811.

O'Brien, S.P. 2010. Crisis Early Warning and Decision Support: Contemporary Approaches and Thoughts on Future Research. *International Studies Review* 12: 87-104.

Pevehouse, J.C. and J.S. Goldstein. 1999. Serbian compliance or defiance in Kosovo? Statistical analysis and real-time predictions. *Journal of Conflict Resolution* 43: 538-546.

Salehyan, I. 2009. *Rebels without borders: transnational insurgencies in world politics.* Ithaca: Cornell University Press.

Sarkees, M.R. and F. Wayman. 2010. *Resort to War: 1816-2007.* Washington, DC: CQ Press.

Schrodt, P.A. and D.J. Gerner. 2000. Using Cluster Analysis to Derive Early Warning Indicators for Political Change in the Middle East, 1979-1996. *American Political Science Review* 94: 803-818.

Schultz, Kenneth A. 2001. *Democracy and Coercive Diplomacy.* New York: Cambridge University Press.

Singer, J.D. and M.D. Wallace. 1979. *To augur well: early warning indicators in world politics.* Beverly Hills: Sage Publications.

Snyder, J. 2000. *From Voting to Violence: Democratization and Nationalist Conflict.* New York: Norton.

Vincent, J.E. 1980. Scientific Prediction versus Crystal Ball Gazing: Can the Unknown be Known? *International Studies Quarterly* 24: 450-454.

Ward, M.D., R.M. Siverson and X. Cao. 2007. Disputes, Democracies, and Dependencies: A Reexamination of the Kantian Peace. *American Journal of Political Science* 51: 583-601.

CHAPTER 44

Improving ICEWS Models: Forecasting SOUTHCOM Events of Interest Using Ensemble Methods

Patrick Bentley, Stephen M. Shellman, Brian Levey

Strategic Analysis Enterprises, Inc.
Williamsburg, VA
prbentley@gmail.com

ABSTRACT

The U.S. government and military need accurate forecasting tools in order to anticipate the outbreak of violent conflict throughout the world. The authors demonstrate the strength of multi-model approaches to forecasting events of interest (EOIs) (e.g., rebellion, insurgency, ethnic-religious conflict, domestic and international crisis) in the SOUTHCOM Area of Responsibility (AOR). Bayesian Model Averaging (BMA) is used to produce ensemble forecasts that outperform any of the constituent models including: logistic time-series models, linear mixed and hierarchical models, classification and regression tree models, and spatial models. Moreover, the authors demonstrate the utility of modeling countries individually before the forecasts are combined for inclusion in the BMA ensemble. The multi-method ensemble approach generates superior results compared to the constituent models, and produces better, more reliable early warnings of impending crises.

Keywords: ICEWS, forecasting, ensemble methods

1 OVERVIEW OF ICEWS PROJECT

The U.S. government and military need accurate forecasting tools in order to anticipate the outbreak of violent conflict throughout the world. The fulfillment of this need was one of the primary objectives of the DARPA-funded Integrated Crisis

Early Warning System (ICEWS) program. Broadly, the goal of this program was to develop a system to monitor, assess, and forecast national, sub-national, and international crises throughout the world. Such a system would allow top-level military commanders to anticipate crises and allocate resources efficiently in order to mitigate them. The primary Events of Interest (EOIs) that ICEWS focused on were: Domestic Political Crisis, Rebellion, Insurgency, Ethnic/Religious Violence, and International Crisis.

To accomplish the objective of forecasting crises, the program sought to integrate the best social scientific modeling approaches to produce more accurate predictions of the Events of Interest (EOIs) than any of these modeling approaches could produce independently. These modeling approaches included agent-based models, geo-spatial network models, and frequentist and Bayesian statistical models. For our part, we employed logistic time-series models, linear mixed and hierarchical models, classification and regression tree models, and spatial models in our attempts to forecast occurrences of the EOIs. We often constructed multiple statistically and theoretically sound models that produced slightly different forecasts. Facing this, we used a Bayesian Model Averaging algorithm to combine the models into a multi-method ensemble that generated superior results compared to the constituent models, and produced better, more reliable early warnings of impending crises.

In what follows, we present our efforts at forecasting Domestic Political Crisis in the SOUTHCOM Area of Responsibility. First, we will define our dependent variable and describe our data. We then describe the process by which we modeled this EOI using a cross-sectional logistic regression model, a cross-sectional mixed effects model, and individual-country logistic regression models. While we do not present the results of these models here, we discuss the relevant social scientific theory that guided our variable selection and model building. As we show, these models have good accuracy statistics, but they do not always yield the same prediction. We then explain how we use the Bayesian Model Averaging tool to produce better, more reliable results than the individual models. Finally, we present the forecasting statistics of these models to confirm the advantages of ensemble modeling. We conclude with a less-detailed presentation of our forecasting results for the International Crisis EOI.

2 FORECASTING DOMESTIC POLITICAL CRISIS

A country experiences a Domestic Political Crisis (DPC) when there is significant opposition to the government that manifests itself as, for example, a power struggle between two political factions involving disruptive strikes or clashes between supporters or nationwide, violent protests calling on the resignation of the government. Moreover, a Domestic Political Crisis does not reach the level of an insurgency or rebellion (O'Brien 2010, 90).

We use monthly-level data on 24 countries from July 2001 through September 2011 giving us 2,952 country-months. The dependent variable in our models is coded '1' when a country experiences a DPC in a given month and '0' otherwise. These data were coded by research assistants at SAE, inc. and Lustick Consulting using a variety of open source news Of those country-months, 486 have an occurrence of DPC with one country experiencing DPC in every month, eleven never experiencing it, and the remaining twelve having some months with a DPC and some without.

Our main independent variables of interest are monthly level political events data that capture the dynamic interaction between actors within each country (e.g. government, opposition, dissident, and religious actors). We generate these data using an automated event data extractor (Jabari) which processes digital news reports and identifies who did what to whom and when (Van Brackle 2012). These events are then aggregated at a monthly level identifying, for instance, the level of hostility or cooperation between various actors with a country. These variables are lagged for a period of six months. In doing so, we can forecast how behavior between actors six months ago precede DPC in the present month.

We also use some annual level structural data that account for the environment that each country is operating in. These data include indicators such as the State Department Political Terror Scale, the ethnic fractionalization of a country, and the CIRI physical integrity score. These variables are each lagged for a period of twelve months. Since we are forecasting in six-month intervals and these indicators are only recorded annually, it is necessary to capture the value of this variable from the previous year.

2.1 Cross-Sectional Logistic Regression Model

Our first forecasting model is a logistic regression model, which is the most common statistical model used to analyze binary dependent variables like Domestic Political Violence. The model performs a maximum-likelihood calculation that estimates parameters that have the highest probability of producing the observed data.

We find that each of our indicators is statistically and substantively significant. The most impactful variables in our model are our event variables. For example, we find that as the government becomes more hostile to its citizens, there is less likely to be a DPC. If the people know that they will be severely repressed for standing up to the government, then they are less likely to do so. This is consistent with our finding on the State Department Political Terror Scale variable. For this variable, we include a squared term, and find that an increase in political terror will increase the likelihood of DPC to a point, but once the political terror gets severe enough, the likelihood of DPC decreases. This is consistent with the inverted-U hypothesis (Hibbs 1974; Denardo 1986) that states that at low levels of repression there is nothing to rebel against, at high levels of repression it is almost virtually impossible to rebel. Thus, medium levels of repression spark dissent.

On the whole, this model performs fairly well correctly predicting 92.14% of the observations as shown in Table 1a (Logit column). It does, however, only predict 69.38% of the occurrences of a DPC correctly. In an effort to improve on these statistics, we estimate a mixed effects model in order to separate cross-sectional effects from individual country effects.

Table 1a SOUTHCOM Domestic Political Crisis, In-Sample Forecasting Accuracy Statistics

	Logit	Mixed Effects	BMA 1	Individual Countries	BMA 2
Percent Correctly Classified	92.14%	93.51%	94.40%	92.39%	96.51%
Sensitivity (% 1s Correct)	69.38%	72.70%	75.63%	84.18%	86.88%
Specificity (% 0s Correct)	96.61%	97.59%	98.08%	95.34%	98.41%

2.2 Cross-Sectional Mixed Effects Model

The mixed effects model allows us to analyze multiple levels of analysis, which is important for making reliable forecasts of both human and nation-state behavior. In this model, a variable can exhibit one of two types of effects on the likelihood of DPC: a fixed effect or a random effect. A variable modeled as a fixed effect has the same effect across each country, whereas a variable that is a random effect will have a different effect on each country.

This model has two fixed effects, which are both statistically and substantively significant. Across all countries, the number of protests against the government increases the probability that there will be a DPC in the future. We also find that the as the degree of political terror that a government imposes on its people increases, the likelihood that a DPC will occur decreases. Both of these results are intuitive.

The random effect in this model is the degree of physical integrity that the people of a country enjoy. For some countries, the effect of this indicator is positive and in others it is negative, but these findings are not at odds with each other. As physical integrity increases in some countries, it makes sense that the likelihood of DPC would decrease because the people experience less repression by the government. At the same time, the people enjoy more rights and have greater freedom to assemble and voice their opposition against the government account for the increase in the likelihood of DPC in some countries. It depends on the various baseline levels of physical integrity represented in each country and whether they are already high or low as well. When teasing out the effects, we find the overall relationships to be consistent with the inverted-U hypotheses discussed above.

This model yields 93.51% overall accuracy and predicts the occurrence of DPC with 72.70% precision as reported in Table 1a, which is a slight improvement over the cross-sectional logistic regression model. In an effort to improve the precision score, we take one more step to extract even more information from these data to improve our forecasts by modeling individual countries using a logistic regression.

2.3 Individual-Country Logistic Regression Models

The cross-sectional logistic regression model was treating every country the same, and the mixed effects model allowed some parameters to vary by country while others treated all countries the same. Our final step is to individually model those countries that experienced a DPC in our temporal period. The variables we use in these models are similar to those we have used in the other two models.

Table 1a shows that these models produce a more than 11% increase in sensitivity over the mixed-effects model and a nearly 15% increase over the logistic regression model to 84.18%, which is a great improvement. Having fit these models, the question remains: how do we know which model is the best or correct model? Generally, we use several criteria for identifying the best model.

First, the choice of an estimation technique must take into consideration the properties of the data and possibility of violating the assumptions of a given model (e.g. autocorrelation, heteroskedasticity, multicollinearity). We were able to eliminate several types of models in this manner, but both the logistic regression and the mixed effects model were suitable for our needs. Second, there is always uncertainty about which variables to include in a given model, so theory offers some insight as to which variables should have an effect on the EOI. Each of these models has different variables, and both are theoretically sound. Finally, after fitting multiple models, we narrow the field down to one by using some criterion or fit statistic (e.g. R-squared, RMSE, AIC, BIC, Vuong test). Each of these criteria has its virtues and flaws, but none is perfect.

Given these criteria, it would be difficult to eliminate any of these models and conclude that one is better than the other. Even if we did, there would still be some uncertainty as to whether the model we chose is best or correct. Ideally, we could retain all models, so perhaps a better question is: how can we leverage the information from each of these models to achieve a better forecast than any offer individually? Rather than assuming that there is one model that best fits the data, it is much more realistic to acknowledge that some models fit some of the data better than others. More analysis reveals that while there is overlap in the two models' predictions, each model predicts an independent set of crises and misses a different set. How can we get the best of both models while leaving behind the worst? In the next section, we combine these two models with a Bayesian Model Averaging algorithm.

2.4 Bayesian Model Averaging

The Bayesian Model Averager combines the best parts of multiple models to generate a single forecast based on weighted component model predictions based on the past performance of the models on test data. In addition to combining multiple predictions into a single forecast, the BMA is agnostic to modeling approach (e.g. statistical, agent-based, systems dynamic, etc.); any model that generates a forecast can be incorporated into the aggregation.

The BMA works as follows to produce a single forecast from multiple predictive models. Assume we have K predictive models $M_1, M_2, \ldots, M_k$ with an outcome of interest $p(y^*|M_k)$. Using Bayes Rule, we can derive the following:

$$E(y^*) = \sum_{k=1}^{K} E(y^*|M_k)p(M_k|y^T)$$

where $E(y^*)$ is the expected probability of an event and $E(y^*|M_k)p(M_k|y^T)$ is the weighted prediction of all component models. See Rafferty (1995) for further discussion of Bayesian Model Averaging.

In order to actually combine the two cross-sectional models and the individual country models, we had to use the BMA algorithm twice because of the nature of modeling countries individually. In order to get an averaged forecast for each country-month, each model needs to produce a forecast for each country month. With the individual countries, we can only model those countries that have experienced both months with DPC and without DPC because the dependent variable must have some variation. Using these models, we can extract forecasts for these countries, but we need forecasts for the other countries as well.

We first combine the two cross-sectional models using the BMA. You can find the results from this model in Table 1a under column BMA 1. We then replace the forecasts in BMA 1 for the countries with an individual model with the forecasts from each respective model leaving the forecasts from the countries we did not model individually intact. Finally, we combine the two cross-sectional models with the individual-country models/BMA 1, which produces the results for BMA 2.

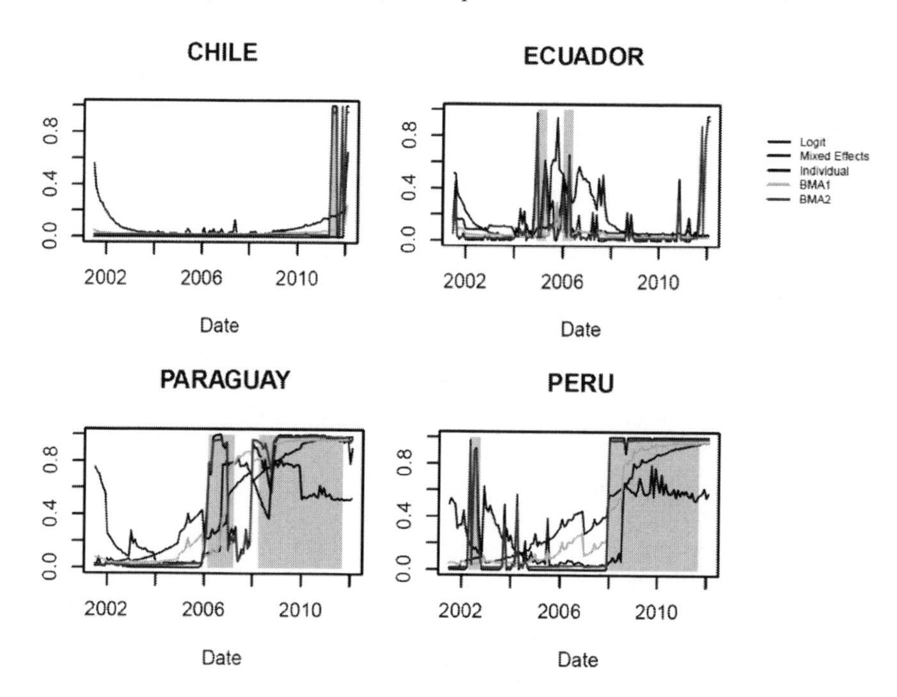

Figure 1: Model Results, Domestic Political Crisis, Selected Countries.

Table 1a shows the results for these two BMAs. BMA 1 produces a slight but noticeable increase in all of the reported statistics over the two cross-sectional models and improves the sensitivity by nearly 3%. The big improvement comes with BMA 2 where we achieve a sensitivity score of 86.88% as well as improvements in overall percentage correctly classified and specificity. In both BMAs we find that the forecasting ability of the combination of the constituent models is superior to any of the models used to produce the BMA.

As mentioned earlier, one of the primary reasons for combining these models using the BMA is because each model predicts an independent set of crises while missing a different set. Figure 1 shows this to be the case and how the BMA is able to harness the information from the individual models to produce a more reliable forecast. Paraguay experiences a DPC beginning in 2006 and concluding in 2007, then another begins in 2008 and continues until the end of the temporal period. The red line shows the final BMA that combines each of the models. Throughout the whole temporal period, this BMA nearly perfectly predicts the occurrence and lack thereof of DPC, while neither the constituent models nor the first BMA (in green) are as successful. Some predict DPC where there is none and fail to predict it when it occurs. Nevertheless, the BMA is able to use these models to successfully predict DPC in Paraguay. The other countries in Figure 1 show a similar pattern with more or less success.

2.5 Forecasting

While it is certainly important that our models be able to produce great in-sample prediction statistics, the real test for forecasting is to see how the models perform in out-of-sample situations. Table 1b shows the forecasting accuracy statistics for these models in three different time periods.

For the first set of statistics, we ran our models on data through December 2009. Using the parameter estimates from these models, we forecasted over the six-month period from January to June of 2010. During this time period, the logistic regression only predicted 58.80% of occurrences of DPC correctly and only 87.50% overall. Furthermore, the mixed effects model did not converge on a result and, thus, did not produce any forecasts. Finally, the individual models preformed well with 89.25% of DPC correctly predicted. As a consequence of the mixed effects model not converging, we were only able to produce forecasts for the countries with individual country models when we combined them with the cross-sectional logistic regression in the BMA. The results of the BMA performed above the metric thresholds set by our research program (80% overall; 70% precision) coming in at 73.53% precision, and almost 94% overall.

In the latter two out of sample temporal periods, the final BMAs predicted over 99% of the observations correctly. In the forecasting period from July to December 2010, we were able to correctly predict every occurrence of DPC correctly. We see a similar pattern in the forecasting period from January to June 2011 being able to predict almost 97% of the occurrences of DPC correctly.

Table 1b SOUTHCOM Domestic Political Crisis, Out-Of-Sample Forecasting Accuracy Statistics

Models run on data through December 2009 with forecasts made for Jan-Jun 2010.					
	Logit	**Mixed Effects**	**BMA 1**	**Individual Countries**	**BMA 2**
Percent Correctly Classified	87.50%	N/A	N/A	95.45%	93.75%
Sensitivity (% 1s Correct)	58.80%	N/A	N/A	89.29%	73.53%
Specificity (% 0s Correct)	96.36%	N/A	N/A	100%	100%
Models run on data through June 2010 with forecasts made for Jul-Dec 2010.					
	Logit	**Mixed Effects**	**BMA 1**	**Individual Countries**	**BMA 2**
Percent Correctly Classified	91.67%	95.83%	98.61%	98.48%	99.31%
Sensitivity (% 1s Correct)	76.47%	100%	100%	100%	100%
Specificity (% 0s Correct)	96.36%	94.73%	98.25%	97.62%	99.12%
Models run on data through December 2010 with forecasts made for Jan-Jun 2011.					
	Logit	**Mixed Effects**	**BMA 1**	**Individual Countries**	**BMA 2**
Percent Correctly Classified	95.14%	99.31%	99.31%	96.97%	99.30%
Sensitivity (% 1s Correct)	77.42%	96.77%	96.77%	96.00%	96.77%
Specificity (% 0s Correct)	100%	100%	100%	97.56%	100%

As these results show, we are able to produce very accurate forecasts of Domestic Political Crisis by combining several different models with a BMA. In the next section, we present our forecasting results for the International Crisis EOI to show that the success of this procedure is not an artifact of modeling this specific EOI.

3 INTERNATIONAL CRISIS

The ICEWS project defines the International Crisis (IC) EOI as "conflict between two or more states or elevated tensions between two or more states that could lead to conflict" (O'Brien 2010, 90). To model IC, we use a dependent variable produced by the same process as DPC. We also draw our independent variables from the events data and structural data discussed above. Table 2a shows the in-sample accuracy statistics.

Table 2a SOUTHCOM International Crisis, In-Sample Forecasting Accuracy Statistics

	Logit	Mixed Effects	BMA 1	Individual Countries	BMA 2
Percent Correctly Classified	98.43%	98.60%	98.70%	96.17%	98.98%
Sensitivity (% 1s Correct)	71.43%	70.54%	73.21%	89.29%	84.82%
Specificity (% 0s Correct)	99.50%	99.72%	99.72%	97.42%	99.54%

As with DPC, the logistic regression and mixed effects models have a sensitivity level slightly above 70% and the individual country models correctly predict the occurrence of IC at 89.29%. Combining these with the BMA greatly increases the sensitivity over the cross-sectional models to 84.82% with 98.98% correctly predicted overall.

Table 2b SOUTHCOM International Crisis, Forecasting Accuracy Statistics

Models run on data through December 2009 with forecasts made for Jan-Jun 2010.					
	Logit	**Mixed Effects**	**BMA 1**	**Individual Countries**	**BMA 2**
Percent Correctly Classified	93.06%	95.14%	92.36%	69.44%	96.53%
Sensitivity (% 1s Correct)	92.31%	92.31%	92.31%	46.15%	92.31%
Specificity (% 0s Correct)	93.13%	95.42%	92.37%	82.61%	96.95%
Models run on data through June 2010 with forecasts made for Jul-Dec 2010.					
	Logit	**Mixed Effects**	**BMA 1**	**Individual Countries**	**BMA 2**
Percent Correctly Classified	95.83%	95.83%	95.83%	83.33%	95.83%
Sensitivity (% 1s Correct)	66.67%	66.67%	66.67%	72.22%	72.22%
Specificity (% 0s Correct)	100%	100%	100%	94.44%	99.21%
Models run on data through December 2010 with forecasts made for Jan-Jun 2011.					
	Logit	**Mixed Effects**	**BMA 1**	**Individual Countries**	**BMA 2**
Percent Correctly Classified	86.11%	85.42%	85.42%	66.67%	90.97%
Sensitivity (% 1s Correct)	100%	0.00%	0.00%	100%	100%
Specificity (% 0s Correct)	85.82%	87.23%	87.23%	63.64%	90.78%

Table 2b shows the out-of sample accuracy statistics for the same temporal periods as presented with DPC. In the first and last forecasting period, we correctly predict the occurrence of IC at 92.31% and 100% respectively. We are not as successful in the second forecasting period only correctly forecasting 72.22% of the occurrences of IC, and in fact, none of the models perform better than this. That said, we're still above the 70% goal that we set prior to beginning this project. The results for IC confirm that combining the individual models with the BMA produces more reliable results than any of the constituent models can produce by themselves.

4 CONCLUSION

The ICEWS program aimed to develop a system to forecast the occurrence of various types of violent political conflict. Using both logistic regression and mixed effects modeling techniques, we were able to achieve acceptable but not great accuracy. We saw improvement by combining these models with a Bayesian Model Averaging algorithm. Military commanders will only be able to use such a system as ICEWS if it is accurate. We conclude that using ensemble modeling techniques provides the best means to achieve that accuracy.

REFERENCES

Hibbs, D. 1973. *Mass Political Violence*. New York: Wiley.

O'Brien, S.P. 2010. Crisis Early Warning and Decision Support: Contemporary Approaches and Thoughts on Future Research. *International Studies Review* 12: 87–104.

Raftery, A.E. 1995. 'Bayesian model selection in social research (with Discussion)' in *Sociological Methodology*, ed. P. V. Marsden, Blackwell, Cambridge, MA, pp. 111-196.

Van Brackle, D. 2012. Improvements in the Jabari event coder. 2nd International Conference on Cross-Cultural Decision Making: Focus 2012.

De Nardo, J. 1985. *Power in Numbers: The Political Strategy of Protest and Rebellion.* Princeton, NJ: Princeton University Press.

ACKNOWLEDGMENTS

The majority of this research has been funded by the Defense Advanced Research Projects Agency (DARPA). Continuing financial support comes from the Office of Naval Research (ONR).

CHAPTER 45

Human Factors in Military Satellite Operations and Space Situational Awareness

Thomas J. Solz Jr., Kathleen Bartlett, Denise Nicholson

MESH Solutions, LLC
Orlando, FL
tsolz@dsci.com, kbartlett@dsci.com

ABSTRACT

From 2004 through 2009, senior leaders in the United States Air Force (USAF) expressed a need to improve overall Air and Space Command and Control (C^2) and Space Situational Awareness (SSA). Satellite Operations is the means by which operators monitor satellite health/data, track satellites, and command satellites in order to operate and configure the vehicle (Telemetry, Tracking and Commanding). SSA provides "the ability to 'see' and understand what is going on in space" and serves as a foundation for all operations in that domain.

Improvements have been made in Satellite Operations and SSA, but these enhancements have primarily focused on doctrine, organization, software, and equipment. Very few steps have been taken to ensure human factors issues are addressed for the operators. There is little evidence of a proper Human Systems Interface (HSI) approach in the design of the majority of the currently fielded space operator interfaces and a lack of standardized human factors practices within the industry across the space mission spectrum. In fact, with few exceptions, the space operations systems lack standardization themselves. Over the past few years, Air Force research organizations have stepped up to aid the space operational commands as they address these shortcomings.

Keywords: Satellite Operations, Satellite C2, Space Situational Awareness (SSA), human factors, Human System Interface (HSI), space operations

1 SPACE OPERATIONS

"There is something more important than any ultimate weapon. That is the ultimate position—the position of total control over Earth that lies somewhere out in space. That is the distant future, though not so distant as we may have thought. Whoever gains that ultimate position gains control, total control, over the Earth, for the purposes of tyranny or for the service of freedom."

— Lyndon B. Johnson, United States Senator, 1958

Space capabilities are a force multiplier when integrated across the range of joint military operations. The military and civilian dependency on the space based medium has increased exponentially since satellites were first placed in orbit over 50 years ago. Military commanders and operators must understand how to exploit and defend a weapon system that operates independent of geographical boundaries (Joint Publication 3-14, 2009). They must be experts in the ground systems and links that command and control such space based assets. Space operators must also take into account the space and terrestrial weather, and foreign and natural threats; they must provide timely, accurate, and relevant mission data and services to U.S. and coalition forces engaged in combat. Space-based assets, unlike aircraft, ships, and ground vehicles, never return to base for down time, maintenances, or physical upgrades. They are on station and performing the mission 24/7. Space operators must be constantly vigilant, as their mission does not end, during time of peace or conflict.

1.1 Satellite Operations

Satellite operations are conducted to maneuver, configure, operate and sustain on-orbit assets. Satellite operations are characterized as spacecraft and payload operations. Spacecraft operations include telemetry, tracking, and commanding, maneuvering, monitoring state-of-health, and maintenance sub-functions (Joint Publication 3-14, 2009). Satellite operations rely on an operations center, ground-based remote tracking stations strategically located around the globe, the command and control links these stations provide, the dedicated communications network, and the space vehicle and mission payload itself.

The various networks combined ensure total C^2 of space resources. Additionally, as a critical and essential link between the satellite operator and joint force, and a significant contributor to SSA, satellite operations include protection mechanisms to assure access to space assets (Joint Publication 3-14, 2009)

1.2 Space Situational Awareness

Although sharing many of the same functions as Satellite C^2, the SSA mission requires the processing and fusion of data from many dissimilar entities. The SSA operators do not directly command and control satellites as do their satellite C2 counterparts. They are instead responsible for all space and space-terrestrial interactions. This includes friendly forces, potential threats, manned space operation protection, satellite collision avoidance, unpredictable and uncooperative space faring foreign nations, space and terrestrial environmental factors, as well as space debris.

SSA provides "the ability to 'see' and understand what is going on in space" and serves as a foundation for all operations in that domain. A primary function of SSA is to detect, track, identify, and catalog all man-made objects in orbiting the earth.

In the Space 2007 Conference and Exposition, Lt. Gen. Michael Hamel, commander of the Space and Missile Systems Center, stated "Truly space has become integral to every aspect of what we do in military operations today. The ability of U.S. forces to know what's happening in space is critical to having assured mission capability. Space situational awareness is now the means by which we intend to knit all those capabilities together" (Chavanne, 2007).

2 HUMAN FACTORS AND SPACE OPERATIONS

"A fundamental tenet of human factors is that the human operator lies at the center of system design, the yardstick by which the form, fit, and function of hardware and software must be gauged." — Charlton and O' Brien, 2002

I began my Air Force career as a human factors engineer at Wright-Patterson Air Force Base, Ohio. During that four year tour of duty I developed a unique perspective regarding military weapon systems design that I would carry with me as a space operations operator for the next 16 years. During those subsequent space operations tours, I served in several different space mission areas in SSA and Satellite C^2. Each time I trained and operated a new space weapon system, I was personally amazed at the lack of human factors consideration in the weapon systems' workstations, layout, interface, and overall design. Despite the fact that space operations was the newest and most "high tech" arm of the military, it certainly did not receive the same Human Systems Integration (HSI) deliberation afforded to other military weapons systems such as aircraft, ships, or ground-based fighting vehicles. Most of the space systems I operated were developed in the late 1970s through the 1980s. The systems lacked Graphical User Interfaces, and utilized monochrome three-letter menu commands with several levels. Operators were required to learn multitudes of these directives to quickly execute mission commands during dynamic real world operations or exercise scenarios. In my experience, the HSI approach for each space weapon system was left to the individual contractors' discretion.

Fortunately, there has been a growing interest in HSI for space operations and weapon systems as a result of the 2004 the Air Force Scientific Advisory Board updates to the National Security Space Acquisition Policy (Rader and Smith, 2008). In recent years the Air Force Research Laboratory (AFRL) has teamed up with Air Force Space Command (AFSPC) and supporting aerospace contractors to conduct several space human factor assessments. Their research efforts are documented in the following sections.

2.1 Military Definition of Human Factors Engineering

Human Factors Engineering is the essential link between system design engineering and end users/operators. The field of human factors engineering uses scientific knowledge about human behavior in specifying the design and use of a human-machine system. The application of human factors engineering will create a human-system interface that will operate within human performance capabilities, meet system functional requirements, and accomplish mission objectives (MIL-HDBK-46855A).

The goal of human factors engineering is to maximize the ability of an individual or crew to operate and maintain a system at required levels by eliminating design-induced difficulty and error. Human factors engineers work with systems engineers throughout all phases of system development to design and evaluate human-system interfaces to ensure they are compatible with the capabilities and limitations of the potential user population. This holds true for space operators and space weapon systems as much as it does for their air, land, and sea brethren.

2.2 Human Factors Issues in Space C^2

Legacy military satellite operation centers and Remote Tracking Stations use outdated display, control, and workstation technology, degrading situational awareness and increasing crew workload and confusion (Ianni, 2003). To add to this, the layout of the Remote Tracking Station satellite C^2 consoles is such that crew members cannot effectively see or speak to one another during high tempo operations. The operations floors are noisy areas and consoles are separated by large pallets of hardware, which operate the antiquated systems used to link to and control the satellites. This places a great strain on crew coordination. From 2003 to 2005 an effort was made by the Air Force to analyze these critical operation centers' problems by forming a HSI Tiger Team, led by the Air Force Research Laboratory Human Effectiveness Directorate (AFRL/HE) (Monk and Popik, 2006).

AFRL/HECP coordinated with AFSPC to conduct research to increase space C^2 crew effectiveness. Their recommendations included three-dimensional displays and voice activated commands to provide the operator with better overall satellite situational awareness and a more natural interface to query and obtain satellite data. AFRL also recommended leveraging earlier Unmanned Aerial Vehicle (UAV) workstation design research performed by AFRL/HECP in 2002, since ground to satellite control had many aspects in common with ground to UAV control (Ianni 2003).

In 2006 the AFRL/HE Tiger Team addressed the human factors design aspects of users' workspace and the facility's habitability. The Tiger Team studied individual work environment, crew arrangements, and operational workstations. The study was conducted in accordance with the Military Standard 1472F, *Human Engineering,* and from the *Human Engineering Guide to Equipment Design.* AFRL/HE provided guidelines for future C^2 operation center design (Monk and Popik, 2006).

The Ground Systems Architecture Workshop (GSAW) provides a forum for the world's spacecraft ground system experts to collaborate with other ground system users, developers, and researchers. Over the past ten years, the GSAW has hosted multiple HSI tutorials, presentations, working groups, and panel discussions on satellite operations issues and solutions. This forum has addressed military and industry HSI standards, design approaches, and contractor developed tools to gain a better foothold on satellite operation human factors issues (Boltz and Andrusyszyn, 2008).

2.3 Human Factors Issues in SSA

As space-based platforms increasingly perform functions that are vital to national security, such as reconnaissance, surveillance, early warning, and communications, the more critical and complex a task it becomes to monitor SSA.

Inconsistency among systems in the wide array of diverse SSA locations creates challenges for human users. The Department of Defense (DoD) has spent millions of dollars in human factors engineering for air, land, and sea weapons systems. Comparatively, very little DoD funding and effort has been directed toward human factors considerations for SSA weapon systems.

AFRL/RH research has led the way to ensure critical SSA information is exploited to the maximum extent through fusion technologies and workflow design. In 2010 AFRL/RH conducted cognitive task analysis of the Joint Space Operations Center (the focal point of U.S. military SSA) and the National Air and Space Intelligence Center to provide a foundation for SSA and space C^2. AFRL's belief is that "human factors technology that facilitates the flow of task could possibly yield a much timelier and more accurate SSA picture" (Aleva et al., 2010). AFRL's current research projects are investigating multi-sensor fusion for space C^2 and computer based technology to address SSA visualization and operator collaboration issues. The results will be provided to AFSPC to leverage.

3 CONCLUSION

"Victory smiles upon those who anticipate the changes in the character of war, not upon those who wait to adapt themselves after the changes occur."
— Giulio Douhet

Foreign military space operations and weapons systems capabilities are rapidly increasing. If the US military is to maintain its space superiority edge in the near future, it must rely on the C2 and SSA cornerstones to fully understand the status of its own systems and the capability and intentions of potential adversaries. Utilizing a proper HSI approach and standardization of space weapon systems, with emphasis on operator considerations, will help ensure our space operators have the tools and optimized interfaces to succeed against the unique challenges encountered in the "High Frontier."

REFERENCES

Aleva, D. and J. McCracken, Dr. 2009. JSpOC Cognitive Task Analysis. *Proceedings of the Advanced Maui Optical and Space Surveillance Technologies Conference*. Maui, Hawaii.

Boltz, L. and J. Andrusyszyn. 2008. Human Factors Engineering in System Design for Operationally Responsive Ground Systems. *In: Ground Systems Architecture Workshop Session 3: Development Strategies*, Redondo Beach, California, 2008.

Charlton, S. and T. O'Brien. eds., 2002. *Handbook of Human Factors Testing and Evaluation*. Mahwah, New Jersey London: Erlbaum Associates, Publishers.

Chavanne, B. 2007. Space Integral To Military, General Says. Aerospace Daily and Defense Report [pdf]. Available from: <http://www.aviationweek.com/aw/generic/story_channel.jsp?channel=space>.

Department of Defense 1996. *Human Engineering Program Process and Procedures*. (Version MIL-HDBK-46855a). Available from: <www.hf.faa.gov/docs/508/docs/46855a.pdf>.

Department of Defense 2009. *Space Operations*. (Joint Publication 3-14) Available from: <www.fas.org/irp/doddir/dod/jp3_14.pdf>

GlobalSecurity.org [online]. 2012. Available from: <http://www.globalsecurity.org/space/systems/ground-spt.htm>.

Greenfield, T. and D. Stambolian. 2004. *Spaceport 1-G Human Factors for Optimal Space Transportation System Design*. Available from: <http://www.spacearchitect.org/pubs/Stambolian-Greenfield-2004.pdf>.

Ianni, J. 2003. Human Interfaces For Space Situational Awareness. [pdf]. Available from: <http://www.spacearchitect.org/pubs/Stambolian-Greenfield-2004.pdf>.

Monk, D. and D. Popik. 2006. Air Force Research Lab Wright-Patterson AFB OH Human Effectiveness Directorate. *Air and Space Operations Center (AOC) Facility Design Guidelines: A Human Factors Engineering Perspective*. Available from: http://handle.dtic.mil/100.2/ADA471733

Rader, S. and D. Smith. 2008. Open Discussion with NASA Exploration Interoperability Team. *In: Ground Systems Architecture Workshop Session 11A*, Redondo Beach, California, 2008.

CHAPTER 46

Data for a Worldwide ICEWS and Ongoing Research

Andrew Shilliday and Jennifer Lautenschlager

Lockheed Martin Advanced Technology Laboratories
Arlington, VA, USA, Kennesaw, GA, USA
{andrew.e.shilliday, jennifer.lautenschlager}@lmco.com

ABSTRACT

The ICEWS Core Data Repository (ICDR) consists of more than 16 million news articles dating back more than a decade, from multiple languages, and compounded from hundreds of individual publication sources. It additionally collects and interpolates nation state variables, such as GDP, population, ethnic fractionalization and polity score; it stores, analyses, and aggregates structured events extracted from news articles by the JabariNLP Event Coder; and manages the database of national and international socio-political players.

ICEWS was initially focused on the US Pacific Command's (USPACOM) Area of Responsibility (AOR) and as such, the ICDR was populated with the data needed to support some 29 focus countries. It was since expanded to support every COCOM, necessitating the generation of like data for nearly every country—175 countries worldwide. This ambitious task required a substantial overhaul to the ICDR and the development of new tools to automate or otherwise assist in building the corpus of background knowledge necessary to understanding and incorporating news, events, and other indicators particular to the 175 nations.

We will describe the process of applying ICEWS technology to the global landscape, highlighting the successes, pitfalls, and lessons learned along the way. We will also describe our strategies for maintaining the high degree of accuracy achieved for the USPACOM-centric data during the expansion, and explain our validation and evaluation methodologies and results. We will also present the ICEWS Dictionary Editor, a new tool enabling teams of researchers to quickly and effectively populate the Actor Dictionary, and bring the vision of a global ICDR (and global ICEWS) into fruition.

1 THE ICEWS CORE DATA REPOSITORY

Central to the operations of all ICEWS capabilities (viz., iCAST, iTRACE, iSENT, and iDIME; see Kettler et al., 2012), the ICEWS Core Data Repository (ICDR) and its integrated data environment are responsible for collecting, processing, analyzing, and storing the copious structured and unstructured data required for and used within the system. Over the course of the four-year research effort, ICEWS data requirements have grown considerably in both volume and complexity, maturing into what is currently a robust, global database consisting of some 16-million news articles, 10 million structural data points, 50k social actors, and 10 million events (and counting). And like so many other large-scale data-driven systems, the transition into a global ICDR has not been a continuously smooth one. In this chapter, we describe challenges we encountered as the ICEWS data needs have changed and expanded, and introduce a number of automated and semi-automated tools developed at Lockheed Martin Advanced Technology Laboratories (LM ATL) in support thereof. We begin by briefly describing the kinds of data used within the ICEWS components, and within the ICDR area of responsibility.

1.1 Data and Data Processing for ICEWS forecasting

As noted in Kettler et al., 2012, ICEWS forecasting capabilities (iCAST) use a mixed-method approach to predicting instability events in foreign nations (e.g., domestic political crisis, insurgency). Within the suite of integrated forecasting models, the statistical models in particular rely on a wide variety of indicator variables collected, processed, and ultimately provisioned as input. The ICDR manages two principal classes of model input variables: structural indicators and aggregated behavioral event data.

Structural data consists of country-level slow-changing variables, such as GDP or level of ethnic fractionalization. Each data point is relative to a specific country and date (at varying intervals). For each of the nearly 10,000 structural indicators, the ICDR stores relevant metadata (e.g., the source of the data, frequency of updates, and interval and units of data points within the dataset), with which it fills data requests from models (Ruvinsky, 2011), interpolating and extrapolating over incomplete datasets where necessary.

1.2 Event Data

Behavioral data consists in aggregations of coded interactions between socio-political actors (i.e., cooperative or hostile actions between individuals, groups, sectors and nation states). Events are automatically identified and extracted from news articles by the JabariNLP Event Coder (Van Brackle, 2012), an advanced derivative of the TABARI Event Coder (Schrodt, 2009), incorporating statistical natural language processing into the otherwise completely rule-based text extraction system. These events are essentially triples consisting of a source actor, an event type (according to the CAMEO taxonomy of events), and a target actor. Geographical-temporal metadata are also extracted and associated with the relevant events within a news article. The resulting event data are used (within ICEWS) by both iCAST and iTRACE. For iCAST, the data are aggregated into about 5000 indicator variables that are, in turn, fed to the statistical forecasting models. Each variable is defined across several dimensions: whether the source and target are from the same country (i.e., whether the event is domestic or international); to which sectors the actors belong (e.g., government, business, dissidents, non-exclusively); and the severity of the action (viz., cooperative, hostile, or neutral). Each indicator variable can be aggregated by the overall sum of events falling within the constraints of the indicator definition (grouped by country and time period), or by the sum or average of the numerical (Goldstein) scores attributed to each class of event (ranging from -10 to 10 with negative values indicating hostility, and positive values representing cooperation).

Event data are also aggregated on the fly for iTRACE, where a user might desire a high-level picture of the activity for a specific country or region. Figure 1, for example, shows a daily intensity view of the events pertaining to India (where either the source or target was associated with India) with the hostile event counts on the negative scale, and the cooperative event counts on the positive. This requirement for rapid, dynamic aggregation over such a large and complex dataset presented a significant challenge for the ICDR. Given the hierarchical nature of both actors and events, event data queries could require expensive recursive calls into the database, that would result in a significant performance hit and poor user experience. One key insight to improving performance was the use of the nested set model (Hillyer, 2005) to represent the actor hierarchy, allowing for fast, non-recursive access to all descendants in the actor membership tree and subsequently all events pertaining to a group (or country) and any of its members.

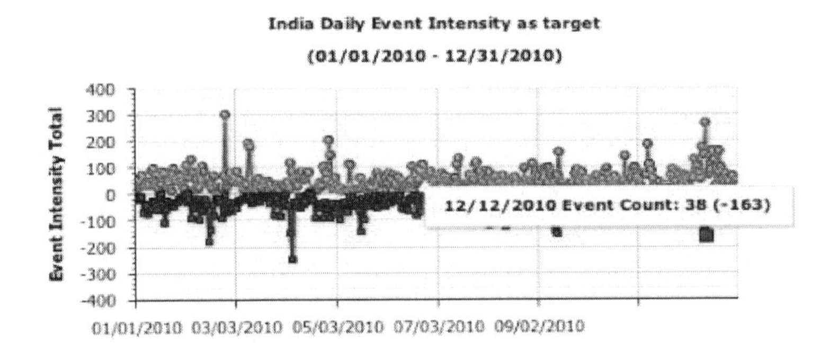

Figure 1 iTRACE daily intensity view for India in 2010

1.3 The Actor and Verb Dictionaries

Although JabariNLP incorporates some NLP components (syntax tree parsing and parts-of-speech tagging), the Event Coder, and its predecessor are essentially rule-based text extraction systems. As such, they require sophisticated dictionaries to govern how structured event data are extracted from unstructured text. (We've also explored purely statistical extraction systems, i.e., BBN's SERIF and SAE's Xenophon event coder, and while a formal small-scale experiment comparing SERIF to TABARI significantly favored SERIF, JabariNLP had already closed the gap between rule-based and statistical extraction systems, and given the time and effort required to train SERIF to the remaining 80% of the CAMEO event types, JabariNLP was ultimately selected. Moreover, SERIF required to be initially trained on the same set of dictionaries, so their development was necessary regardless.)

There are three kinds of dictionaries used by JabariNLP: The Actor Dictionary maps the names of individuals, groups, and countries to CAMEO Actor Codes, indicating the country and sectors within that country with which the actor is associated. The Verb Dictionary maps verb phrases to CAMEO Event Codes; it uses a special syntax to inform which are the verbs and relevant modifiers that allow the system to select the best interpretation of a sentence. The verb rules also sometimes dictate where, relative to the verb, the coder should look for the source actor and target actor (i.e., does the source come before the verb as in "President Barack Obama urged China to … ," or after verb as in "Iraqi security forces were attacked by al-Qaeda militants … ." Finally, the Agent Dictionary is similar to the Actor Dictionary, in that it informs the coder of words or phrases to be treated as actors, but the Agent Dictionary describes generic terms that are associated with specific roles or sectors (i.e., "police," "prime minister," "armed rebels").

In Phase 1 of the ICEWS effort, these dictionaries were plain text files that were fed directly into the TABARI event coder. Given these dictionaries and a sentence, TABARI would generate one or more triples consisting of source and target actor codes, and an event code. This proved sufficient for the purposes of modeling, where events are aggregated into indicator variables (Section 1.2), but having only the generic actor codes, we lost a great deal of specificity in our event data. In Phase 2, we implemented the actor dictionary as a database in the ICDR, whereby individuals and groups, and the time-bound, hierarchical membership relationships between them were captured and events could be mapped directly to the actual actors involved. This increased level of detail in our event data kindled government interest in a tool to browse and analyze the data directly (see iTRACE in Kettler et al., 2012).

2 DATA PROCESSING CHALLENGES IN EARLY ICEWS

In late 2009, a successful iTRACE demo to PACOM put a greater emphasis on the quality of the event data as users could now view the coded event triples and the stories from which they were extracted. In January 2010, we established TABARI's performance in a series of evaluations at 58% precision. In January 2010, we established TABARI's performance in a series of evaluations at 58 percent precision, sufficient to produce signal for models to generate accurate forecasts, but not as a human accessible base for analysis. This estimate was calculated by manually evaluating a sampling of event data against their original news stories. Precision was calculated as the percentage of event triples that were correct. Recall is another key dimension of coding accuracy, but because recall requires the existence of a (manually) built ground truth, we used the number of events generated from a set of stories as a proxy for recall: i.e., given a relatively high level of precision (and thus a low level of false positives), a higher number of events generated should imply a higher recall. This was deemed too low for iTRACE. Thus, an LM ATL team, led by Dr. Shilliday, was tasked to improve the event data accuracy, setting a target quality level of 80% precision, considered to be peak as human coding performance.

Thus, an LM ATL team, led by Dr. Shilliday, was tasked to improve the event data accuracy, setting a target quality level of 80 percent precision, considered to be peak as human coding performance. (Interestingly, this assessment of human performance (King and Lowe, 2003) evaluated accuracy using the event type as the only metric, whereas ICEWS accuracy judgments are made on both the accuracy of the event and the source/target actors involved. Our own evaluations show human coding performance is closer to 40 percent accuracy given issues of cross-coder consistency, fatigue, etc.)

In February 2010 we completed initial development of Jabari, our Java version of TABARI. This involved extensive testing and debugging (of both TABARI and Jabari). Both systems ultimately produced identical results on a test data set, and thus Jabari became the new baseline coder. In addition, we built a number of tools to support and improve the event data ingest/coding process, including statistical, pre-coding story filters (to remove unwanted stories—e.g., sports, entertainment, etc.); sentence delineation; post-coding event filtering, etc. We leveraged Java-based open source tools/packages including Apache OpenNLP (2011) and the Univ. of Massachusetts MALLET (McCallum, 2002). To improve its performance from the ~57 percent baseline, we identified potential sources of coding errors in the event coding process including gaps in news coverage; changes in news story format; coding of unwanted news stories (e.g., entertainment news); story parsing errors; incorrect event coding (wrong source, target, or event type); and post-coding filtering errors. We then systematically evaluated these to determine the causes of error (where possible); fixes to the pre- and post-processing tools (e.g., filters); actor/verb dictionary errors/omissions; and bug fixes/enhancements to the event coder itself.

JabariNLP (Van Brackle, 2011) was conceived as a means to help guide the application of patterns through the use of additional information—namely syntactic structure of sentences obtained from the OpenNLP parser. Significant improvement in accuracy (precision) was demonstrated with JabariNLP vs. Jabari/TABARI, but carried a sizable hit in recall. Additional enhancements were made to improve recall without sacrificing precision, primarily focusing on ways in which the coder identified actors, i.e., pronoun dereferencing and an improved agent inference engine that used context from elsewhere in the text to determine the country or group associated with generic agents (e.g., Jabari/TABARI could only recognize the agent term "Police" to mean the "Australian Police" if the term "Australia" was nearly adjacent to the term, whereas JabariNLP could now infer such associations by recognizing that the story was about Australia, and that any such references should by default be associated thusly). In 2010, JabariNLP was selected as the baseline event coder going forward into Phase 3, having achieved a precision of 71 percent, a marked improvement over its predecessors.

3 EXPANDING TO A GLOBAL ICEWS

Building on the previous successes, and due to overwhelming enthusiasm by USCOCOMs, Phase 3 of the ICEWS executed a transition of capabilities to the USSTRATCOM ISPAN program of record (Starz, 2012), necessitating a six-fold expansion of the system's initial focus area (29 countries in the USPACOM AOI) to 175 countries worldwide across all six COCOMs. The effort involved acquiring news stories for all 175 countries (our current 10-year archive consisted of articles explicitly relevant to the 29 PACOM countries), and generating actor dictionaries for the remaining 146 countries.

3.1 Expanding the Source Data

A major change made to support this "coding for the world" was the switch from LexisNexis (LN) to Dow Jones Factiva as the story provider. Factiva provided ICEWS tools with easier access to news stories via an application programming interface (API) (versus using potentially brittle web-scraping scripts to access LN stories) and automatic downloading enabled by standard FTP/HTTP/HTTPS access. Factiva news stories were provided in an easily parseable XML format with delineated metadata, unlike LexisNexis' HTML documents. Metadata included index terms that helped identify relevant stories. Factiva provided high quality technical support, working with LM to reduce irrelevant or duplicate stories (approximately 1/3 of the LN data set), and identify supplemental news sources as needed. Data set quality – roughly measured by the number of events generated per story downloaded (as an indicator of data relevancy)—increased from 0.35 events/story for LN to 0.96 for Factiva (in both cases, coding only the first six sentences of each story). As part of expanding to 175 countries, we increased the number of news sources from 75 to more than 250.

With addition of SOUTHCOM as a pilot site for ICEWS in 2011, the need to mine stories in Spanish and Portuguese became evident as a means to augment the comparatively scarce volume of South American news articles. Working with Factiva, we obtained additional English online-only stories and Spanish/Portuguese stories. As of February 2012, of the approximately 16 million stories in ICEWS, roughly 2.2 million are in foreign languages (1.5 million in Spanish, 700K in Portuguese). We considered a number of commercial off-the-shelf machine translation tools, and executed an evaluation of SysTran and SDL's Language Weaver, comparing both against a baseline using Google Translate. Performance was determined according to the precision of extracted events over translated documents (as measured against the English translation, but with an eye to original foreign text), and the ratio of generated events to processed stories as a proxy for recall. Both systems achieved comparable precision scores, but Language Weaver significantly outperformed SysTran in recall indicating, if nothing else, that the resulting English text was a closer match to the English journalist writing style, for which the verb dictionary was tuned. Empirically, the Language Weaver translations appeared to be superior to those of SysTran, and the tool was ultimately selected and incorporated into the ICDR event coding process. The event data extracted from these translated documents are less accurate than those extracted from native English documents (67 percent for Portuguese, and 65.5 percent for Spanish). Furthermore, the ratio of events to stories is quite a bit lower for Spanish/ Portuguese documents than for English. Both of these results are to be expected given these translated documents tend to have more grammatical mistakes. In the end, however, the translated Spanish and Portuguese texts proved to be sufficient to bring the SOUTHCOM story and event volume to within an acceptable range.

3.2 Expanding the Actor Dictionary

The expansion of ICDR Actor Dictionaries to cover all 175 countries (as well as all relevant international entities) constituted a significant collaboration effort between Strategic Analysis Enterprises, Inc. (SAE), Pennsylvania State University (PSU) and LM ATL, executed over six months in 2011. The dictionary generation process is largely a manual effort wherein social science experts must not only determine who constitutes a relevant socio-political player, but classify each player by the groups and sectors to which they belong, and the time period during which the classification holds. We used homegrown automatic entity extraction, data mining, and web scraping tools to populate first-pass lists of possibly relevant actors from a variety of data sources (Janes, CIA Factbook, Wikipedia, Rulers.org, etc.).

In parallel, improvements were made to how actors were represented as "first class entities" so that they could have multiple types (e.g., group/sector membership—such as "government" or "insurgent") and properties (e.g., birth date, death date, gender). LM ATL developed the Actor Dictionary Editor tool, shown in Figure 2, to enable easier and more consistent editing of dictionaries. The Dictionary Editor enables the users (SAE, PSU, and LM ATL) to edit the sector hierarchy, add new per-country dictionaries, and edit the actors contained therein. For each actor, the user entered the name and provided a list of candidate patterns for use in JabariNLP (particularly useful if the actor had alternate spellings or aliases). Additionally, they provided membership details associating the actor with one or more countries or groups, and within each country, the sector to which the actor belongs and the role he or she plays within that group or sector. Each membership relationship is optionally bound by a start and end date. The ICDR uses this information for querying and aggregating over the event data (e.g., so that one could query for all of the negative events targeting the North Korean administration, and any event specifically involving Kim Jon-Il between 10/8/1997 and 12/17/2011 would be included). Within the Dictionary Editor, a user can also define new classes of aggregate indicator variables for use by new forecasting models.

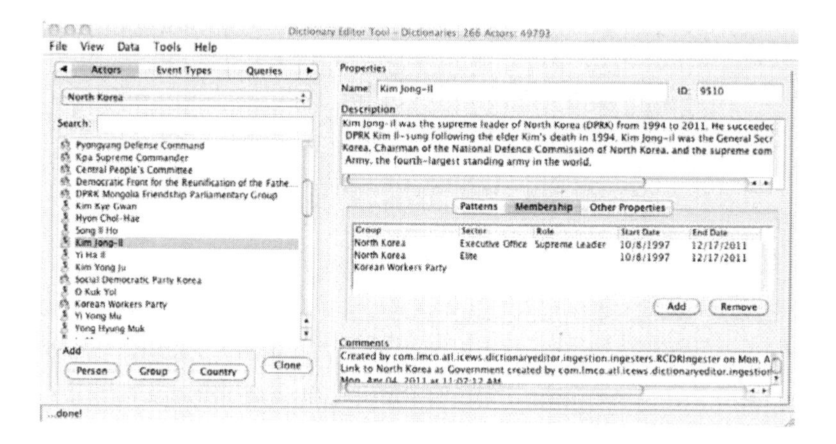

Figure 2 The ICEWS Actor Dictionary Editor

3.3 The ICEWS Data Tool

The Dictionary Editor operated as a stand-alone tool, where dictionaries modified within are saved as a series of XML actor dictionary files. The ICEWS Data Tool is used to bring the resulting dictionaries into the ICDR for use during event processing. The tool was originally developed during Phase 2 to support a small number of recurring data tasks, and was subsequently reengineered in Phase 3 as the administrator's interface to the database, handling nearly every facet of data upkeep that occurs in the ICDR. To name a few of its functions, the tool is responsible for importing and processing newly downloaded news articles, executing JabariNLP and processing/filtering the results, and pushing new data out to remote installations at SOUTHCOM, PACOM, and STRATCOM. The Data Tool operates via the command line, so is easily scripted into batch operations and executed automatically as recurring scheduled jobs.

3.4 Event Data Evaluations

Given the speed at which dictionaries were being generated, quality assurance was a significant concern. We conducted several inter-coder reliability studies (comparing results from multiple dictionary developers over the same set of actors to obtain a measure of consistency). Furthermore, we established a quality gate using previously developed dictionaries forcing that each developer precisely understood the process and desired data. LM ATL also executed several per-COCOM event data evaluations to ensure consistent levels of accuracy across the newly-supported and existing country dictionaries. As previously mentioned, event data were judged based on the correctness of the selected event type, as well as the correctness of the selected source and target actors. Each human judge was provided a randomly selected set of events, each with its source news documents highlighting the sentence from which it was drawn.

The judges were provided guidance as to how each dimension of an event should be graded, and under what conditions they could allow a certain amount of latitude with "good enough" events. If an event was extracted having a source actor that was correct, though slightly more general than what was offered in the text (e.g., the system selected Australian Prime Minister whereas the sentence indicated Australian Prime Minister John Howard), then the judge could mark it as correct. Furthermore, JabariNLP often will extract multiple events from a given sentence, and as the events are selected randomly, it was not guaranteed that the sentence's entire set of events was represented in a sample. Thus, the evaluators were instructed to count an event as correct if the event was true for the given sentence, even where the event was not the central point of the text. Evaluation sets were crafted to have some percentage of shared events between two or more evaluators to obtain a measure of inter-evaluator reliability.

In October 2011, we performed the final evaluations for 10 years worth of event data using the completed 175 country actor dictionaries. Out of the 15.6 million stories processed, over 10.9 million events were generated (yielding a 0.7 events/story ratio, already a huge improvement over Phase 2), and an overall precision of 75.42 percent with a 3.10 percent confidence interval. A subsequent blind operational evaluation at PACOM confirmed these findings (and in fact generated a 2 percent higher accuracy than our own internal evaluations).

4 BEYOND ICEWS

Over Phases 2 and 3 of the effort, ICEWS has been subject to dramatic expansion in size and complexity of the data, as well as an overall broadening of the scope and application thereof. These changes have brought with them a number of challenges in generating and managing ICEWS data and ensuring it is of sufficient quality to satisfy their new requirements. We have met these and other challenges, building new tools and redesigning existing ones as needed, and established and executed an evaluation methodology demonstrating a significantly improved overall data quality. As we move into the next phase of ICEWS research (now W-ICEWS), we will be expanding the ICEWS *forecasting* capabilities to include all 177 countries worldwide. We also plan to further improve upon our data quality to achieve an overall accuracy above 80 percent.

4.1 Acccessing the Data

At the time of this writing, the Office of Naval Research (ONR) is pursuing approval for open release of the ICEWS event data set for academic/research use. Please contact the author regarding any interest in exploring the current availability of these data.

ACKNOWLEDGMENTS

Core team members included Duke University (led by Dr. Michael Ward); Innovative Decisions, Inc. (led by Drs. Suzanne Mahoney and Dennis Buede), Lustick Consulting, Inc. (led by Dr. Ian Lustick), the Pennsylvania State University (led by Prof. Phil Schrodt), and Strategic Analysis Enterprises, Inc. (led by Dr. Steve Shellman). ICEWS was supported by the Defense Advanced Research Projects Agency (DARPA). We especially thank Drs. Sean O'Brien and Philippe Loustaunau for their vision in creating and guiding the ICEWS Program and the ICEWS operational champions at SOUTHCOM, PACOM, and STRATCOM. Future work is being supported by Worldwide ICEWS contract through the Office of Naval Research.

REFERENCES

Apache OpenNLP Development Community. 2011. Apache OpenNLP Developer Documentation. Online, http://incubator.apache.org/opennlp/documentation/1.5.2-incubating/manual/opennlp.html.

Hillyer, M. 2005. Managing Hierarchical Data in MySQL. Online, http://mikehillyer.com/articles/managing-hierarchical-data-in-mysql/.

Kettler, B. and Hoffman, H. 2012. Lessons Learned in Instability Modeling, Forecasting, and Mitigation from the DARPA Integrated Crisis Early Warning System (ICEWS) Program. *2nd International Conference on Cross-Cultural Decision Making: Focus 2012.*

King, G. and Lowe, W. 2003. An Automated Information Extraction Tool for International Conflict Data with Performance as Good as Human Coders: A Rare Events Evaluation Design. *International Organization* 57(3): 617–642

McCallum, A.K. 2002. MALLET: A Machine Learning for Language Toolkit. http://mallet cs.umass.edu

Ruvinsky, A., Shilliday, A., Wedgwood, J., and Welsh, J. 2011. Purpose-Driven Metadata for Verification and Validation of HSCB Models. Abstract presented at the *Human Social Culture Behavior Modeling Focus 2011.*

Schrodt, Philip A. 2008. Tabari : Textual Analysis By Augmented Replacement Instructions. PDF file, http://web.ku.edu/~keds/tabari.dir/tabari.manual.0.6.3b7.pdf.

Starz, J., Hoffman, M., Roberts, J., Losco, J., Spivey, K., and Lautenschlager, J. 2012. Supporting situation understanding (past, present, and implications on the future) in the STRATCOM ISPAN program of record. *2nd International Conference on Cross-Cultural Decision Making: Focus 2012.*

Van Brackle, D. and Wedgwood, J. 2011. Event Coding for HSCB Modeling: Challenges and Approaches. Abstract presented at the *Human Social Culture Behavior Modeling Focus 2011.*

Van Brackle, D. and Haglich, P. 2012. Improvements in the Jabari event coder. *2nd International Conference on Cross-Cultural Decision Making: Focus 2012.*

Index